Arrow-Pushing in Organic Chemistry

Arrow-Pushing in Organic Chemistry

An Easy Approach to Understanding Reaction Mechanisms

Second Edition

Daniel E. Levy

WILEY

This edition first published 2017
© 2017 John Wiley & Sons, Inc.

The right of Daniel E Levy to be identified as the author of this work has been asserted in accordance with law.

Registered Offices
John Wiley & Sons, Inc., 111 River Street, Hoboken, NJ 07030, USA

Editorial Office
111 River Street, Hoboken, NJ 07030, USA

For details of our global editorial offices, customer services, and more information about Wiley products visit us at www.wiley.com.

Wiley also publishes its books in a variety of electronic formats and by print-on-demand. Some content that appears in standard print versions of this book may not be available in other formats.

Library of Congress Cataloguing-in-Publication Data

Names: Levy, D. E. (Daniel E.)
Title: Arrow-pushing in organic chemistry : an easy approach to understanding reaction mechanisms / Daniel E. Levy.
Other titles: Arrow pushing in organic chemistry
Description: Second edition. | Hoboken, NJ : John Wiley & Sons, Inc., 2017. | Includes bibliographical references and index. | Includes index.
Identifiers: LCCN 2016043334| ISBN 9781118991329 (pbk.) | ISBN 9781118991206 (epub)
Subjects: LCSH: Chemistry, Organic–Textbooks. | Reaction mechanisms (Chemistry)–Textbooks.
Classification: LCC QD253.2 .L48 2017 | DDC 547/.2–dc23
LC record available at https://lccn.loc.gov/2016043334

Cover image: Wiley
Cover design: Courtesy of Daniel E. Levy; (Background) © 0shut0/Gettyimages

Set in 10/12pt Times by SPi Global, Pondicherry, India

Printed in the United States of America

10 9 8 7 6 5 4 3 2 1

Dedicated to the memory of Henry Rapoport (1918–2002)
Professor Emeritus of Chemistry
University of California–Berkeley
a true teacher and mentor

Contents

Preface

Organic chemistry is a general requirement for most students pursuing degrees in the fields of biology, physiology, medicine, chemical engineering, biochemistry, and chemistry. Consequently, many of the students studying organic chemistry initially do so out of obligations to required curriculum rather than out of genuine interest in the subject. This is, in fact, alright and expected as almost all college students find themselves enrolling in classes in which they either have no interest or cannot foresee application of the subject to their future vocation. Alternatively, there are students who are intrigued with the potential application of organic chemistry to fields including pharmaceuticals, polymers, pesticides, food science, and energy. However, whichever group represents the individual students, there is always a common subset of each that tenuously approaches the study of organic chemistry due to rumors or preconceived notions that the subject is extremely difficult and requires extensive memorization. Having personally studied organic chemistry and tutored many students in the subject, I assure you that this is not the case.

When first presented with organic chemistry course material, one can easily be caught up in the size of the book, the encyclopedic presentation of reactions, and the self-questioning of how one can ever decipher the subject. These students frequently compile endless sets of flash cards listing specific chemical reactions and their associated names. Like many of my classmates, I began to approach the subject in this manner. However, this strategy did not work for me as I quickly realized that memorization of reactions did not provide any deductive or predictive insight into the progression of starting materials to products and by what mechanisms the transformations occurred. In fact, the fundamental fault in the "memorization strategy" is that in order to be effective, the student must memorize not only all chemical reactions and associated reaction names but also all associated reaction mechanisms and potential competing processes. It wasn't until I abandoned the "memorization strategy" that I began to do well in organic chemistry and develop a true appreciation for the subject and how the science benefits society.

The presumption that introductory organic chemistry entails very little memorization is valid and simplifies the subject provided the student adheres to the philosophy that the study of organic chemistry can be reduced to the study of interactions between organic acids and bases. From this perspective, organic chemistry students can learn to determine the most acidic proton in a given molecule, determine the most reactive site (for nucleo-philic attack), and determine the best reactants (nucleophiles and electrophiles) and how to predict reaction products. In learning to predict these components of organic reactions, the beginning organic chemist will be able to deduce reasonable routes from starting materials to products using the basic mechanistic types involved in introductory organic chemistry. Furthermore, through an understanding of how electrons move, extrapolations from ionic or heterolytic mechanisms can be used to explain free radical and pericyclic processes. Finally, by utilizing the principles discussed in this book, the student will gain a better understanding of how to approach the more advanced reaction types discussed as the introductory organic chemistry course progresses.

The goal of this book is not to present a comprehensive treatment of organic chemistry. Furthermore, this book is not intended to be a replacement for organic chemistry texts or to serve as a stand-alone presentation of the subject. This book is intended to supplement organic chemistry textbooks by presenting a simplified strategy to the study of the subject in the absence of extensive lists of organic reactions. Through application of the principles presented herein, including new chapters covering free radicals, carbenes, and pericyclic reactions, it is my hope that this second edition, when used as intended, will aid the beginning student in approaching organic chemistry as I did—with little memorization and much understanding.

DANIEL E. LEVY, PH.D.

Acknowledgements

I would like to express my deepest appreciation to my wife, Jennifer, and to my children, Aaron, Joshua, and Dahlia, for their patience and support while writing this book. I would also like to express special thanks to Dr. Lane Clizbe for his editorial contributions and to Professor James S. Nowick for his suggestions regarding content for the second edition.

About the Author

Daniel E. Levy received his Bachelor of Science in 1987 from the University of California at Berkeley where, under the direction of Professor Henry Rapoport, he studied the preparation of 4-amino-4-deoxy sugars and novel analogs of pilocarpine. Following his undergraduate studies, Dr. Levy pursued his Ph.D. at the Massachusetts Institute of Technology. Under the direction of Professor Satoru Masamune, he studied sugar modifications of amphotericin B, the total synthesis of calyculin A, and the use of chiral isoxazolidines as chiral auxiliaries. In 1992, Dr. Levy completed his Ph.D. and has since worked on various projects involving the design and synthesis of novel organic compounds. These compounds include glycomimetic inhibitors of fucosyltransferases and cell adhesion molecules, peptidomimetic matrix metalloproteinase inhibitors, carbocyclic AMP analogs as inhibitors of type V adenylyl cyclase, heterocyclic ADP receptor antagonists, inhibitors of calmodulin-dependent kinase, and nanoparticle delivery vehicles for siRNA-based therapeutics. In 2010, Dr. Levy founded DEL BioPharma LLC—a consulting firm providing research and development services to emerging pharmaceutical companies.

Arrow-Pushing in Organic Chemistry is Dr. Levy's third book—the first edition having been published in 2008. In 1995, Dr. Levy coauthored a book entitled *The Chemistry of C-Glycosides* (1995, Elsevier Sciences). Collaborating with Dr. Péter Fügedi, Dr. Levy developed and presented short courses entitled "Modern Synthetic Carbohydrate Chemistry" and "The Organic Chemistry of Sugars," which were offered by the American Chemical Society Continuing Education Department. With Dr. Fügedi, Dr. Levy coedited his second book entitled *The Organic Chemistry of Sugars* (2005, CRC Press).

Chapter *1*

Introduction

The study of organic chemistry focuses on the chemistry of elements and materials essential for the existence of life. In addition to carbon, the most common elements present in organic molecules are hydrogen, oxygen, nitrogen, sulfur, and various halogens. Through the study of organic chemistry, our understanding of the forces binding these elements to one another and how these bonds can be manipulated are explored. In general, our ability to manipulate organic molecules is influenced by several factors that include the nature of **functional groups** near sites of reaction, the nature of **reagents** utilized in reactions, and the nature of potential **leaving groups**. In addition, these three factors impart further variables that influence the course of organic reactions. For example, the nature of the **reagents** used in given reactions can influence the reaction **mechanisms** and ultimately the reaction **products**. By recognizing the interplay between these factors and by applying principles of **arrow-pushing**, which in reality represents bookkeeping of electrons, reasonable predictions of organic mechanisms and products can be realized without the burden of committing to memory the wealth of organic reactions studied in introductory courses. In this chapter, the concept of **arrow-pushing** is defined in context with various **reaction types**, **functional groups**, **mechanism types**, **reagents/nucleophiles**, and **leaving groups**.

1.1 DEFINITION OF ARROW-PUSHING

Organic chemistry is generally presented through a treatment of how organic chemicals are converted from starting materials to products. For example, the **Wittig reaction** (Scheme 1.1) is used for the conversion of **aldehydes** and **ketones** to **olefins**, and the **Diels–Alder reaction** (Scheme 1.2) is used for the formation of six-membered ring systems and treatment of alkyl halides with reagents such as tributyltin hydride (Scheme 1.3),

Arrow-Pushing in Organic Chemistry: An Easy Approach to Understanding Reaction Mechanisms,
Second Edition. Daniel E. Levy.
© 2017 John Wiley & Sons, Inc. Published 2017 by John Wiley & Sons, Inc.

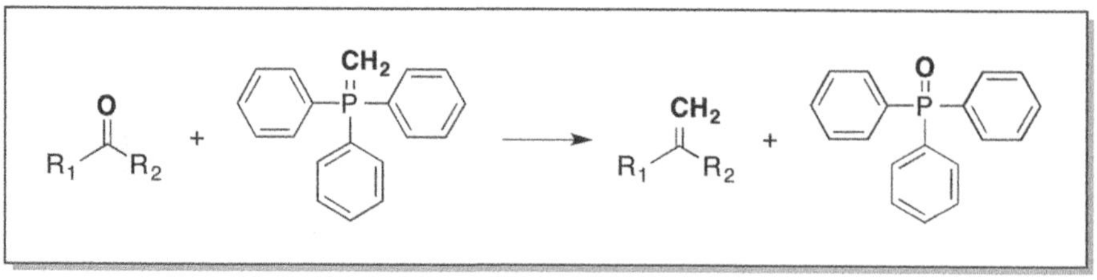

Scheme 1.1 *Example of the Wittig reaction.*

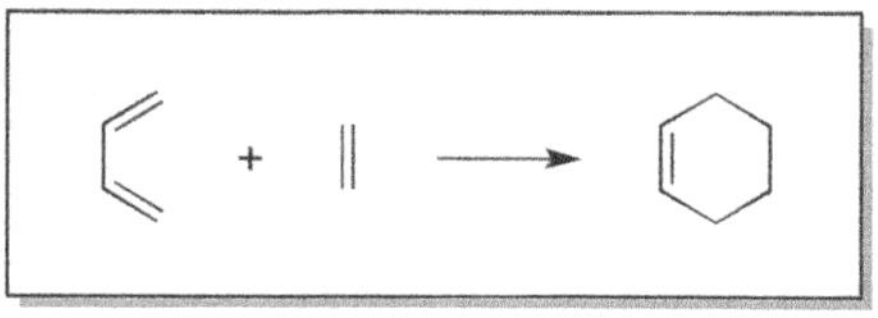

Scheme 1.2 *Example of the Diels–Alder reaction.*

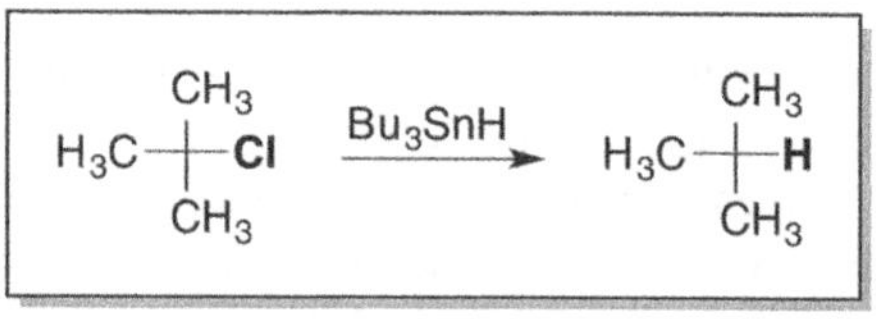

Scheme 1.3 *Example of a tin hydride dehalogenation.*

resulting in removal of the associated halides. However, by presenting these reactions as illustrated in Schemes 1.1, 1.2, and 1.3, no explanation is provided as to how the starting materials end up as their respective products.

By definition, the outcome of any chemical reaction is the result of a process resulting in the breaking and formation of chemical bonds. Referring to material covered in most general chemistry courses, bonds between atoms are defined by sets of two electrons. Specifically, a single bond between two atoms is made of two electrons, a double bond between atoms is made of two sets of two electrons, and a triple bond between atoms is made of three sets of two electrons. These types of bonds can generally be represented by **Lewis structures** using pairs of dots to illustrate the presence of an electron pair. In organic chemistry, these dots are most commonly replaced with lines. Figure 1.1 illustrates several types of chemical bonds using both electron dot notation and line notation. The list of bond types shown in Figure 1.1 is not intended to be exhaustive with respect to functional groups or potential combinations of atoms.

While chemical bonds are represented by lines connecting atoms, electron dot notation is commonly used to represent **lone pairs** (nonbonding pairs) of electrons. **Lone pairs** are found on **heteroatoms** (atoms other than carbon or hydrogen) that do not require bonds with additional atoms to fill their valence shell of eight electrons. For example, atomic **carbon** possesses four valence electrons. In order for **carbon** to achieve a full complement of eight valence electrons in its outer shell, it must form four chemical bonds, leaving no electrons as **lone pairs**. Atomic **nitrogen**, on the other hand, possesses five valence electrons. In order for **nitrogen** to achieve a full complement of eight valence electrons, it must form three chemical bonds, leaving two electrons as a **lone pair**. Similarly, atomic **oxygen**

Single Bonds		Double Bonds		Triple Bonds	
Electron Dots	Lines	Electron Dots	Lines	Electron Dots	Lines
$H_3C:CH_3$	H_3C-CH_3	$O::O$	$O=O$	$N:::N$	$N \equiv N$
$H_3C:Cl$	H_3C-Cl	$H_2C::CH_2$	$H_2C=CH_2$	$HC:::CH$	$HC \equiv CH$
$H_3C:NH_2$	H_3C-NH_2	$H_2C::NH$	$H_2C=NH$	$HC:::N$	$HC \equiv N$
$H_3C:OH$	H_3C-OH	$H_2C::O$	$H_2C=O$		
$H_3C:SH$	H_3C-SH	$H_2C::S$	$H_2C=S$		

Figure 1.1 *Examples of chemical bonds.*

Single Bonds		Double Bonds		Triple Bonds	
Electron Dots	Lines	Electron Dots	Lines	Electron Dots	Lines
$H_3C:CH_3$	H_3C-CH_3	$\overset{\cdot\cdot}{O}::\overset{\cdot\cdot}{O}$	$\overset{\cdot\cdot}{O}=\overset{\cdot\cdot}{O}$	$:N:::N:$	$:N \equiv N:$
$H_3C:\overset{\cdot\cdot}{\underset{\cdot\cdot}{C}l}:$	$H_3C-\overset{\cdot\cdot}{\underset{\cdot\cdot}{C}l}:$	$H_2C::CH_2$	$H_2C=CH_2$	$HC:::CH$	$HC \equiv CH$
$H_3C:\overset{\cdot\cdot}{N}H_2$	$H_3C-\overset{\cdot\cdot}{N}H_2$	$H_2C::\overset{\cdot\cdot}{N}H$	$H_2C=\overset{\cdot\cdot}{N}H$	$HC:::N:$	$HC \equiv N:$
$H_3C:\overset{\cdot\cdot}{O}H$	$H_3C-\overset{\cdot\cdot}{O}H$	$H_2C::\overset{\cdot\cdot}{O}:$	$H_2C=\overset{\cdot\cdot}{O}:$		
$H_3C:\overset{\cdot\cdot}{S}H$	$H_3C-\overset{\cdot\cdot}{S}H$	$H_2C::\overset{\cdot\cdot}{S}:$	$H_2C=\overset{\cdot\cdot}{S}:$		

Figure 1.2 *Examples of chemical bonds and lone pairs.*

possesses six valence electrons. In order for **oxygen** to achieve a full complement of eight valence electrons, it must form two chemical bonds, leaving four electrons as two sets of **lone pairs**. In the examples of chemical bonds shown in Figure 1.1, **lone pairs** are not represented in order to focus on the bonds themselves. In Figure 1.2 the missing **lone pairs** are added where appropriate. **Lone pairs** are extremely important in understanding organic mechanisms because they frequently provide the sources of **electron density** necessary to drive reactions as will be discussed throughout this book.

As organic reactions proceed through the breaking and subsequent formation of chemical bonds, it is now important to understand the various ways in which atomic bonds can be broken. In general, there are three ways in which this process can be initiated. As shown in Scheme 1.4, the first is simple separation of a single bond where one electron from the bond resides on one atom and the other electron resides on the other atom. This type of bond cleavage is known as **homolytic cleavage** because the electron density is equally shared between the separate fragments and no charged species are generated. It is this process that leads to **free radical** reactions.

Unlike homolytic cleavage, **heterolytic cleavage** (Scheme 1.5) of a chemical bond results in one species retaining both electrons from the bond and one species retaining no electrons from the bond. In general, this also results in the formation of **ionic** species where

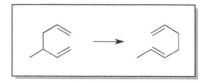

Scheme 1.4   *Illustration of homolytic cleavage.*

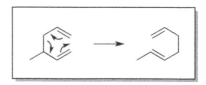

Scheme 1.5   *Illustration of heterolytic cleavage.*

Scheme 1.6   *Illustration of a concerted reaction (Cope rearrangement).*

Scheme 1.7   *Illustration of arrow-pushing applied to the Cope rearrangement.*

the fragment retaining the electrons from the bond becomes **negatively charged** while the other fragment becomes **positively charged**. These charged species then become available to participate in ion-based transformations governed by the electronic nature of reactants or adjacent functional groups.

Having introduced homolytic cleavage and heterolytic cleavage as the first two ways in which bonds are broken at the initiation of organic reactions, attention must be drawn to the possibility that bonds can rearrange into lower energy configurations through **concerted mechanisms** where bonds are simultaneously broken and formed. This third process, associated with **pericyclic reactions**, is illustrated in Scheme 1.6 using the **Cope rearrangement** and does not involve **free radicals** or **ions** as intermediates. Instead, it relies on the overlap of **atomic orbitals**, thus allowing the transfer of electron density that drives the conversion from starting material to product. Regardless, whether reactions rely on **free radicals**, **ions**, or **concerted mechanisms**, all can be explained and/or predicted using the principles of **arrow-pushing**.

Arrow-pushing is a term used to define the process of using **arrows** to conceptually move **electrons** in order to describe the mechanistic steps involved in the transition of **starting materials** to **products**. An example of **arrow-pushing** is illustrated in Scheme 1.7 as applied to the **Cope rearrangement** introduced in Scheme 1.6. As the **Cope rearrangement** proceeds through a **concerted mechanism**, the movement of electrons is shown in a

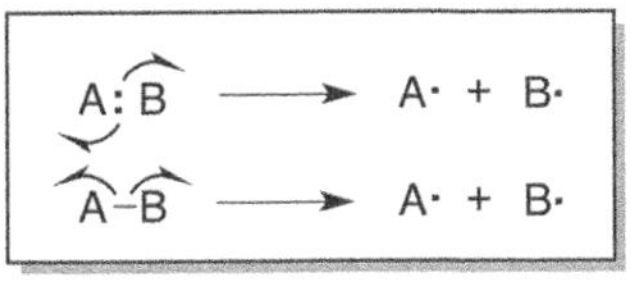

Scheme 1.8 *Application of arrow-pushing to homolytic cleavage using single-barbed arrows.*

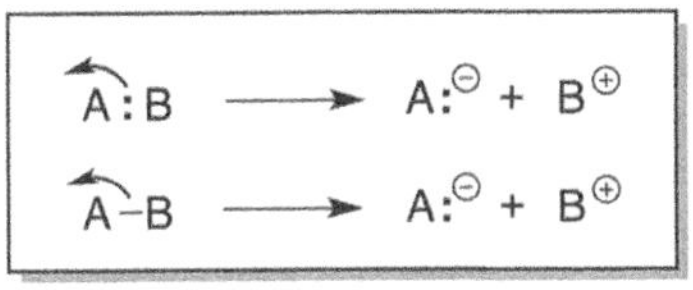

Scheme 1.9 *Application of arrow-pushing to heterolytic cleavage using double-barbed arrows.*

single step. As will become apparent, **arrow-pushing** is broadly useful to explain even very complex and multistep mechanisms. However, while **arrow-pushing** is useful to explain and describe diverse mechanistic types, it is important to note that different types of arrows are used depending on the type of bond cleavage involved in a given reaction. Specifically, when **homolytic cleavage** is involved in the reaction mechanism, **single-barbed arrows** are used to signify the movement of **single electrons**. Alternatively, when **heterolytic cleavage** or **concerted** steps are involved in the reaction mechanism, **double-barbed arrows** are used to signify the movement of **electron pairs**. Schemes 1.8 and 1.9 illustrate the use of appropriate arrows applied to **homolytic cleavage** and **heterolytic cleavage**, respectively.

1.2 FUNCTIONAL GROUPS

Having presented the concept of **arrow-pushing** in context of the steps that initiate chemical reactions, some factors impacting the flow of electrons leading from starting materials to products can now be explored.

As a general rule, **electrons** will flow from atomic centers **high in electron density** to atomic centers **low in electron density**. This dependence on **polarity** is similar to the way that electricity flows in an electrical circuit. If there is no difference in **electrical potential** between the ends of a wire, electricity will not flow. However, if a **charge** is applied to one end of the wire then the wire becomes **polarized** and electricity flows. If we imagine a simple **hydrocarbon** such as ethane, we can analogously relate this system to a **non-polarized wire**. Both carbon atoms possess the same density of electrons and thus ethane has no polarity. However, if functionality is added to ethane through the introduction of groups bearing **heteroatoms**, the **polarity** changes and electron flow can be used to induce chemical reactions. These heteroatom-bearing groups are known as **functional groups** and serve to donate or withdraw electron density.

While **functional groups** can be either **electron donating** or **electron withdrawing**, these properties rely upon the specific heteroatoms the functional group is composed of and the configuration of these heteroatoms relative to one another. With respect to the specific heteroatoms, **electronegativity** of the heteroatoms is the driving force influencing **polarity**. Thus, the more **electronegative** the atom, the greater the affinity the atom has for electrons. As a calibration for **electronegativity**, the **Periodic Table of the Elements** serves as an excellent resource. Specifically, moving from left to right and from bottom to

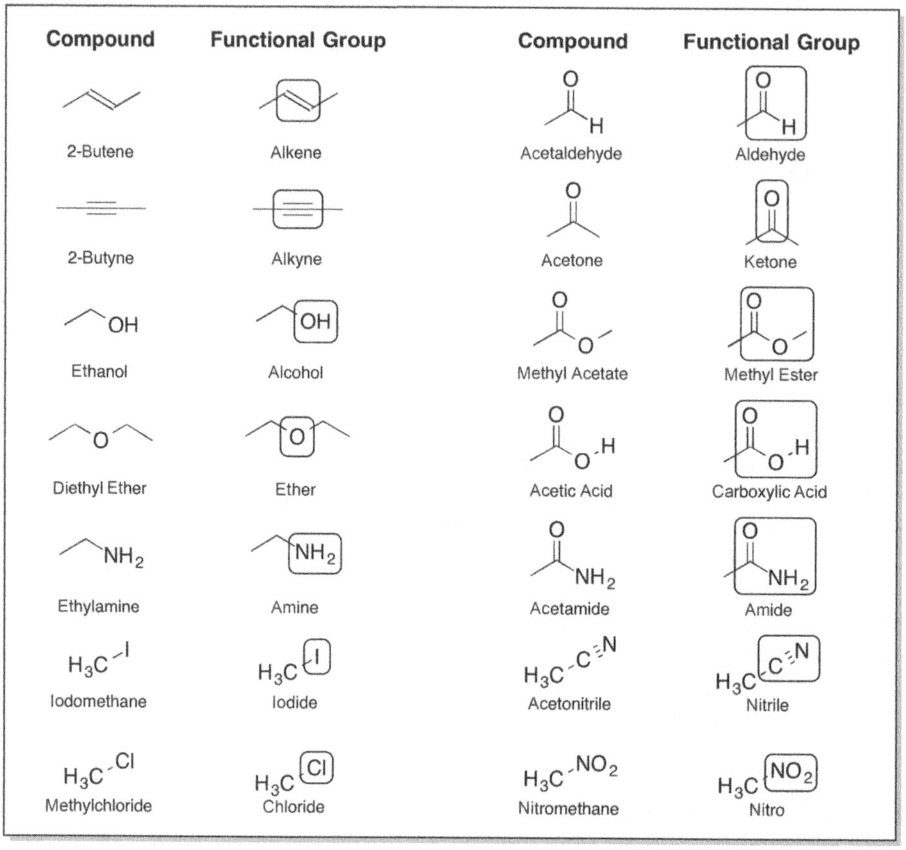

Figure 1.3 *Common organic functional groups.*

top, electronegativity increases. For example, nitrogen is more electronegative than carbon, and oxygen is more electronegative than nitrogen. Likewise, fluorine is more electronegative than chlorine, and chlorine is more electronegative than bromine. It is important to note that the influence of **electronegativity** on **polarity** is so strong that simply replacing a carbon atom with a **heteroatom** is enough to impart strong changes in **polarity** compared to the parent structure. Figure 1.3 illustrates common organic **functional groups** as components of common organic molecules.

Polarity in organic molecules is generally represented as **partial positive** (δ^+) charges and **partial negative** (δ^-) charges. These **partial charges** are induced based on the presence of **heteroatoms** either by themselves or in groups. These **heteroatoms**, as described in the previous paragraph and in Figure 1.3, define the various **functional groups**. Returning to the example of ethane as a nonpolar parent, Figure 1.4 illustrates how polarity changes as influenced by the introduction of **heteroatoms** and **functional groups**. As shown, **heteroatoms** such as nitrogen, oxygen, and halogens, due to their increased **electronegativities** compared to carbon, adopt **partial negative charges**. This causes **adjacent carbon atoms** to take on **partial positive** characteristics. As illustrated in Figure 1.4, charges on carbon atoms are not limited to positive. In fact, when a carbon atom is adjacent to a positive or partial positive center, it can adopt partial negative characteristics. This ability to control the charge characteristics of carbon atoms leads to the ability to create reactive centers with a

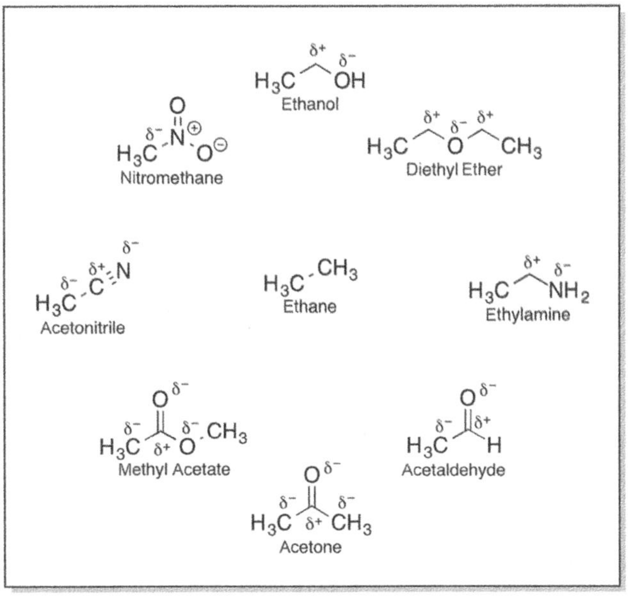

Figure 1.4 *How functional groups influence polarity.*

diverse array of properties. By taking advantage of this phenomenon of **induced polarity**, we are able to employ a multitude of chemical transformations, allowing for the creation of exotic and useful substances relevant to fields ranging from material science to food science to agriculture to pharmaceuticals.

1.3 NUCLEOPHILES AND LEAVING GROUPS

As discussed in Section 1.2, **polarity** is the key to the ability to initiate most **chemical reactions**. However, this is not the only factor influencing the ability to initiate reactions. In fact, the type of reaction on a given molecule is often dependent on the nature of the **solvent** and the **reagents** used. For example, solvent **polarity** can influence the **reaction rate** and the **reaction mechanism**. Furthermore, the nature of the chemical **reagents** used can affect the **reaction mechanism** and the identity of the final product. The following definitions will be key to understanding the terminology used in the following chapters.

Nucleophiles are reagents that have an affinity for **positively charged species** or **electrophiles**. In organic reactions, **nucleophiles** form **chemical bonds** at sites of **partial positive charge** through donation of their electrons. This generally results in the need for the starting compound to release a **leaving group**. An example of a **nucleophilic reaction** is shown in Scheme 1.10 where Nu: represents the nucleophile and L: represents the leaving group. **Arrow-pushing** is used to illustrate the movement of the electron pairs.

Leaving groups are the components of chemical reactions which detach from the starting material. Referring to Scheme 1.10, the **leaving group**, L:, ends up separate from the product while the **nucleophile**, Nu:, becomes incorporated into the product. Furthermore, while an initial evaluation of the material covered in an introductory organic chemistry course may seem overwhelming, the majority of the material covered can be reduced to the principles illustrated in the single reaction shown in Scheme 1.10.

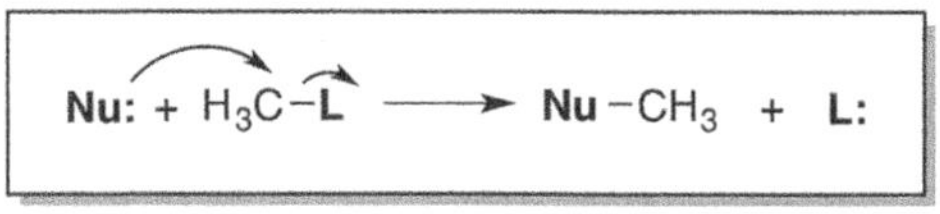

Scheme 1.10 *Example of a nucleophilic reaction.*

1.4 SUMMARY

In this chapter, the basic principle of **arrow-pushing** was introduced in the context of organic reactions driven by **homolytic cleavage, heterolytic cleavage,** or **concerted mechanisms.** Furthermore, the concept of **polarity** was introduced using **heteroatoms** and common organic **functional groups.** This discussion led to the definitions of **nucleophiles** and **leaving groups** in the context of simple **nucleophilic reactions.** Finally, by pulling these ideas together, the concept of approaching the study of mechanistic organic chemistry from a simplified perspective of understanding the principles of **arrow-pushing** was introduced.

While characteristics such as **homolytic cleavage, heterolytic cleavage,** and **concerted mechanisms** were discussed, the principles of **arrow-pushing** apply equally to all. However, with respect to **heterolytic cleavage,** an understanding of the properties of organic acids and bases is essential in order to understand underlying organic mechanisms. These concepts are introduced in Chapters 3 and 4.

PROBLEMS

1. Add arrow-pushing to explain the following reactions:

   a. $N\equiv C^{\ominus}$ + H_3C-I ⟶ $N\equiv C-CH_3$ + $I^{\ominus}$

   b.

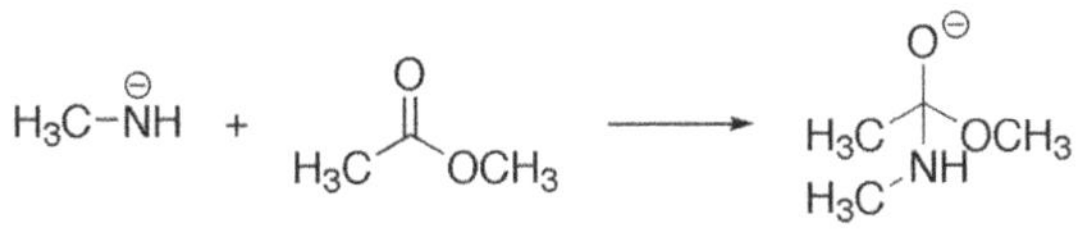

   c.

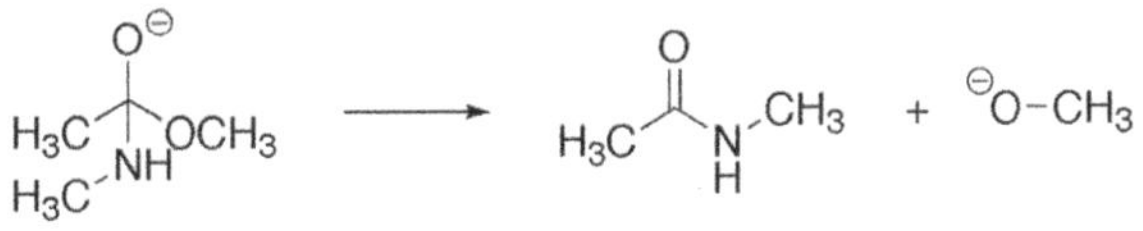

d. $H_3C-\overset{\cdot\cdot}{N}H_2$ + $H_3C\diagup\diagdown Cl$ $\longrightarrow$ $H_3C\diagup\diagdown\overset{H_2}{\underset{\oplus}{N}}-CH_3$ + $Cl^\ominus$

e. $H_3C-\overset{O}{\overset{\|}{C}}-\overset{\ominus}{C}H_2$ + $\overset{O}{\overset{\|}{H}}\diagup C-CH_3$ $\longrightarrow$ $H_3C-\overset{O}{\overset{\|}{C}}-\overset{}{\underset{H_2}{C}}-\overset{\overset{O^\ominus}{|}}{\underset{H}{C}}-CH_3$

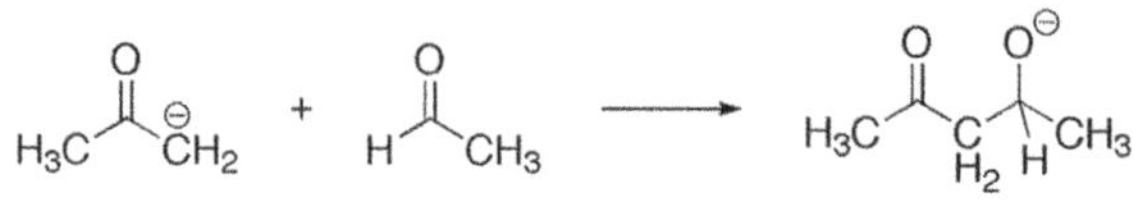

f. $\overset{H_3C\quad CH_3}{\underset{H_3C}{\diagdown}\overset{|}{C}\diagup}\overset{}{\underset{:\overset{\cdot\cdot}{O}:}{}}\diagdown^H$ + $H^\oplus$ $\longrightarrow$ $\overset{H_3C\quad CH_3}{\underset{H_3C}{\diagdown}\overset{|}{C}\diagup}\overset{}{\underset{:\overset{\oplus}{O}}{}}\overset{\diagdown^H}{\diagdown_H}$

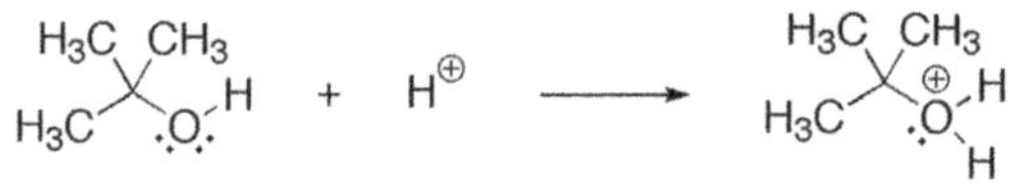

g.

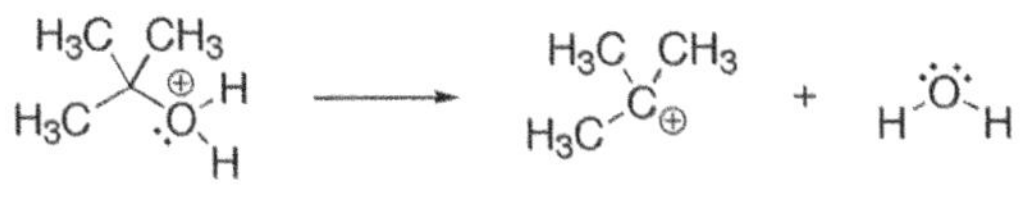

h.

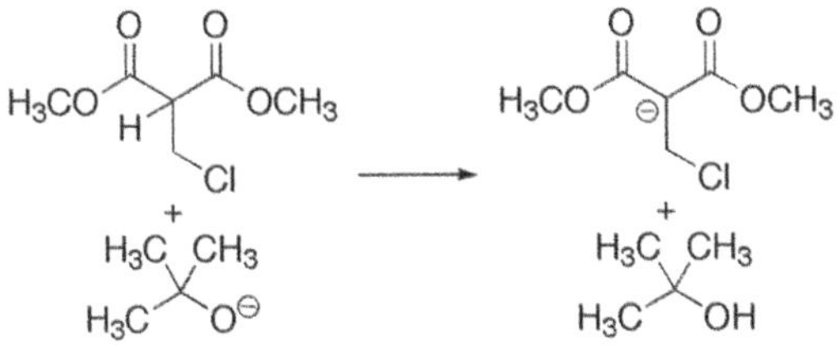

i.

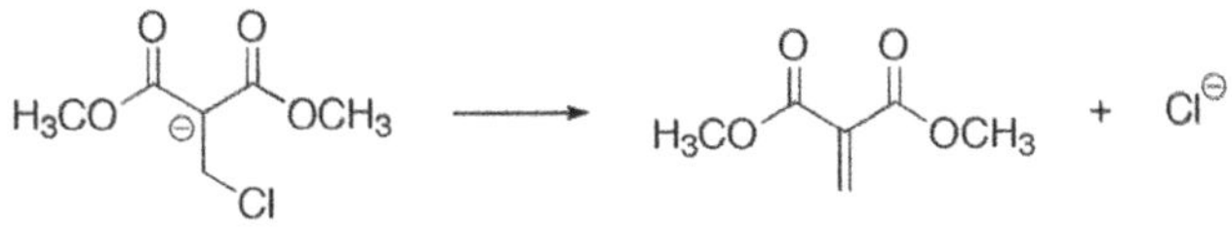

j. Br—Br ⟶ Br· + Br·

k. 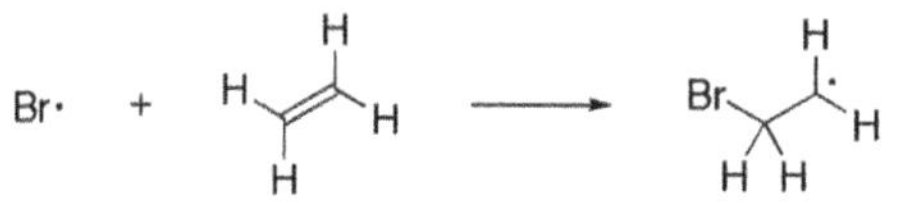 Br· + [ethene] ⟶ [BrCH₂CH₂·]

l. 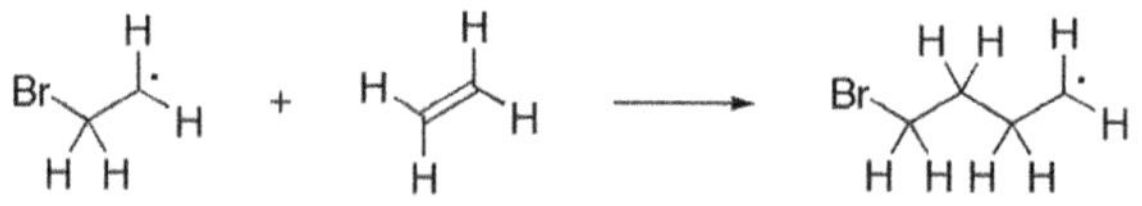 [BrCH₂CH₂·] + [ethene] ⟶ [BrCH₂CH₂CH₂CH₂·]

m.

n.

2. Place the partial charges on the following molecules:

a.

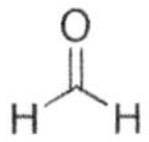

b.

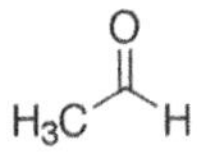

c.

d.

e.

f.

g.

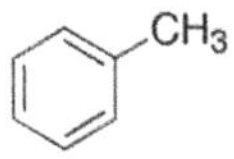

h.

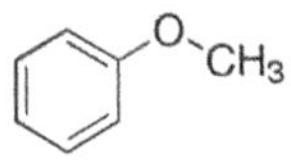

i.

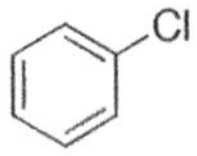

j.

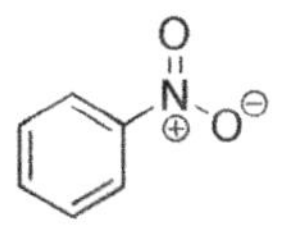

k.

l.

m.

n.

o.

p.

q.

r.

Chapter 2

Free Radicals

Imagine two balls connected by a rubber tether. If the balls are pulled apart, strain is introduced to the stretched tether. With enough energy, the balls can be completely separated and the rubber tether breaks. This frequently used analogy conceptually describes the nature of a bond joining together two atoms. When **bonds** between atoms are broken and created in specific sequences, **chemical reactions** become defined. As introduced in Chapter 1, **atomic bonds** can change as a result of **homolytic cleavage, heterolytic cleavage**, and **concerted** reaction mechanisms. However, the chemistry of **free radicals** depends upon **homolytic cleavage**. In this chapter, the chemistry of **free radicals** is discussed in the context of their formation and associated **chemical reactions**.

2.1 WHAT ARE FREE RADICALS?

Consider the tethered ball analogy discussed earlier. As shown in Figure 2.1, there is no strain on the tether at rest. However, when the balls are pulled apart, significant strain is placed on the tether. With enough strain, this tether can either separate from one of the balls or break evenly. As illustrated in Figure 2.2, if one ball separates from the tether, the model represents **heterolytic cleavage** where the tether represents the bonding pair of electrons. However, if the tether breaks in the middle, the model represents **homolytic cleavage** where one electron remains with each of the formerly bound atoms. As shown in Figure 2.3, the broken tether model can be translated to atoms where the full tether represents an **electron pair** and the tether fragment represents a **single electron**. When **heterolytic cleavage** occurs, the result is the formation of a **cation** and an **anion**. However, **free radicals** are the result of **homolytic cleavage** where an atom (either single or part of a larger molecule) contains a single **unpaired electron**. Finally, while Figure 2.3 may imply that free radicals

Arrow-Pushing in Organic Chemistry: An Easy Approach to Understanding Reaction Mechanisms,
Second Edition. Daniel E. Levy.

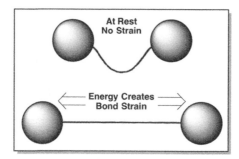

Figure 2.1 *Tethered ball model for bond strain.*

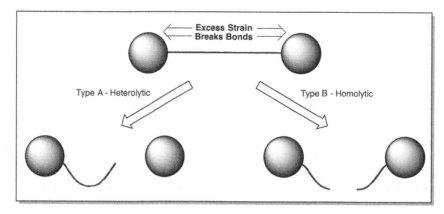

Figure 2.2 *Tethered ball model for breaking bonds.*

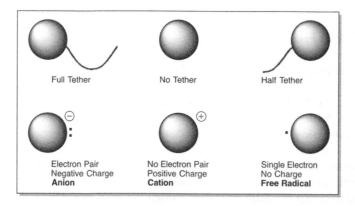

Figure 2.3 *Translation of tethered ball model to anions, cations, and free radicals.*

have no charge, this is not always the case. Section 2.2 introduces the concepts of **radical anions** and **radical cations**.

Free radicals can be represented using the same type of **electron dot structures** (**Lewis structures**) or lines that are used to represent atomic bonds and/or **lone pairs** of electrons (see Figs. 1.1 and 1.2). However, the location of the **free radical** is represented by a single **electron dot**. For simplicity, **free radicals** are generally represented in an

Name	Electron Dots	Abbreviated
Hydrogen atom	H·	H·
Fluorine atom	:F̤·	F·
Chlorine atom	:C̤l·	Cl·
Bromine atom	:B̤r·	Br·
Methyl radical	H:C̤· (with H above and H below)	H₃C·
Trifluoromethyl radical	:F̤: :F̤:C̤· :F̤:	F₃C·
Methoxy radical	H:C̤:Ö· (with H above and H below)	H₃CO·
Nitric oxide	·N̈::Ö:	·NO
(2,2,6,6-Tetramethylpiperidin-1-yl)oxyl Radical TEMPO		

Figure 2.4 *Common free radicals.*

abbreviated form where only the **radical** (and sometimes an adjacent lone pair) is specified. Common **free radicals**, along with their **Lewis dot** and abbreviated structures, are shown in Figure 2.4.

2.2 HOW ARE FREE RADICALS FORMED?

As stated earlier, **free radicals** are the result of **homolytic cleavage** of a single bond joining two atoms. Referring to the tethered ball model (Figs. 2.1 and 2.2), such bond cleavage is the result of energy inducing strain on the bond as the distance between the atoms increases. Most commonly, **thermal** and **photolytic** methods are used to induce **homolytic cleavage** for the formation of **free radicals**. In both cases, **free radical initiators** are typically required, and the **free radicals** formed have no charge (are neutral).

In addition to the use of **free radical initiators**, **electron transfer** methods can be used to generate **free radicals**. Under these conditions, the **free radicals** formed are **ionic** in nature. Specifically, if a single electron is added to a molecule, the newly formed free radical takes on a **negative charge**. Conversely, if a single electron is removed from a molecule, the newly formed free radical takes on a **positive charge**. Negatively charged free radicals are called **radical anions**, and positively charged free radicals are called **radical cations**. Figure 2.5 illustrates **electron transfer** in the formation of **radical ions**.

$$M \xrightarrow{\ +\,e^-\ } M^{\bullet -}$$

$$M \xrightarrow{\ -\,e^-\ } M^{\bullet +}$$

Figure 2.5 *Formation of radical ions via electron transfer.*

2.2.1 Free Radical Initiators

A **free radical initiator** is a chemical containing a single bond that is susceptible to **homolytic cleavage** on exposure to heat (**thermal conditions**) or light (**photolytic conditions**). One such reagent is *N*-**bromosuccinimide**. As illustrated in Scheme 2.1, *N*-**bromo-succinimide** undergoes **homolytic cleavage** forming a **bromine atom** and a **succinimide radical**. The **homolytic cleavage** mechanism is illustrated using **single-barbed arrows** to represent the movement of single electrons (see Scheme 1.8). Common **free radical initiators** are illustrated in Figure 2.6.

Free radical initiators function because they contain **weak bonds** that can be **homolytically cleaved** under relatively mild conditions. Consequently, most **free radical initiators** are unstable, thus requiring storage at low temperatures and protection from light.

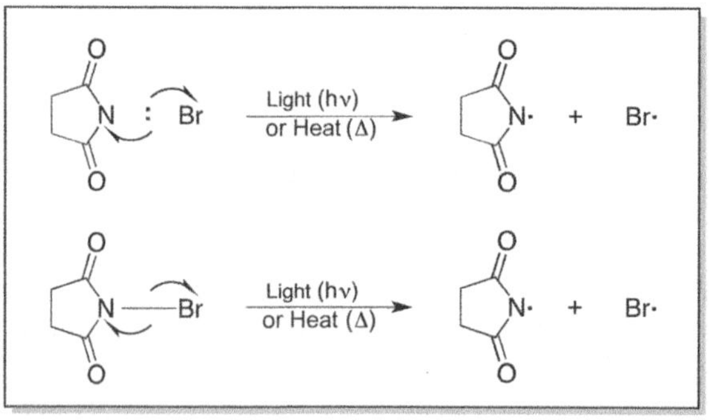

Scheme 2.1 *Homolytic cleavage of N-bromosuccinimide.*

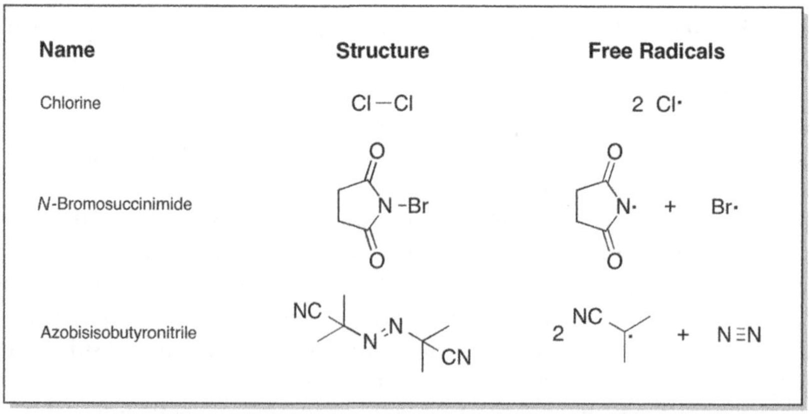

Figure 2.6 *Common free radical initiators.*

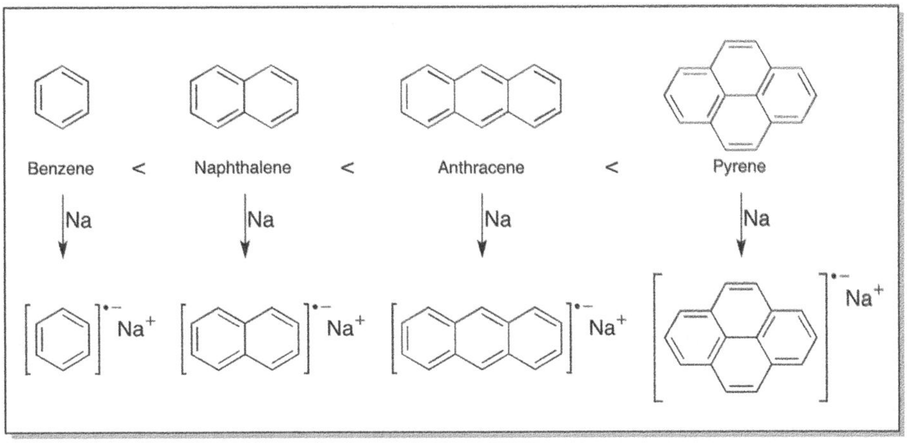

Figure 2.7 *Conjugated aromatic ring systems form radical anions more readily.*

However, it is this same instability that renders **free radical initiators** so valuable in many chemical processes. If a given reaction requires **homolytic cleavage**, the reaction is much more efficient if the initial **homolytic cleavage** is a low energy event.

2.2.2 Electron Transfer

Electron transfer reactions either add or remove electrons from a molecular system. If an electron is added to a molecule, the result is a **reduction** reaction. If an electron is removed from a molecule, the result is an **oxidation** reaction. Some **reductions** forming **free radicals** are the result of treating **aromatic** ring systems with **alkali metals** (Li, Na, and K). As illustrated in Figure 2.7, the ease of **radical anion** formation is related to the extent of **conjugation** in the **aromatic** ring system. For example, **benzene** will react with **sodium metal** more slowly than **naphthalene**. Finally, it should be noted that while the transfer of an electron from an **alkali metal** to an **aromatic** molecule **reduces** the **aromatic** molecule to a **radical anion**, the **alkali metal** becomes **oxidized** forming a **cation**.

Oxidations forming **free radicals** are most relevant to **mass spectrometry** and to the chemistry of **conducting polymers**. Being beyond the scope of this book, **radical cations** will not be discussed further.

2.3 FREE RADICAL STABILITY

By nature, **free radicals** are highly reactive molecular species since they have a deficiency of electrons in the outer shell, thus disobeying the **octet rule**. This level of reactivity is due to **unpaired electrons** preferring to be paired. As illustrated in Scheme 2.2, this pairing of **free radicals** results in the formation of new **covalent bonds**. In situations where different **free radical** species are present, the result is a mixture of products. In the example illustrated in Scheme 2.2, a **methyl radical** and a **chlorine atom** combine to form a mixture of **chlorine**, **chloromethane**, and **ethane**. The pairing of the illustrated free radicals is shown using **arrow-pushing**.

Scheme 2.2 image and equation:

$$H_3C\cdot \; + \; Cl\cdot \; \longrightarrow \; \begin{bmatrix} Cl\cdot \curvearrowright \cdot Cl \\ H_3C\cdot \curvearrowright \cdot Cl \\ H_3C\cdot \curvearrowright \cdot CH_3 \end{bmatrix} \; \longrightarrow \; Cl-Cl \; + \; H_3C-Cl \; + \; H_3C-CH_3$$

Chlorine   Chloromethane   Ethane

Scheme 2.2 *Free radicals readily pair forming covalent bonds.*

Understanding that **free radicals** are highly reactive, there are factors that influence the stability of various **free radicals**. Among these factors is **conjugation**. **Conjugation** is the result of a sequence of two or more consecutive double (or triple) bonds allowing **electrons** to flow throughout a given molecule. Consider **graphite** and **diamond** (Fig. 2.8). **Graphite** is well known to conduct electricity while **diamond** is an excellent insulator. While both of these substances are made entirely of carbon, it is the conjugated nature of **graphite** that imparts its conductive properties. There are no double or triple bonds present in **diamond**.

Referring back to Figure 2.7 regarding the ease of formation of **free radicals**, we recognize that **free radicals** form more easily as the degree of **conjugation** increases. This trend directly relates to the **stability of free radicals**. Specifically, those **free radicals** that are more easily formed are more stable. Thus, we observe that the stability trend of the radical anions illustrated in Figure 2.7 is pyrene > anthracene > naphthalene > benzene.

In addition to **conjugation**, **free radicals** are stabilized by substitutions or **lone pairs** of electrons on adjacent atoms, since these groups **inductively donate** electron density to **electron-deficient sites**. As shown in Figure 2.9, this trend translates to a **tertiary free radical** having greater stability compared to a **secondary free radical**. Likewise, a **secondary free radical** has greater stability compared to a **primary free radical**. Finally, a **free radical** with an adjacent nitrogen or oxygen atom has greater stability compared to a **free radical** with an adjacent carbon atom due to **resonance stabilization**.

The stability trends illustrated in Figure 2.9 are explained by an effect known as **hyperconjugation**. As illustrated in Figure 2.10, **hyperconjugation** relates to the ability of a

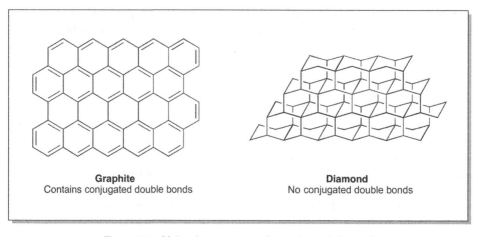

Graphite
Contains conjugated double bonds

Diamond
No conjugated double bonds

Figure 2.8 *Molecular structures of graphite and diamond.*

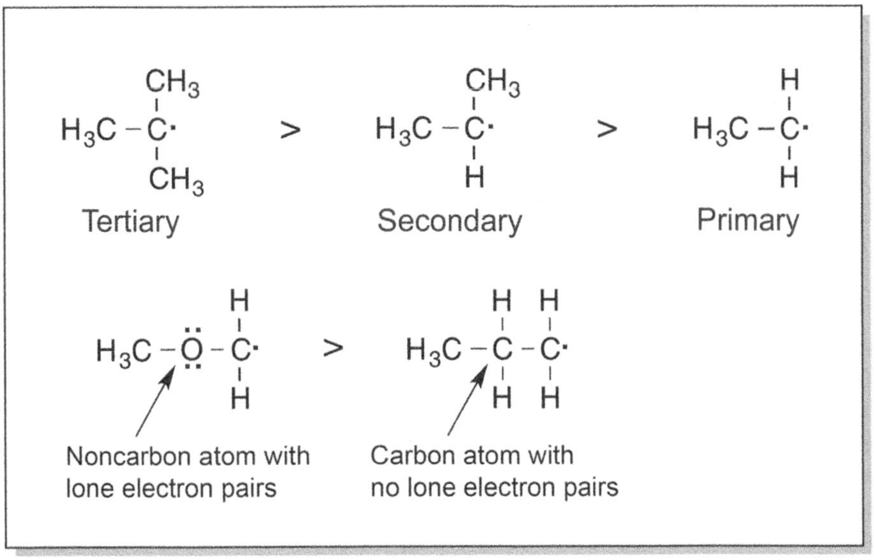

$$H_3C - \overset{\overset{\displaystyle CH_3}{|}}{\underset{\underset{\displaystyle CH_3}{|}}{C}} \cdot \quad > \quad H_3C - \overset{\overset{\displaystyle CH_3}{|}}{\underset{\underset{\displaystyle H}{|}}{C}} \cdot \quad > \quad H_3C - \overset{\overset{\displaystyle H}{|}}{\underset{\underset{\displaystyle H}{|}}{C}} \cdot$$

Tertiary   Secondary   Primary

$$H_3C - \overset{..}{\underset{..}{O}} - \overset{\overset{\displaystyle H}{|}}{\underset{\underset{\displaystyle H}{|}}{C}} \cdot \quad > \quad H_3C - \overset{\overset{\displaystyle H}{|}}{\underset{\underset{\displaystyle H}{|}}{C}} - \overset{\overset{\displaystyle H}{|}}{\underset{\underset{\displaystyle H}{|}}{C}} \cdot$$

Noncarbon atom with   Carbon atom with
lone electron pairs   no lone electron pairs

Figure 2.9 *Order of free radical stability.*

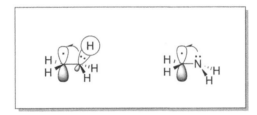

Figure 2.10 *Hydrogen atom s orbitals can donate electron density to adjacent centers of electron deficiency as can heteroatoms bearing lone electron pairs.*

hydrogen atom to donate **electron density** from its *s*-**orbital** to sites of neighboring **electron deficiency**. This same effect is observed when **lone pairs** of electrons are present and adjacent to sites of **electron deficiency**. By recognizing the trends of **free radical stability**, the outcomes of free radical reactions can be predicted.

2.4   WHAT TYPES OF REACTIONS INVOLVE FREE RADICALS?

While **free radicals** are highly reactive species, their reactive properties can be utilized in a variety of useful chemical transformations. Among these are introduction of halogens to organic molecules, formation of polymers, and oxidation/reduction processes. These examples, presented in the following sections (Sections 2.4.1–2.4.3), are not intended to be exhaustive relative to the breadth of available **free radical** reactions. They are, however, intended to provide mechanistic insight into how single electrons flow in **free radical-** dependent **reaction mechanisms**.

2.4.1 Halogenation Reactions

Halogenation reactions are among the most important reactions in organic chemistry because of their utility in adding reactive functionality to unreactive centers. Scheme 2.3 illustrates various **bromination** reactions applied to **alkanes**, **allylic/benzylic** compounds, and **carboxylic acids**. It is important to note that while **bromination** reactions are illustrated, similar reactions are available for the formation of **chlorides** and **iodides**. Finally, as is discussed throughout this book, there are many reactions that generate similar products through different mechanisms. Regarding Scheme 2.3, the illustrated **alkyl bromination**, **allylic bromination**, and **benzylic bromination** all proceed via **free radical** mechanisms. The **bromination** of **carboxylic acids** utilizing the **Hell–Volhard–Zelinsky** reaction does not involve **free radicals** and is included solely to illustrate the variety of nonreactive sites that can be activated utilizing appropriate **halogenation** methodologies.

    Free radical bromination of **alkanes** proceeds when an **alkane** is treated with **bromine** under conditions appropriate for the **homolytic** cleavage of **bromine**. In general, **homolytic cleavage** is initiated utilizing either heat or light. Once **bromine** undergoes **homolysis** forming two **bromine atoms**, a **bromine atom** extracts a **hydrogen atom** from the **alkane** forming **hydrogen bromide** and an **alkyl radical**. The **alkyl radical** then pairs with a **bromine atom** forming an **alkyl bromide**. Applied to **methane**, this reaction is illustrated in Scheme 2.4 utilizing **arrow-pushing**. It is important to note that this reaction will form a mixture of products as described in Scheme 2.2.

    Allylic and **benzylic halogenation** reactions are very similar in that the initially formed **allylic** or **benzylic free radical** is stabilized due to **conjugation**. As shown in Scheme 2.5,

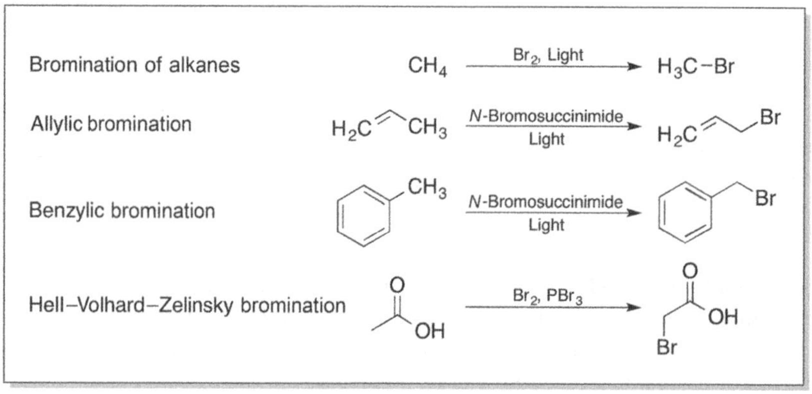

Scheme 2.3   *Examples of bromination reactions.*

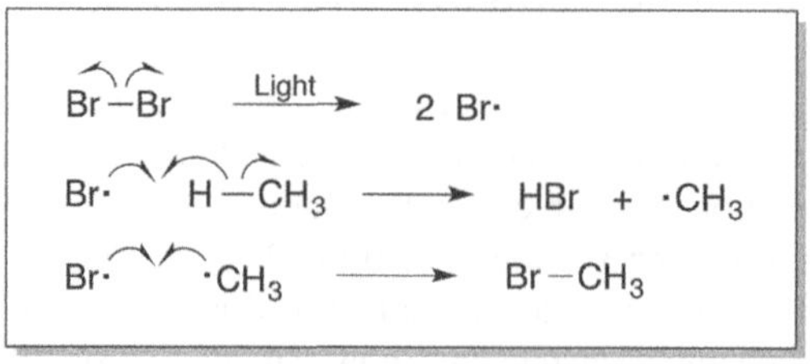

Scheme 2.4   *Arrow-pushing mechanism for bromination of methane.*

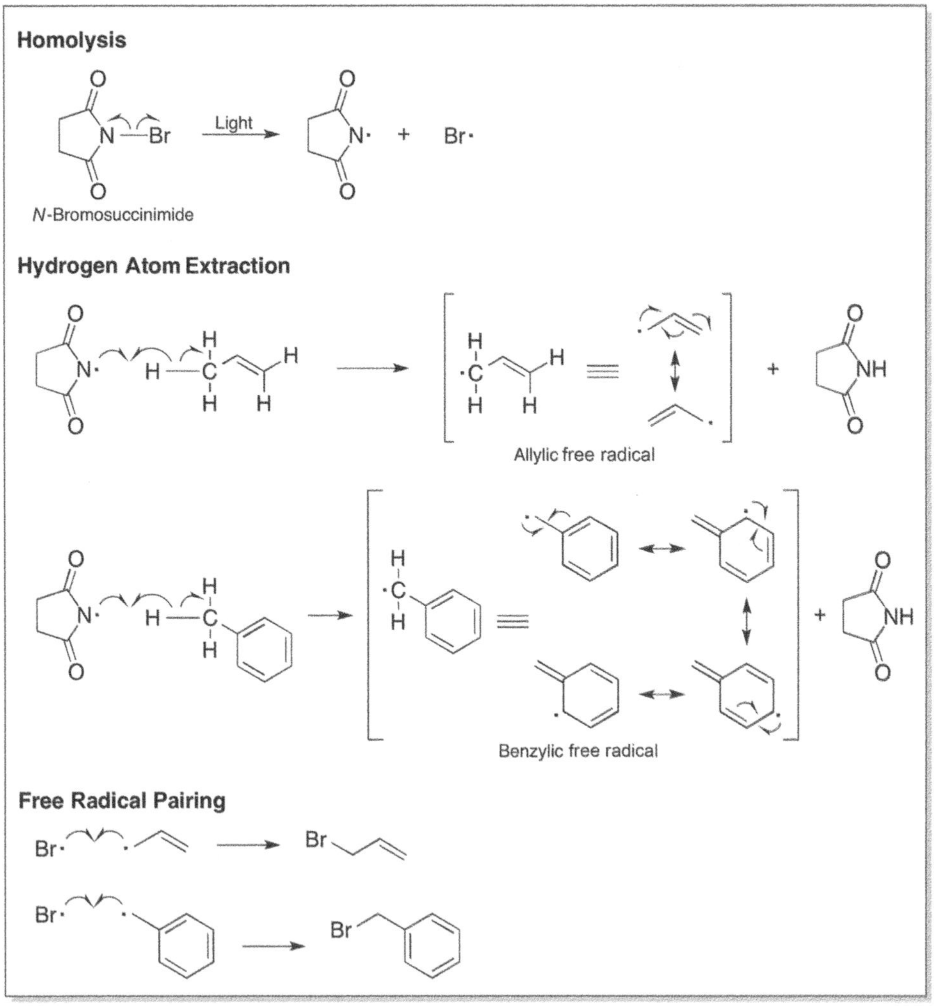

Homolysis

N-Bromosuccinimide

Hydrogen Atom Extraction

Allylic free radical

Benzylic free radical

Free Radical Pairing

Scheme 2.5 *Allylic and benzylic bromination (halogenation).*

when *N*-bromosuccinimide is exposed to light, **homolysis** results in formation of a **succinimide free radical** and a **bromine atom**. The **succinimide free radical** then extracts a **hydrogen atom** from the **allylic** or **benzylic** position forming the stabilized **allylic** or **benzylic free radical**. Once formed, the **allylic** or **benzylic free radical** pairs with a **bromine atom** forming the **bromination** product.

Conjugation refers to the **delocalization** of **electrons** onto different carbon atoms. This effect is illustrated in Scheme 2.5 using **arrow-pushing** to show the movement of electrons and **double-headed arrows** to show the resulting **resonance structures**. The **free radical resonance structures** for **butene** and **toluene**, derived as illustrated in Scheme 2.5 for **propene** and **toluene**, are shown in Scheme 2.6. These **resonance structures** are shown pairing with a **bromine atom** along with their respective products. As illustrated, in the case of an **allylic free radical**, **conjugation** creates the possibility of multiple products forming during the reaction when the **allyl group** is part of a larger structure. This is not

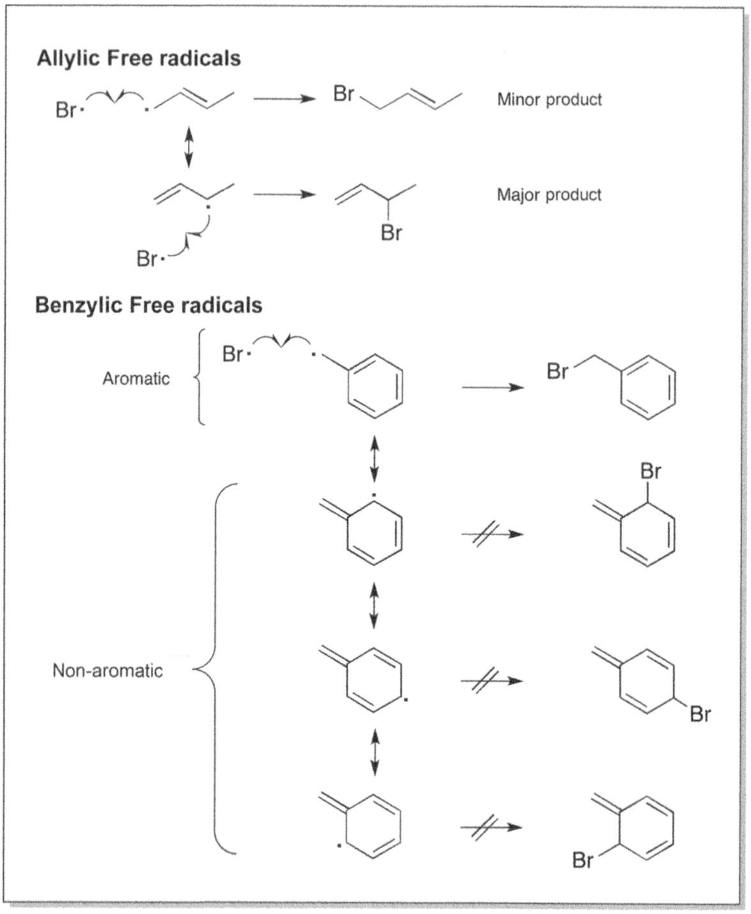

Allylic Free radicals

Minor product

Major product

Benzylic Free radicals

Aromatic

Non-aromatic

Scheme 2.6 *Free radicals stabilized by conjugation can form multiple products.*

the case with **benzylic halogenation** reactions due to the interruption of **aromaticity** should **halogenation** occur on any of the ring carbon atoms. Finally, it is important to remember that the major products of **free radical reactions** will reflect the stability of the **free radicals** formed during the reactions. This is illustrated in Scheme 2.6 where the major **bromination** product of **butene** is the **secondary bromide** while the minor product is the **primary bromide**. Reasons for this selectivity were presented in Figure 2.9 and explained by **hyperconjugation**.

2.4.2 Polymerization Reactions

Formation of **polymers** is the result of joining multiple small structures (**monomers**) together to form long linear or branched structures composed of identifiable repeating units. Common **organic polymers** found in everyday products, along with their respective **monomer** units, are shown in Figure 2.11.

While there are many methods used in the formation of **polymeric** substances, **free radical polymerization** is among the most common and productive processes. Such reactions

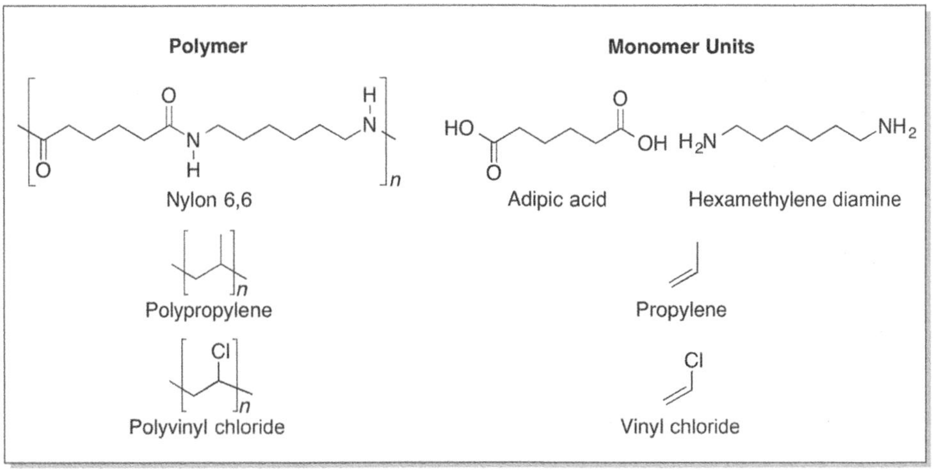

Figure 2.11 Common organic polymers.

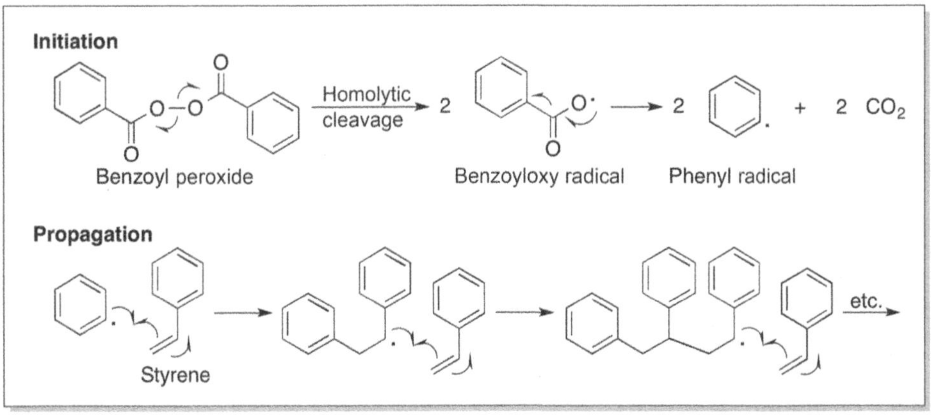

Scheme 2.7 Free radical formation of polystyrene.

typically involve combining a **monomeric** unit with a **free radical initiator**. In order to **propagate** the **polymerization** reaction, it is essential that the terminus of the growing **polymer** chain possesses a **free radical** after each reaction with a **monomer** molecule. The formation of **polystyrene** results from the **free radical polymerization** illustrated in Scheme 2.7 using **benzoyl peroxide** as the **free radical initiator**. As shown using **arrow-pushing**, **benzoyl peroxide** undergoes **homolytic cleavage** to form two **benzoyloxy radicals**, which then further degrade forming **carbon dioxide** and **phenyl radicals**. A **phenyl radical** then reacts with **styrene** initiating formation of the **polystyrene** chain. **Propagation** is the result of the terminal **free radical** of the growing chain reacting with subsequent **styrene** units. **Termination** of the **free radical polymerization** process occurs when two growing chains combine, an initiator phenyl radical reacts with the growing chain, or hydrogen is abstracted from one polymer chain to another (**radical disproportionation**). These **termination** reactions are illustrated in Scheme 2.8.

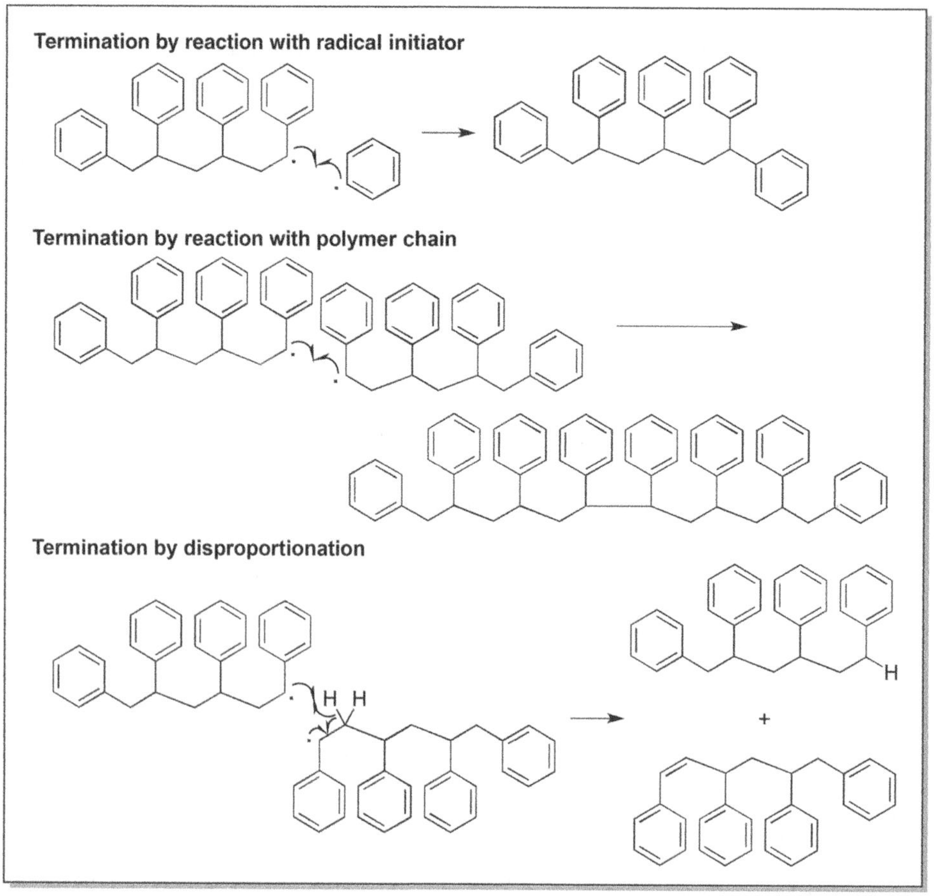

Termination by reaction with radical initiator

Termination by reaction with polymer chain

Termination by disproportionation

Scheme 2.8 *Termination of polystyrene free radical polymerization.*

2.4.3 Oxidation Reactions

The ability to alter the **oxidation state** of organic molecules in a controlled manner is among the most important tools available to organic chemists. Using this chemistry, **alcohols** can be oxidized to **aldehydes** and **ketones**. Additionally, **aldehydes** can be oxidized to **carboxylic acids**. Similarly, **sulfides** can be oxidized to **sulfoxides**, and **sulfoxides** can be oxidized to **sulfones**. Scheme 2.9 generally illustrates these oxidative functional group transformations.

While most **oxidation reactions** proceed under **heterolytic** reaction conditions, many are mediated by **free radicals**. Among the most important of these is the conversion of a **hydrocarbon radical** to a **peroxide** via reaction with **oxygen**. This "autoxidation" reaction is a critical process in initial **functionalization of alkanes** to form more functionalized and useful materials.

The **autoxidation** reaction mechanism is similar to that described for **free radical polymerizations**. As shown in Scheme 2.10, in the presence of a **radical initiator**, a **hydrocarbon** undergoes **hydrogen atom extraction** forming a **hydrocarbon radical**. The **hydrocarbon radical** then reacts with molecular **oxygen** forming a **peroxide radical**. The **peroxide radical** then propagates the **autoxidation** by **hydrogen atom extraction**

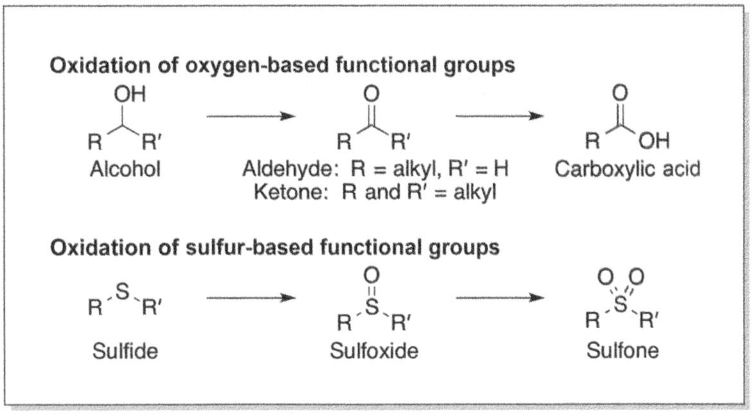

Scheme 2.9 *Examples of oxidative functional group transformations.*

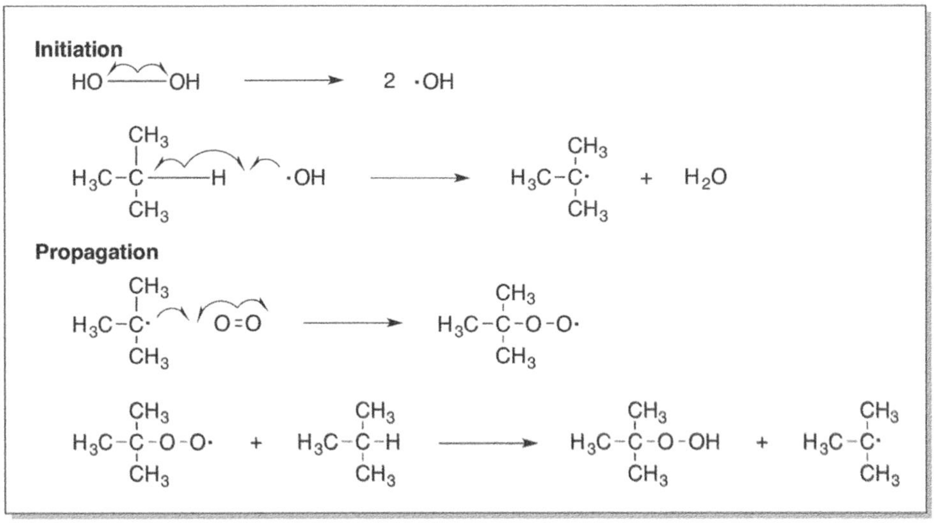

Scheme 2.10 *Mechanism for autoxidation of isobutane.*

from another **hydrocarbon** molecule. **Termination** of the **autoxidation** process proceeds similarly to the **termination** of all **free radical** reactions via **free radical combination** or **disproportionation**.

2.5 SUMMARY

In this chapter, **free radicals** were defined and methods of forming **free radicals** were introduced. Furthermore, factors related to **free radical stability** were explored. Reactions involving **free radicals** were illustrated through **homolytic** bond cleavage and the movement of single electrons utilizing **arrow-pushing** and **single-barbed arrows**. Through the information presented in this chapter, more advanced **free radical** reactions can be understood and their mechanisms can be predicted. The key, as will be apparent throughout this book, is to recognize the weakest bonds and to identify the atoms with the lowest **electron density**.

PROBLEMS

1. Arrange the free radicals in each group from highest to lowest stability:

a.

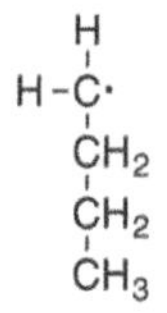

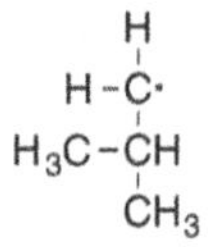

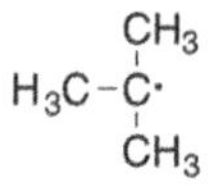

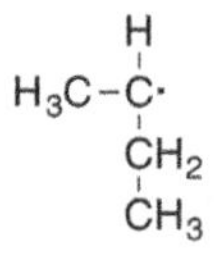

H–C·
|
CH₂
|
CH₂
|
CH₃

H–C·
|
H₃C–CH
|
CH₃

CH₃
|
H₃C–C·
|
CH₃

H
|
H₃C–C·
|
CH₂
|
CH₃

b.

CH₃
|
H₃C–C·
|
N–CH₃
|
CH₃

CH₃
|
H₃C–C·
|
O
|
CH₃

CH₃
|
H₃C–C·
|
CH₂
|
CH₃

c.

CH₃
|
H₃C–C·
|
CH₂
|
CH₃

CH₃
|
H₃C–C·

CH₃
|
H₃C–C·
|
O
|
CH₃

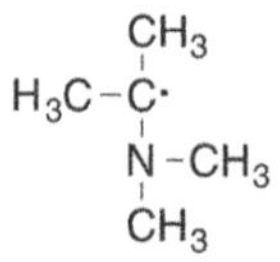

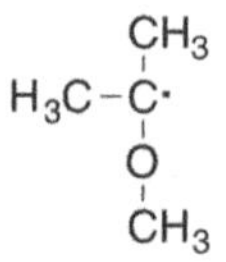

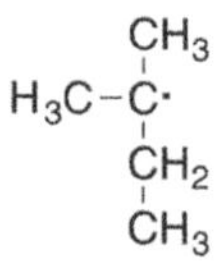

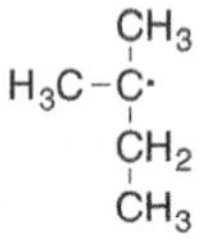

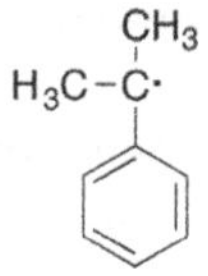

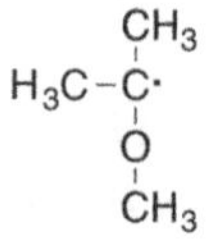

2. Figure 2.6 listed several common free radical initiators including azobisisobutyronitrile (AIBN). Using arrow-pushing, explain the mechanism of AIBN decomposition.

3. Using arrow-pushing, explain how AIBN as a free radical initiator assists reactions using tributyltin hydride—$(C_4H_9)_3SnH$.

4. Predict the major bromination product for each compound when reacted with *N*-bromo-succinimide. Explain your answers.

a.

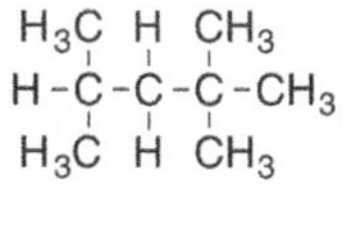

b.

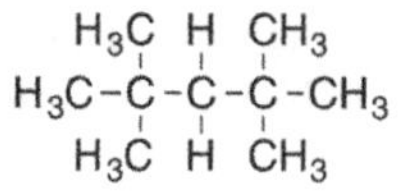

$$\text{H}_3\text{C}-\overset{\overset{\displaystyle \text{H}_3\text{C}}{|}}{\underset{\underset{\displaystyle \text{H}_3\text{C}}{|}}{\text{C}}}-\overset{\overset{\displaystyle \text{H}}{|}}{\underset{\underset{\displaystyle \text{H}}{|}}{\text{C}}}-\overset{\overset{\displaystyle \text{CH}_3}{|}}{\underset{\underset{\displaystyle \text{CH}_3}{|}}{\text{C}}}-\text{CH}_3$$

c.

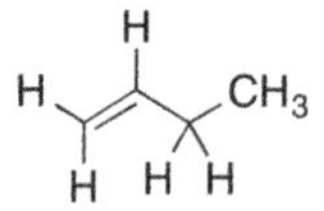

d.

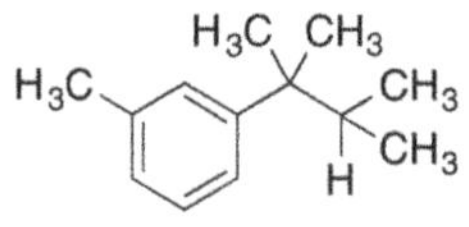

5. Predict the products resulting from free radical "degradation" of each structure. Show your work using arrow-pushing.

a.

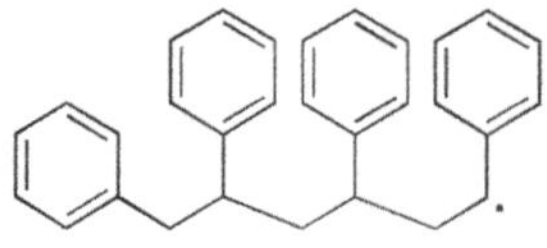

b.

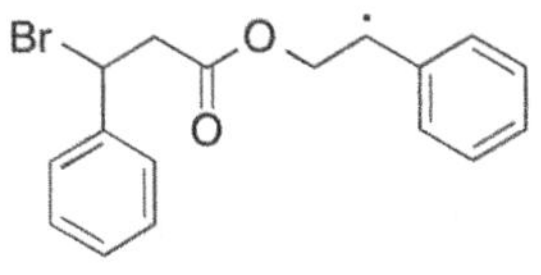

6. Sodium metal reacts with benzophenone to form a deep blue colored radical anion called a ketyl via an electron transfer process. Predict the structure of sodium benzophenone ketyl.

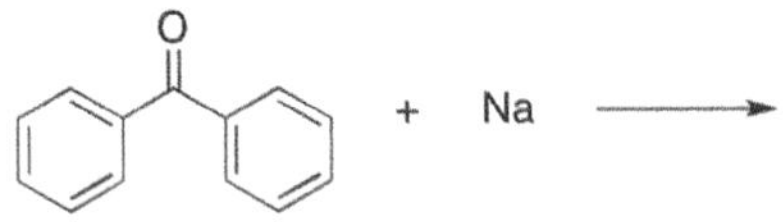

7. Most free radicals are highly reactive species with very short half-lives. Some free radicals are stable and can be stored for long periods of time without the concern of dimerization or degradation. Explain why the (2,2,6,6-tetramethylpiperidin-1-yl)oxyl radical (TEMPO) is stable.

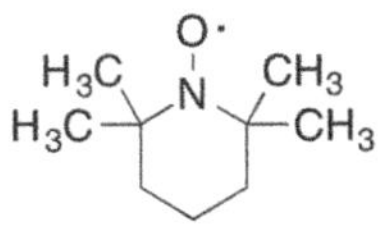

TEMPO

8. Chlorofluorocarbon-mediated ozone depletion is a significant environmental concern. The sequence begins with the photochemical formation of free radicals in the upper atmosphere and continues with reaction of the free radicals with ozone. Using arrow-pushing, show the mechanisms for each step associated with ozone depletion. Account for all electrons.

a. $Cl-CCl_2F \longrightarrow \cdot CCl_2F + Cl\cdot$

b.  $Cl\cdot + O:\overset{\oplus}{O}\cdot O^{\ominus} \longrightarrow Cl-O-\overset{\oplus}{\underset{\cdot}{O}}-O^{\ominus}$

c.  $Cl-O-\overset{\oplus}{\underset{\cdot}{O}}-O^{\ominus} \longrightarrow Cl-O\cdot + O{=}O$

d.  $Cl-O\cdot + O:\overset{\oplus}{O}\cdot O^{\ominus} \longrightarrow Cl\cdot + 2\ O{=}O$

Chapter *3*

Acids

As mentioned at the end of Chapter 1, an understanding of heterolytic reaction mechanisms must be accompanied by an understanding of the properties of organic acids and bases. Through this understanding, an ability to predict the reactive species in organic reactions and the reactive sites in organic molecules will evolve. Therefore, this chapter focuses on the properties of acids, the dissociation constants, and the relative acidities observed for protons in different environments.

3.1 WHAT ARE ACIDS?

The most general description of an **acid** is a molecule that liberates **hydrogen ions**. Therefore, if we consider a molecule, HA, this molecule is said to be an acid if it dissociates as shown in Scheme 3.1. It is important to note that any **acid dissociation** is an **equilibrium process**. Through this equilibrium process, two species, a **proton** (hydrogen **cation**) and an **anion**, are liberated. Furthermore, because this dissociation results in the formation of two ionic (charged) species, it is important to consider why this would be favorable as compared to the neutral state of undissociated HA. The answer to this question lies in the stability of the anion, A$^-$, itself.

Regarding **anionic stability**, there are many relevant factors. Among these are external influences such as **solvent effects** (Fig. 3.1). Specifically, a **polar solvent** has the ability to **stabilize** ionic species through **charge–charge interactions** or **charge–heteroatom interactions**. Conversely, a **non-polar solvent** generally **inhibits** formation of charged species because it cannot interact with the ions. Figure 3.2 lists common polar and non-polar organic solvents. While solvent **polarity** is an important factor in the progression and the rate of reactions, its role applied to **arrow-pushing** relates more to mechanistic determination

---

Arrow-Pushing in Organic Chemistry: An Easy Approach to Understanding Reaction Mechanisms,
Second Edition. Daniel E. Levy.

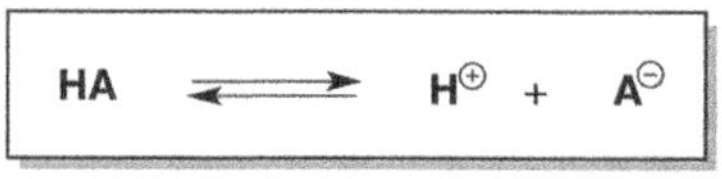

Scheme 3.1   *General representation of acid dissociation.*

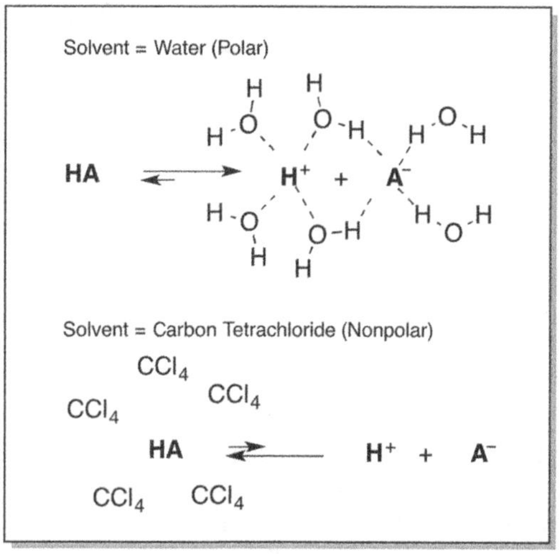

Figure 3.1   *Solvent effects on acid dissociation.*

than to how electrons move. Therefore, solvent **polarity** will not be addressed further in this chapter and will be revisited in the context of various mechanistic types.

In addition to external factors such as **solvent effects**, there are internal factors that influence anionic stability. Among these are **inductive effects** (how do electron-donating or electron-withdrawing substituents affect a molecule?) and **resonance effects** (is the charge localized or delocalized?). As **inductive effects** generally work in concert with **resonance effects**, our primary focus will be on the resonance effects themselves.

3.2   WHAT IS RESONANCE?

When a given molecule or ion can exist with multiple configurations of double/triple bonds or multiple sites bearing positive/negative charges, the molecule or ion is said to possess **resonance** forms. These **resonance** forms can be represented by drawings where the changes in electronic configuration are rationalized using **arrow-pushing**. Furthermore, these changes in electronic configuration occur with no alterations to the connectivity of the individual atoms. For example, as shown in Scheme 3.2, a **carboxylic acid** dissociates into a **proton** and a **carboxylate anion**. As shown in Scheme 3.3, this **carboxylate anion** possesses two **resonance** structures. These **resonance** structures, illustrated using a **double-headed arrow**, are easily explained using **arrow-pushing** to move the electrons associated with the negative charge from one oxygen atom to the other (Scheme 3.4).

Although **carboxylic acids** exist in **equilibrium** with their resonance-stabilized **carboxylate anions**, it is important to understand that resonance stabilization alone will not induce carboxylate anions to form. In fact, when resonance stabilization is not enough

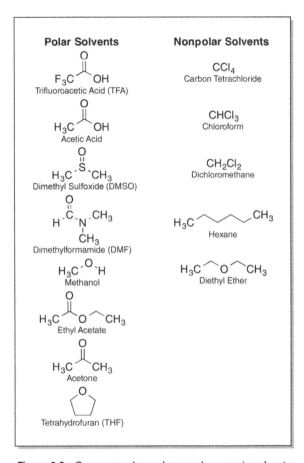

Figure 3.2   *Common polar and non-polar organic solvents.*

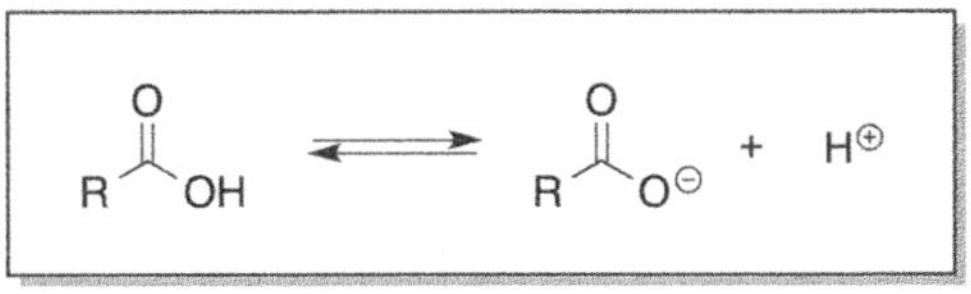

Scheme 3.2   *Dissociation of a carboxylic acid forming a proton and a carboxylate anion.*

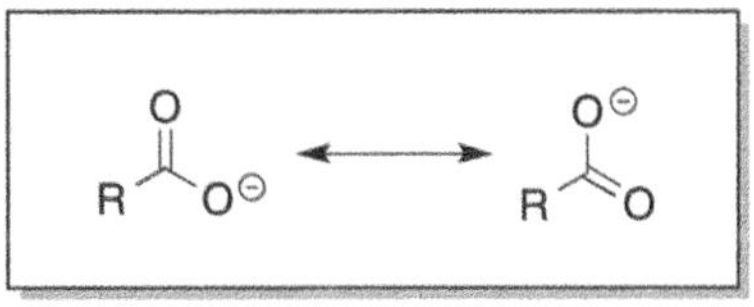

Scheme 3.3   *Resonance forms of the carboxylate anion.*

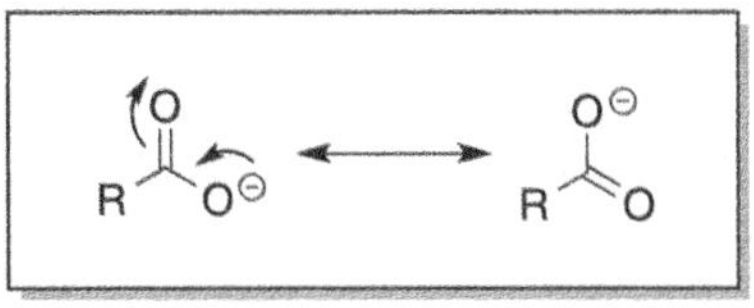

Scheme 3.4   *Rationalization of the carboxylate anion resonance forms using arrow-pushing.*

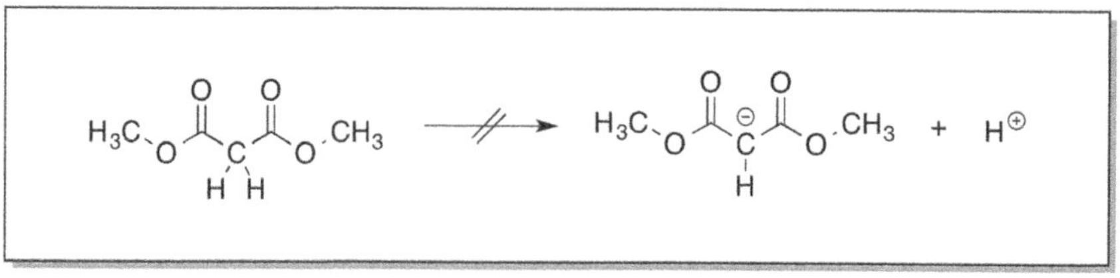

Scheme 3.5   *Dimethyl malonate does not spontaneously liberate malonate anions.*

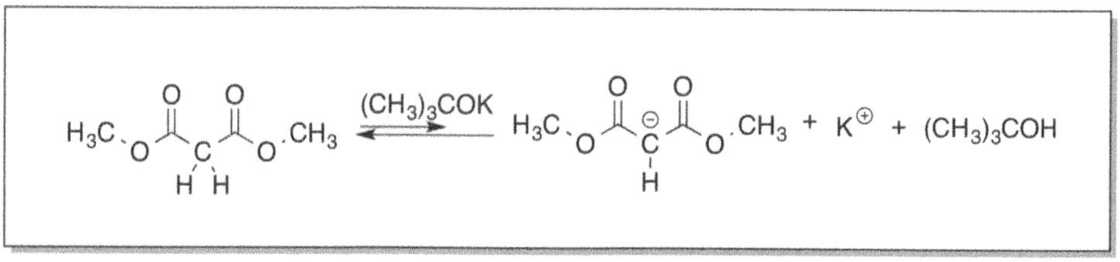

Scheme 3.6   *Potassium tert-butoxide partially deprotonates dimethyl malonate.*

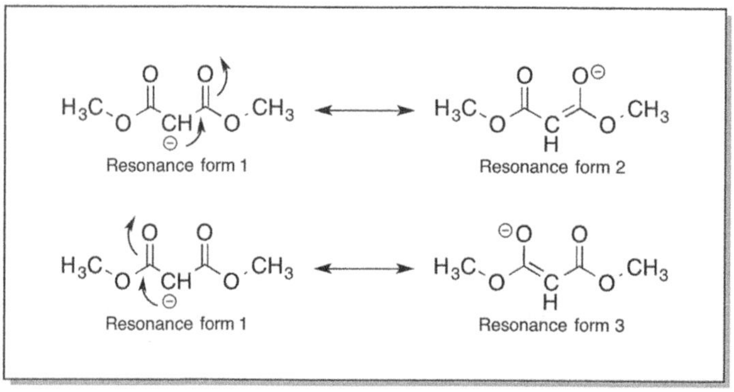

Scheme 3.7   *Resonance forms of the malonate anion rationalized using arrow-pushing.*

to induce formation of a carboxylate anion, addition of a base generally will accomplish this task. For example, considering **dimethyl malonate**, there is **no dissociation** of any **protons** liberating **malonate anions** (Scheme 3.5). The equilibrium lies entirely in favor of neutral dimethyl malonate. However, with the addition of a **base** such as **potassium *tert*-butoxide**, a proton is readily extracted generating malonate anions, potassium cations, and *tert*-butyl alcohol (Scheme 3.6). The three resonance forms of the malonate anion, described using **arrow-pushing**, are illustrated in Scheme 3.7. While **deprotonation** under these conditions does not proceed to completion, the **equilibrium** is such that malonate anions are available in sufficient quantities to react as required.

As illustrated in Schemes 3.2 and 3.5, different organic anions form under different conditions. Some, as illustrated in Scheme 3.2, form through spontaneous dissociation of an acid, while others, as illustrated in Scheme 3.5, require bases to extract protons and liberate anions. Since, as illustrated in Scheme 3.1, acids are defined as substances that liberate protons and since the anions illustrated in Schemes 3.2 and 3.5 form upon liberation of their respective protons, both carboxylic acids and substances such as dimethyl malonate must be defined as acids.

Although carboxylic acids and substances such as dimethyl malonate can both be classified as acids, their relative acidities are clearly very different as illustrated by the different conditions required to liberate protons and anions. In order to understand this phenomenon, it is essential to first understand how acidities are measured.

3.3 HOW IS ACIDITY MEASURED?

Before discussing how acidity is measured, the definition of the **equilibrium constant** should be reviewed. Referring to Scheme 3.1, which illustrates that an **acid** is in **equilibrium** with its dissociated ions, the degree of this **dissociation** as compared to the undissociated acid is measured according to its **equilibrium constant** (K_{eq}). From general chemistry coursework, we remember that K_{eq} is the product of the ion concentrations divided by the concentration of the undissociated acid (Fig. 3.3).

Since Figure 3.3 represents how equilibrium constants are calculated and since we are specifically studying dissociation of acids, K_{eq} can be redefined for acids as the **acid dissociation constant** (K_a) illustrated in Figure 3.4.

Again referring to general chemistry coursework, the degree of acidity of a solution is measured according to the concentration of hydrogen ions present. The **pH** of a solution is defined as the negative logarithm of the hydrogen ion concentration (Fig. 3.5). Similarly, if K_a is converted to its negative logarithm, we calculate the **pK_a** (Fig. 3.6). It is, in fact, the **pK_a** that is used to represent the **acidity** associated with the various hydrogen atoms present in organic molecules.

$$K_{eq} = \frac{[H^{\oplus}][A^{\ominus}]}{[HA]}$$

Figure 3.3 *Definition of the equilibrium constant (K_{eq}).*

$$K_{eq} = \frac{[H^{\oplus}][A^{\ominus}]}{[HA]} = K_a$$

Figure 3.4 *K_a is the K_{eq} specifically related to dissociation of acids.*

$$pH = -\log[H^{\oplus}]$$

Figure 3.5 *Definition of pH.*

$$pK_a = -\log K_a$$

Figure 3.6 *Definition of pK_a.*

$$pK_a = -\log K_a = -\log\left\{\frac{[H^{\oplus}][A^{\ominus}]}{[HA]}\right\} = -\log[H^{\oplus}] + \left\{-\log\frac{[A^{\ominus}]}{[HA]}\right\} = pH + \left\{-\log\frac{[A^{\ominus}]}{[HA]}\right\}$$

Figure 3.7 pK_a values are related to pH.

$$pK_a = pH - \log\left\{\frac{[A^{\ominus}]}{[HA]}\right\}$$

Figure 3.8   The Henderson–Hasselbalch equation.

$$\boxed{\text{When } [A^{\ominus}] = [HA]}$$
$$pK_a = pH - \log\left\{\frac{[A^{\ominus}]}{[HA]}\right\} = pH - \log 1 = pH - 0 = \mathbf{pH}$$

Figure 3.9   In a perfect equilibrium, $pK_a = pH$.

When calculating the **pK_a** of a given hydrogen atom, it is important to remember that the **pK_a** is related to the **pH** of the solution. This relationship is represented in Figure 3.7. As illustrated, the **pH** value can be mathematically separated from the negative logarithm of the ratio of the anion concentration to the concentration of the undissociated acid. Taking this relationship to its final derivation, we find the **Henderson–Hasselbalch equation** (Fig. 3.8) that provides us with the key to determining **pK_a** values and relative **acidities**. An important result derived from the **Henderson–Hasselbalch equation** is that in a **perfect equilibrium** (a system where there is an equal amount of dissociated and undissociated acids), the **pK_a** is equal to the **pH**. This arises from the fact that in a perfect equilibrium, the concentration of HA is equal to the concentration of A⁻ making the ratio of [A⁻] to [HA] equal to 1. Since the log of 1 is 0, this term drops out of the equation (Fig. 3.9).

3.4   RELATIVE ACIDITIES

Having explored the relationships between solution **pH** and **pK_a** values, we can now explore the **relative acidities** of various hydrogen atoms and how these values are influenced by neighboring **functional groups** and **heteroatoms**. In this arena, it is important to remember that how a reaction proceeds is largely dependent upon the *relative acidities* of protons (hydrogen atoms) compared to one another and not on the *absolute acidity* of a given proton.

Considering a compound that produces a solution with **pH** greater than 7, that compound is generally referred to as **basic**. However, a proton of interest (proton A) on this compound may be considered **acidic** compared to another proton (proton B) if the **pK_a** of proton A is lower than the **pK_a** of proton B. In other words, the lower the **pK_a** for a given proton, the more **acidic** that proton is. Consequently, in order to predict the mechanistic course of a given organic reaction, it is *extremely* important to be able to recognize the **most acidic proton** in a given molecule.

As previously stated, **acidities** of various protons are dependent upon their associated **functional groups** and nearby **heteroatoms**. Furthermore, these protons may be either a component of the relevant functional groups or adjacent to relevant functional groups. Figure 3.10 illustrates examples of compounds possessing acidic protons associated with representative functional groups, and Figure 3.11 illustrates examples of compounds possessing acidic protons adjacent to representative functional groups. In both figures, the acidic protons are highlighted in bold.

In studying the relationships between **functional groups** and proton **acidities**, we will first look at **carboxylic acids**. As illustrated in Scheme 3.2, **carboxylic acids** dissociate to form **protons** and **carboxylate anions**. Furthermore, as shown in Scheme 3.3, the **carboxylate anion** is stabilized through two **resonance** forms. It is this **resonance stabilization** that serves as the primary driving force behind the **acidic** nature of **carboxylic acids**. Further evidence of the relationship between **resonance stabilization** of **anions** and **acidity** can be seen when comparing the **pK_a** values of **carboxylic acids** to the **pK_a** values of **alcohols**. As shown in Figure 3.10, **carboxylic acids** (compounds that dissociate into resonance-stabilized carboxylate anions) have **pK_a** values ranging from 4 to 6. However, **alcohols** (compounds that dissociate into alkoxy anions possessing no resonance stabilization) have **pK_a** values ranging from 15 to 19. A comparison of **resonance** capabilities of **carboxylic acids** and **alcohols** is illustrated in Scheme 3.8. As shown, the **resonance**

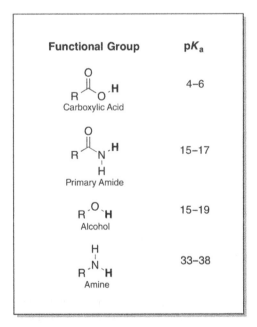

Figure 3.10 *Representative functional groups with associated acidic protons.*

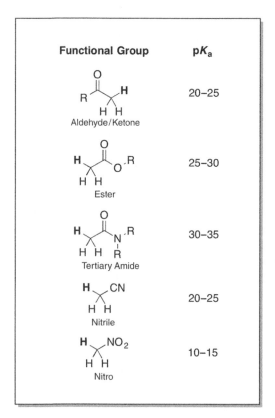

Figure 3.11 *Representative functional groups with adjacent acidic protons.*

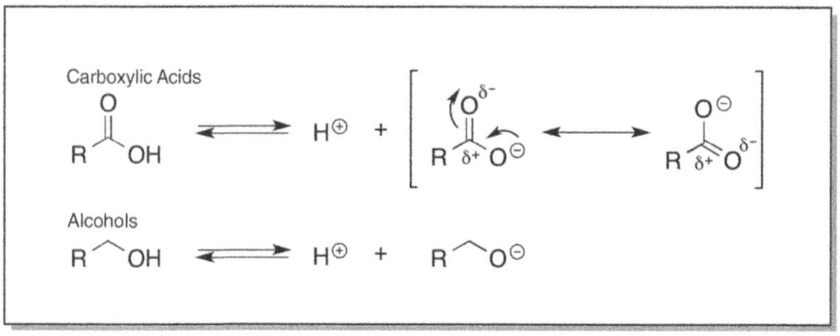

Scheme 3.8 *Resonance capabilities of carboxylic acids compared to alcohols.*

capabilities of the **carboxylate anion** are due to the presence of a **carbonyl** group adjacent to the OH. This **carbonyl** group imparts additional **partial charges** that attract the **negative charge** and distribute it over **multiple centers**. In the case of an **alcohol**, the deprotonated **alkoxide anion** has no place to distribute its charge, and the charge remains entirely on the oxygen. Because **alcohols** have no **resonance** capabilities, their **pK_a** values are higher than those of **carboxylic acids**. Nevertheless, both **carboxylic acids** and **alcohols** are considered organic **acids**.

   Having illustrated how **resonance** effects influence the **relative acidities** of different **functional groups**, it is important to understand why the same **functional groups** in different compounds can possess different **acidities**. In order to address this, we must move from the general representation of compounds presented earlier to a treatment of specific compounds. For example, the **carboxylic acid** generally represented in Scheme 3.2 possesses an "R" group. In organic chemistry, "R" groups are commonly used to represent regions of a compound, which are variable. To illustrate this point, Figure 3.12 lists several common carboxylic acids and their respective **pK_a** values.

   As illustrated in Figure 3.12, **formic acid** (R=H) has a **pK_a** of 3.75. However, if R is changed to an electron-withdrawing group such as CF_3, the anion resulting from dissociation becomes more stabilized resulting in a lower **pK_a** compared to **formic acid**. Alternately, if R is changed to an electron-donating group such as CH_3, the anion resulting from dissociation becomes less stable as illustrated by a higher **pK_a** compared to that of formic acid.

   The changes in the **pK_a** values associated with different carboxylic acids are the result of **inductive effects**. **Inductive effects** occur when **electronegative** groups pull **electron density** away from acidic centers rendering these centers more acidic. Conversely, **inductive effects** also occur when **electropositive** groups push electron density toward acidic centers rendering these centers less acidic. The concept of **electronegativity** was introduced in Section 1.2 and referred to the **periodic table of elements** as a resource for calibration. In comparing **formic acid** to **acetic acid**, the CH_3 group of acetic acid is electron donating

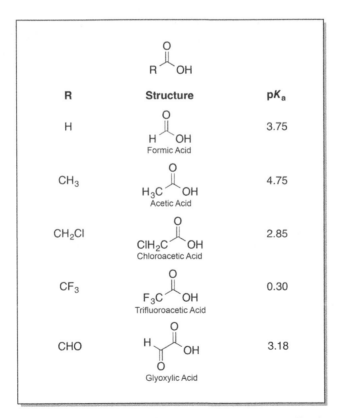

Figure 3.12 *Common carboxylic acids and their respective pK_a values.*

while the H of formic acid is not. This means that there is greater electron density present in the carboxylate anion of acetic acid than in the carboxylate anion of formic acid. An **increase** in **electron density** associated with a carboxylate anion **lowers** the **stability** of the anion and raises the **pK_a**. Thus, as mentioned at the end of Section 3.1, these observations demonstrate that **inductive effects** work in concert with **resonance effects** to alter pK_a values.

While **carboxylic acids** are among the most **acidic** of all organic compounds, we are more frequently interested in removing protons that are not directly associated with the **carboxylic acid functional group**. Additionally, reliance upon removal of **protons** from molecules containing **functional groups** other than **carboxylic acids** is common. In this context, **esters** represent the next functional group we will study. **Esters**, as **functional groups**, are simply oxygen-alkylated **carboxylic acids** (see Fig. 1.3). As such, we cannot remove a proton from the oxygen, as we are able to do with **carboxylic acids**. However, as shown in Scheme 3.9, the proton in the position α (adjacent) to the ester carbonyl can be removed under **basic** conditions. Furthermore, the rationalization of the **acidity** of this proton is represented in Scheme 3.10 using arrow-pushing and the placement of partial charges. The **pK_a** is approximately 25 for the illustrated ester when R is hydrogen.

As with carboxylic acids, **pK_a** values associated with **esters** will change accordingly when the molecule is changed. Specifically, the associated **pK_a** values are subject to the influence of **resonance effects**. An excellent example brings us back to **dimethyl malonate** illustrated in Schemes 3.6 and 3.7. If we consider the second ester group in dimethyl malonate as another electron-withdrawing group as illustrated by the additional resonance forms of the anion, we can understand that, compared to simple esters, the malonate diester will be more acidic. In fact, the **pK_a** associated with **deprotonation** of **dimethyl malonate** is approximately 12. Thus, considering **resonance effects** of neighboring groups, we can predict relative acidities for most organic compounds.

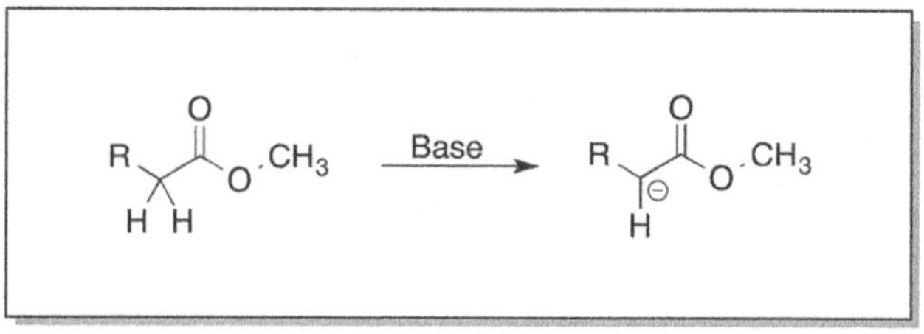

Scheme 3.9   *Esters can be deprotonated α to ester carbonyls.*

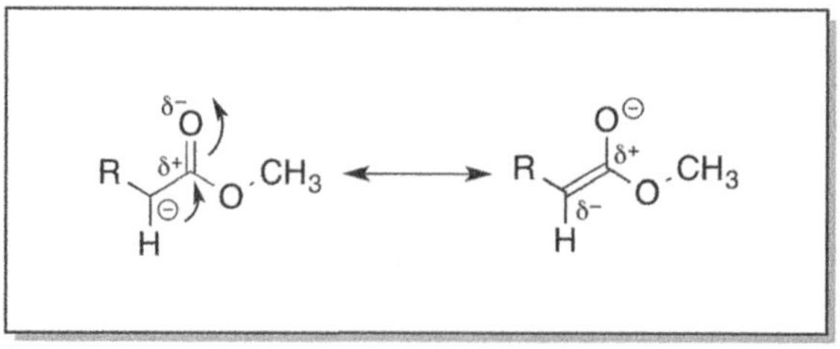

Scheme 3.10   *Rationalization of the acidity of protons α to ester carbonyls.*

3.5 INDUCTIVE EFFECTS

In addition to **resonance effects, inductive effects** influence the acidities of organic molecules. As alluded to through the **pK_a** values of the selected **carboxylic acids** shown in Figure 3.12, **inductive effects** can be either **electron-donating** or **electron-withdrawing**. Specifically, an **electron-donating** inductive effect will result from incorporation of an **electron-donating group**. Similarly, an **electron-withdrawing** inductive effect will result from incorporation of an **electron-withdrawing group**. When an **electron-donating group** is incorporated adjacent to an acidic proton, the **acidity** decreases and the **pK_a** increases. Similarly, when an **electron-withdrawing group** is incorporated adjacent to an acidic proton, the acidity increases and the **pK_a** decreases. Thus, it is imperative for those studying organic chemistry to fully understand what constitutes **electron-donating groups** and **electron-withdrawing groups**.

As illustrated in Scheme 3.11, **electron-withdrawing groups** are readily recognized when the group places either a partial or formal positive charge adjacent to an acidic center. This placement of a partial **positive charge** allows greater **delocalization** of the **negative charge** that develops when the acidic proton is removed. Through this increased **delocalization** of the developing **negative charge**, the stability of the developing anion increases, thus increasing the acidity of the target proton.

As illustrated in Scheme 3.12, **electron-donating groups** are readily recognized when the group places either a partial or formal negative charge adjacent to an acidic center.

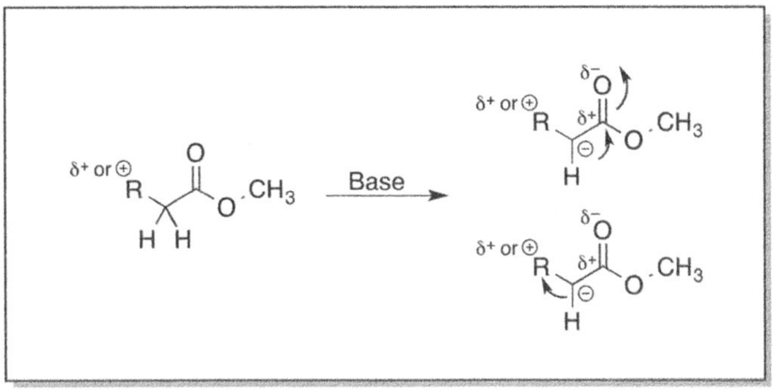

Scheme 3.11 *Electron-withdrawing groups increase acidity by increasing anionic stability.*

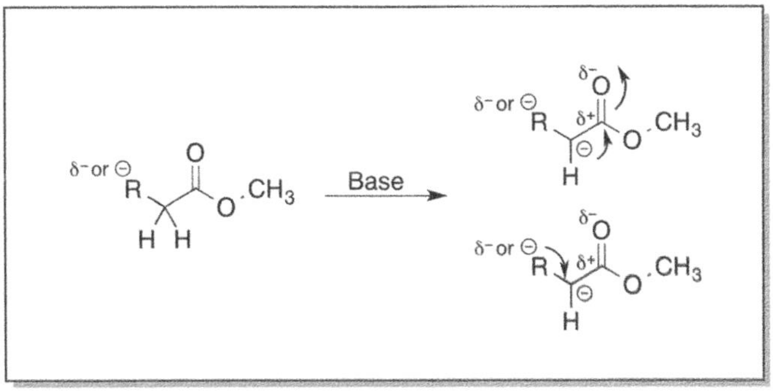

Scheme 3.12 *Electron-donating groups decrease acidity by decreasing anionic stability.*

This placement of a **negative charge** forces destabilization of the **negative charge** that develops when the acidic proton is removed. To illustrate, imagine trying to force two magnets to meet at their negative poles. As the negative poles get closer, the repulsive forces between the magnets increase. As with magnets, two negative charges on adjacent atoms results in a destabilizing situation. By decreasing the stability of a developing **negative charge**, the stability of a developing anion decreases, thus decreasing the acidity of the target proton.

Functional groups were defined and discussed in Chapter 1 (Section 1.2). In that discussion, the concept was presented that functional groups can be either **electron-withdrawing groups** or **electron-donating groups**. In fact, all **inductive effects** result from the introduction of **functional groups** to organic molecules. Furthermore, through an understanding of the characteristics of the various functional groups, one can predict whether these **functional groups** are **electron-donating groups** or **electron-withdrawing groups**. In general, if a **functional group** is capable of absorbing electron density through **delocalization**, it will act as an **electron-withdrawing group**. Such groups include carbonyl-based groups, nitro groups, and nitriles. On the other hand, if a **functional group** possesses free lone pairs of electrons, this **functional group** will act as an **electron-donating group** regardless of the **electronegativity** of the specific atom involved. Such groups include alcohols, ethers, amines, and halogens. From the group of halogens, fluorine is the exception and serves as an **electron-withdrawing group** due to its high **electronegativity**. Finally, while inductive effects thus far have been tied to heteroatoms, it is important to note that alkyl groups are weak **electron-donating groups** and will impact **pK_a** values, as we will soon discuss. Common **electron-donating groups** and **electron-withdrawing groups** are listed in Figure 3.13.

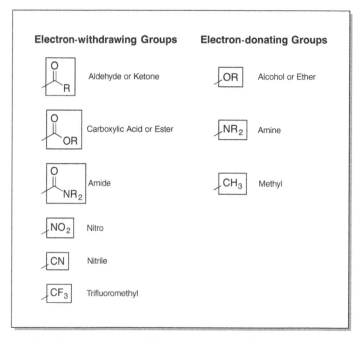

Figure 3.13   *Common electron-withdrawing groups and electron-donating groups.*

3.6 INDUCTIVE EFFECTS AND RELATIVE ACIDITIES

In Section 3.4, the concept of **relative acidities** was presented without fully defining the concept of **inductive effects**. In Section 3.5, the concept of **inductive effects** was defined and linked to **functional groups**. In this section, the concept of **relative acidities** is presented in conjunction with discussions of associated **functional groups** and their respective **inductive effects**. Since a treatment of this subject was already introduced in Section 3.4 as applied to carboxylic acids and esters, this section focuses on oxygen- and nitrogen-containing functional groups lacking carbonyl components. Specifically, the relative acidities of **alcohols** and **amines** are discussed.

Beginning with alcohols, the simplest is methanol with a **pK_a** of 15. As illustrated in Figure 3.14, as **branching** of the alkyl group increases, so does the **pK_a**. This observation is a direct result of the **inductive effect** associated with **alkyl groups** such as methyl (CH_3). As mentioned in Section 3.5, the **inductive effects** of **methyl groups** result in donation of electron density. This increase in electron density in the vicinity of the oxygen atom destabilizes the anion resulting from deprotonation, thus increasing the **pK_a**. The more methyl groups adjacent to the OH, the greater the effect as illustrated in the case of *tert*-butanol with a **pK_a** of 18–19.

While our discussion of how inductive effects alter alcohol **pK_a** values has focused primarily on **electron-donating groups**, we cannot ignore the effect resulting from incorporation of **electron-withdrawing groups**. As one would expect, if **electron-donating groups** adjacent to alcohol **functional groups** increase **pK_a** values through destabilization of anions resulting from **deprotonation**, the opposite effect should be observed when **electron-withdrawing groups** are used. This is in fact the case as supported by the entry for 2,2,2-trifluoroethanol in Figure 3.14. The electron-withdrawing

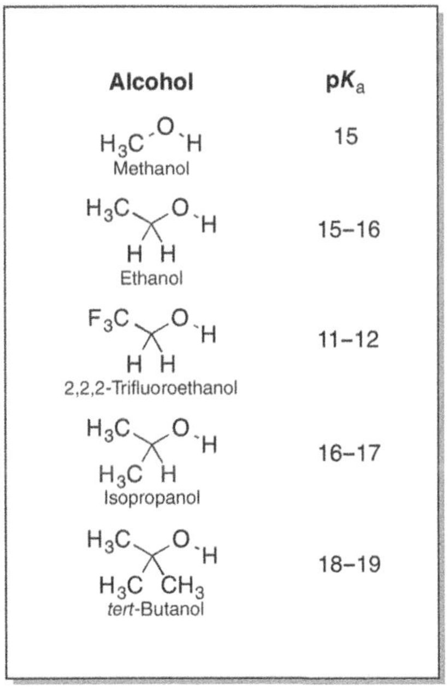

Figure 3.14 *pK_a values associated with alcohols increase as alkyl branching increases.*

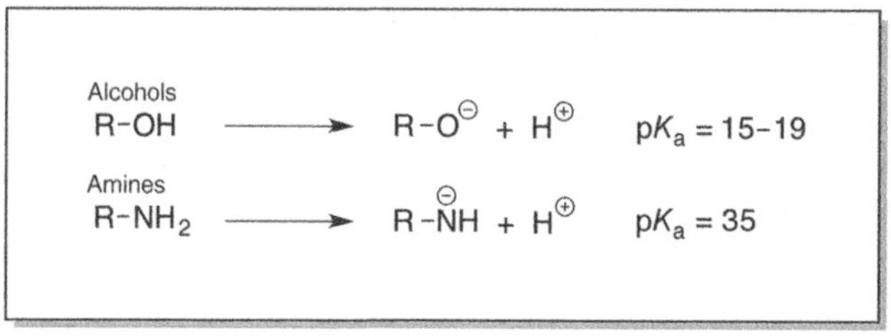

Alcohols
$$R-OH \longrightarrow R-O^{\ominus} + H^{\oplus} \qquad pK_a = 15\text{--}19$$

Amines
$$R-NH_2 \longrightarrow R-\overset{\ominus}{NH} + H^{\oplus} \qquad pK_a = 35$$

Scheme 3.13   *Amines and alcohols can both be deprotonated.*

nature of the trifluoromethyl group adds stabilization to the anion resulting from **deprotonation** and reduces the **pK_a** compared to ethanol.

Amines are similar to **alcohols** in that they can be deprotonated under basic conditions to generate anions (Scheme 3.13). However, because nitrogen is less **electronegative** than oxygen, considerably stronger bases are required to affect these deprotonations. This is further supported by the significantly higher **pK_a** values measured for amines as compared to alcohols (Scheme 3.13). These observations should not indicate in any way that the rules regarding **inductive effects** are different. In fact, they are exactly the same, and **amines** possessing **electron-donating groups** will exhibit higher **pK_a** values, while **amines** possessing **electron-withdrawing groups** will exhibit lower **pK_a** values.

3.7   RELATIVE ACIDITIES OF HYDROCARBONS

Any discussion surrounding **pK_a** values would not be complete without addressing the **deprotonation** of **hydrocarbons**. Since hydrocarbons generally refer to organic molecules made up only of hydrogen and carbon, **inductive effects** resulting from introduction of **functional groups** are not relevant, and we do not usually consider these compounds to be **acidic**. However, **pK_a** values of various protons associated with hydrocarbons have been measured. As illustrated in Scheme 3.14, saturated hydrocarbons are the least acidic, while olefinic and acetylenic protons have acidities on the same order of magnitude as those associated with esters and amines.

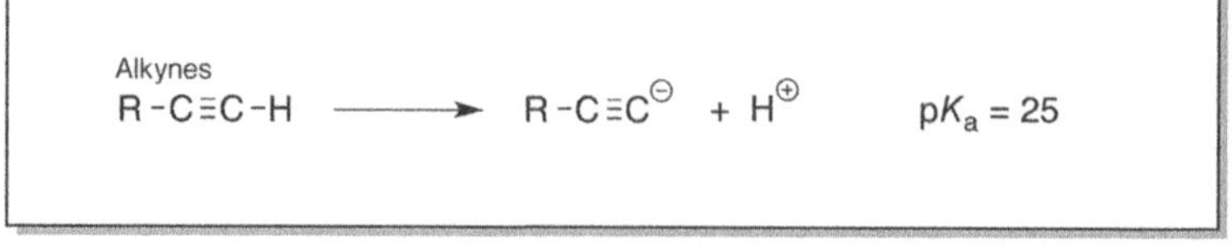

Alkanes
$$CH_4 \longrightarrow \overset{\ominus}{CH_3} + H^{\oplus} \qquad pK_a = 50\text{--}75$$

Alkenes
$$\underset{R}{\overset{R}{}}C=C\underset{R}{\overset{H}{}} \longrightarrow \underset{R}{\overset{R}{}}C=\overset{\ominus}{C}\underset{R}{} + H^{\oplus} \qquad pK_a = 35\text{--}40$$

Alkynes
$$R-C\equiv C-H \longrightarrow R-C\equiv C^{\ominus} + H^{\oplus} \qquad pK_a = 25$$

Scheme 3.14   *Hydrocarbons can be deprotonated and have measurable pK_a values.*

3.8 SUMMARY

This chapter focused on the **definition** of **acids** as applied to **organic molecules**. Furthermore, the impacts of **electronegativities** and **functional groups** on the **acidities** of various types of protons were rationalized in the context of **inductive effects**. As discussions advance throughout this book and through organic chemistry coursework, a clear understanding of the various pK_a values associated with different environments and how they relate to one another will be beneficial. Consequently, a complete familiarization of the pK_a values presented in this chapter is essential. For convenience, all the pK_a values discussed in this chapter are listed in Appendix 1 and **THEIR GENERAL MAGNITUDES WITH RESPECT TO ONE ANOTHER SHOULD BE COMMITTED TO MEMORY**. This is the only recommended memorization task associated with this treatment of **arrow-pushing** and will greatly facilitate the development of skills enabling the prediction of the mechanistic progression of organic reactions.

PROBLEMS

1. Explain how the Henderson–Hasselbalch equation can be used, in conjunction with a titration curve, to determine a pK_a.

2. What is the pH of a solution of acetic acid (pK_a=4.75) that has been titrated with 1/4 an equivalent of NaOH?

3. Draw the resonance structures of the following charged molecules.

   a.

   b.

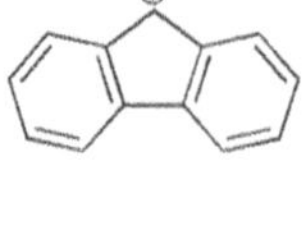

c.

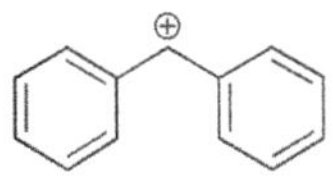

d.

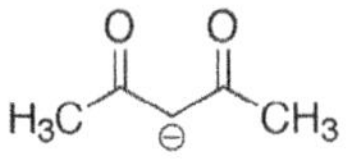

e.

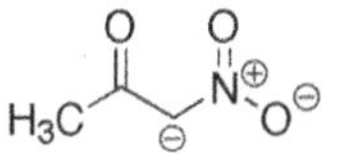

f.

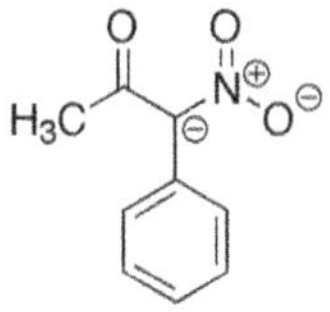

g.

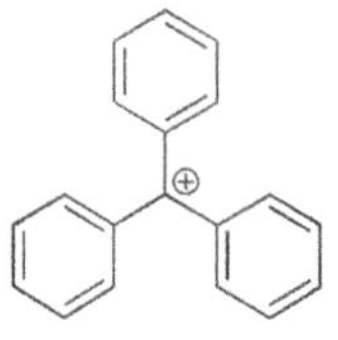

h.

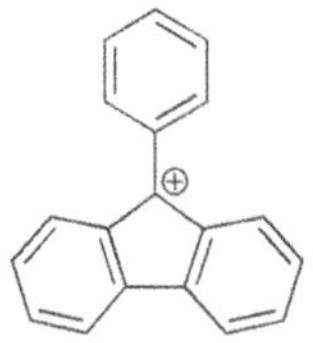

4. Which cation from Problem 3 is more stable, 3g or 3h? Explain using partial charges.

5. How will the following substituents affect the pK_a of benzoic acid (raise, lower, or no change)? Explain using partial charges to illustrate inductive effects. Remember, *o* refers to *ortho* positions, *m* refers to *meta* positions, and *p* refers to the *para* position. **In addressing these problems, assume that the acidity of the carboxylic acid is influenced solely by the partial charges induced by additional ring substituents.**

   a.   *o*-NO$_2$

   b.   *p*-NO$_2$

c.   *m*-**NO$_2$**

d.   *p*-**OH**

e.   *m*-**OH**

f.   *p*-**NH$_2$**

g.   *m*-**CH$_3$**

h.   *p*-**CH$_3$**

i.   *m*-**CHO**

j.   *p*-**OCH$_3$**

k.   *o*-**NO**

    l.   *p*-**Cl**

    m.   *m*-**Cl**

6. Arrange the following groups of molecules in order of increasing acidity. Explain your results using partial charges and inductive effects.

7. Predict pK_a values for the highlighted protons in following molecules. Rationalize your answers.

    a.

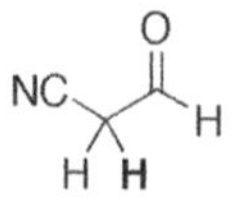

    b.

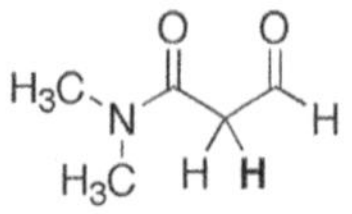

c.

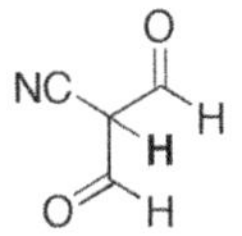

d.

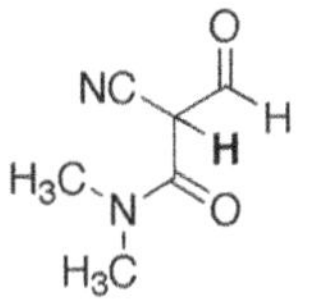

8. Predict the order of deprotonation of the various protons on the following molecules. Back up your answers with appropriate pK_a values.

a.

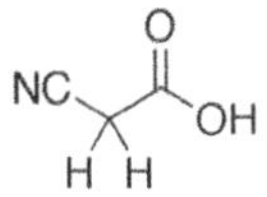

b.

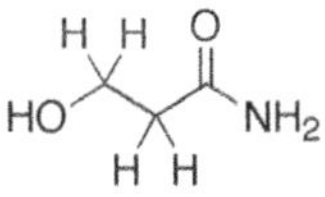

c.

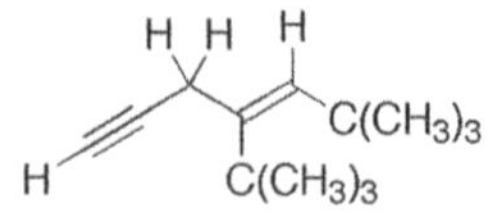

d.

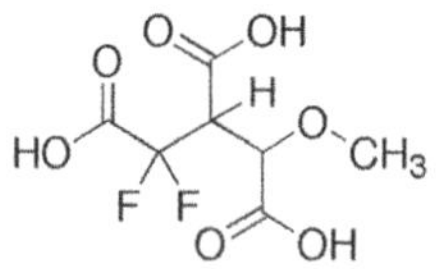

9. Which proton is the most acidic? Rationalize your answer.

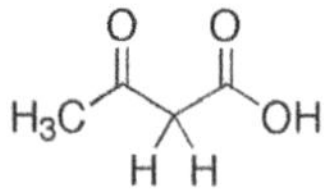

10. Using the pK_a values given in Appendix 1, calculate the equilibrium constants for the following reactions.

a.

b.

c.

$$HCl + Br^{\ominus} \rightleftharpoons HBr + Cl^{\ominus}$$

d.

Chapter *4*

Bases and Nucleophiles

In Chapter 3, the concept of **acidity** was introduced and discussed as related to organic molecules. Additionally, various **functional groups** were presented along with the concepts of **resonance effects** and **inductive effects**. Moving forward, **resonance effects** and **inductive effects** were applied to rationalize variations in **pK_a** values that exist among compounds bearing similar **functional groups**. All of these factors were described using **arrow-pushing**.

While relative acidities are extremely important to our abilities to accurately predict the mechanistic courses of organic reactions, we must recognize that in addition to **acids**, most **heterolytic** reactions involve **bases** as well. Furthermore, as will soon be discussed, **bases**, in organic chemistry, generally are able to function as **nucleophiles**. Therefore, this chapter serves as an introduction to the types of **bases** and **nucleophiles** that drive mechanistic organic chemistry.

4.1   WHAT ARE BASES?

The most general definition of a **base** is a molecule that has an affinity for protons. For example, if we consider a molecule, B, or an anion, B⁻, these species are said to be **bases** if they react with an **acid**, HA, as shown in Scheme 4.1. As with the **dissociation** of **acids**, discussed in Chapter 3, the reaction of a **base** with an **acid** is an **equilibrium** process which produces an anionic species, A⁻, and, depending on the charged nature of the base, either a cationic or a neutral species. Furthermore, considering the species formed when a base reacts with an acid, the anionic component, A⁻, is said to be the **conjugate base** of the starting acid, HA. Likewise, the cationic or neutral species formed, BH⁺ or BH, is said to be the **conjugate acid** of the starting base, B⁻ or B. Common **base** classes and examples of

Arrow-Pushing in Organic Chemistry: An Easy Approach to Understanding Reaction Mechanisms,
Second Edition. Daniel E. Levy.
© 2017 John Wiley & Sons, Inc. Published 2017 by John Wiley & Sons, Inc.

Scheme 4.1 *General representation of bases reacting with acids.*

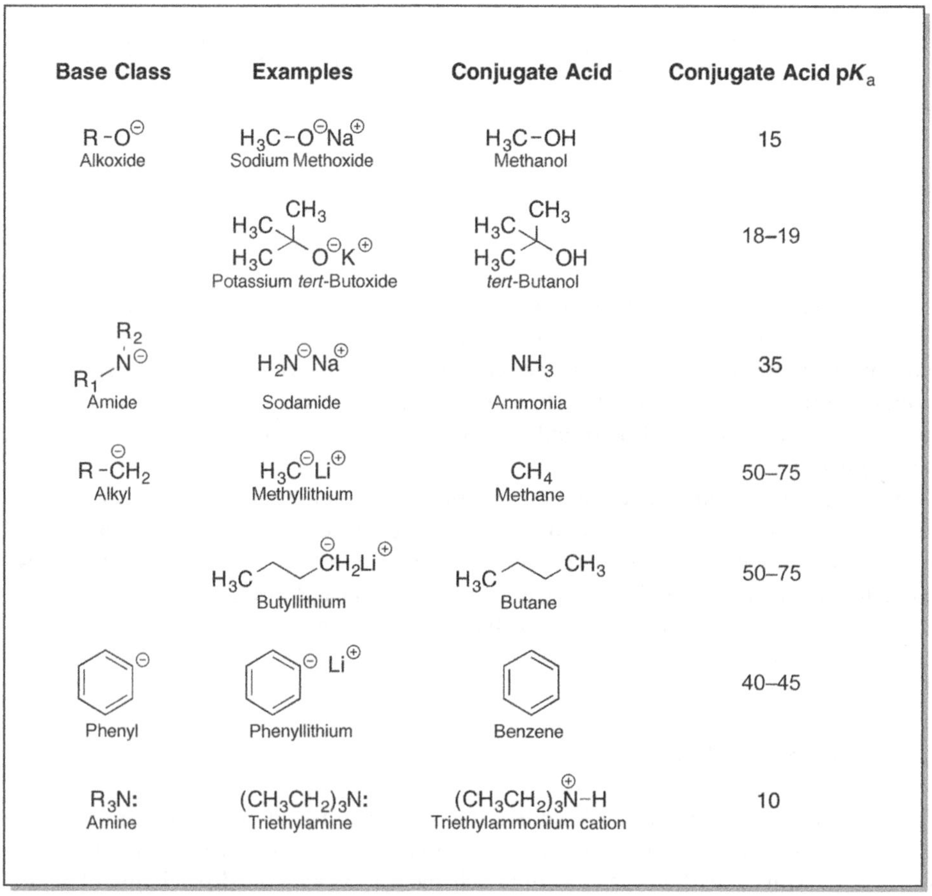

Figure 4.1 *Common bases used in organic chemistry.*

bases used in organic chemistry are shown in Figure 4.1 along with their **conjugate acids** and associated **pK_a** values.

In looking at the **conjugate acid pK_a** values listed in Figure 4.1, we realize that in order for the reactions represented in Scheme 4.1 to occur, the **conjugate acid** of a given **base** must have a **pK_a** value that is higher than the **pK_a** value associated with a proton of interest. For example, as shown in Scheme 4.2, we would not expect triethylamine to effectively deprotonate methyl acetate because the **pK_a** of methyl acetate is 15 **pK_a** units higher than the **pK_a** of the triethylammonium cation. Since each **pK_a** unit represents a factor of 10 (see

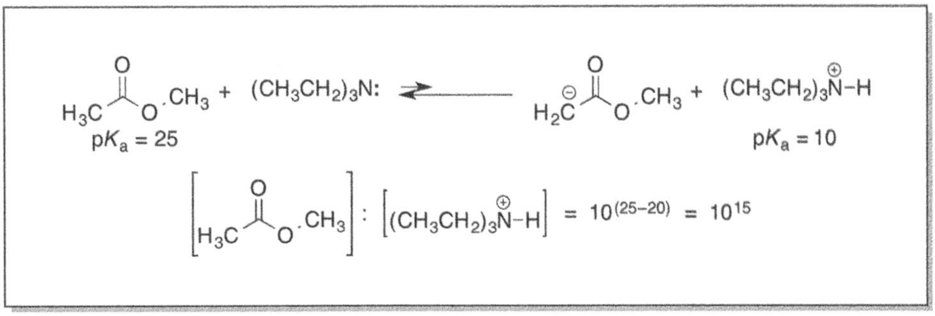

Scheme 4.2 *Equilibrium between methyl acetate and triethylamine.*

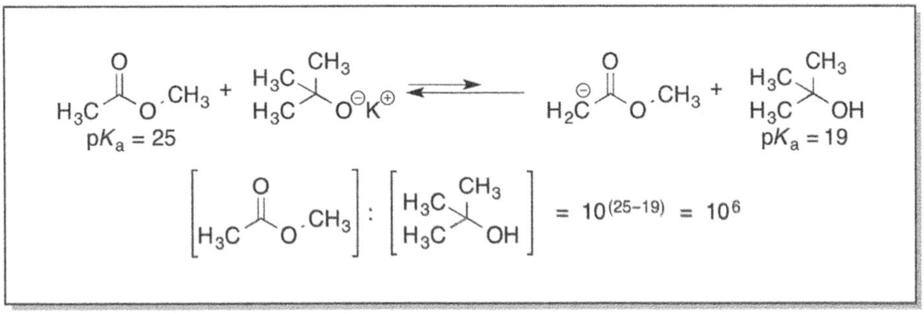

Scheme 4.3 *Equilibrium between methyl acetate and potassium tert-butoxide.*

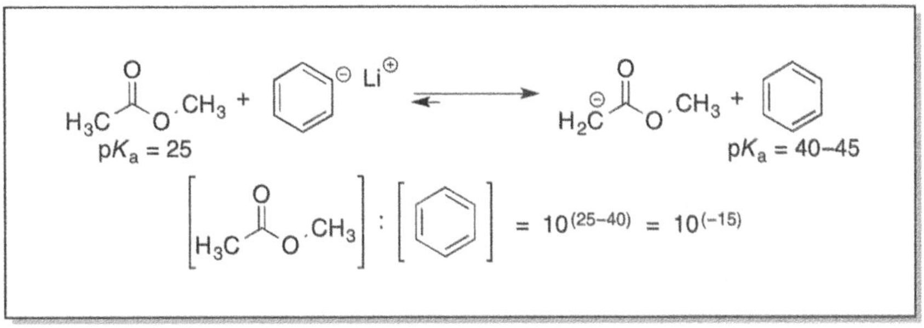

Scheme 4.4 *Equilibrium between methyl acetate and phenyllithium.*

the definition of **pK$_a$** described in Chapter 3), this differential indicates that for each 10^{15} molecules of methyl acetate, only one will be deprotonated. As shown in Scheme 4.3, the more basic potassium *tert*-butoxide will have a greater effect on this equilibrium because the **pK$_a$** differential between that of methyl acetate and that of *tert*-butanol is 6. By comparison, for every 10^6 molecules of methyl acetate, one will be deprotonated. Finally, in the absence of reactions other than deprotonation, Scheme 4.4 illustrates that very strong bases such as phenyllithium will affect essentially complete deprotonation. This is due to the **pK$_a$** differential between methyl acetate and benzene being -15 indicating that for every molecule of methyl acetate, there will be 10^{15} molecules of benzene and 10^{15} molecules of deprotonated methyl acetate. Thus, it is important to understand the **relative acidities** of all

$$(CH_3CH_2)_3N\text{:} \quad + \quad H^{\oplus} \quad \rightleftharpoons \quad (CHCH_2)_3\overset{\oplus}{N}\text{---H}$$
$$pK_a = 10$$

Scheme 4.5 *Amine basicity is related to the nitrogen lone pair.*

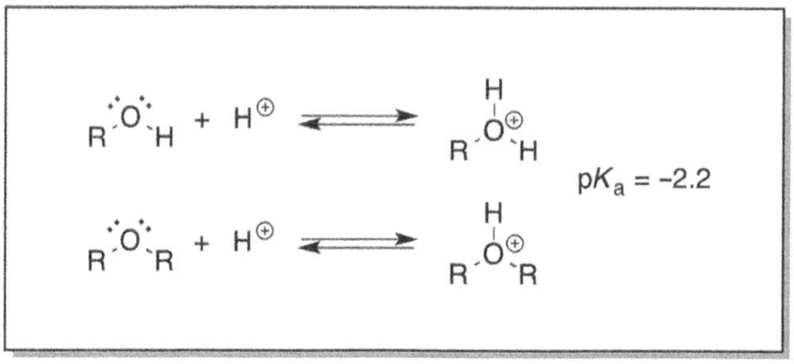

$$pK_a = -2.2$$

Scheme 4.6 *Alcohol and ether oxygens can be protonated.*

components involved in organic reactions in order to predict the direction and outcome from a mechanistic perspective.

Again referring to the bases listed in Figure 4.1, triethylamine stands out because it is the only base listed that does not rely on a negative charge to impose basicity. In fact, the basicity of amines such as triethylamine is attributable, in part, to the **lone pair** associated with nitrogen (Scheme 4.5). Expanding upon the ability of **lone pairs** to act as bases and attract protons, we can expect that atoms and functional groups that possess lone pairs will also have measurable basicity and that their corresponding **conjugate acids** will have measurable **pK$_a$** values.

Moving beyond amines in our discussion of the acidity of conjugate acids of neutral bases, let us now consider alcohols and ethers. The protonation of these functional groups, illustrated in Scheme 4.6, results in positively charged trivalent oxygen atoms referred to as **oxonium ions**. While this protonation is possible due to oxygen possessing two nonbonded electron pairs, it is not surprising that oxygen is far less basic than nitrogen due to its increased electronegativity making the lone pairs less available for protonation. Consequently, **oxonium ions** are far more acidic than their corresponding **ammonium ions** and exhibit **pK$_a$** values around −2.2.

Extending beyond alcohols and ethers, **conjugate acids** of carbonyl-based functional groups are known. Specifically referring to **carboxylic acids and esters**, the corresponding **conjugate acids**, illustrated in Scheme 4.7, have **pK$_a$** values around −6. Furthermore, **protonated aldehydes and ketones**, illustrated in Scheme 4.8, have **pK$_a$** values ranging from −7 to −9.

While not practical as **bases**, as demonstrated by **pK$_a$** values, the protonation of carbonyl-based functional groups is important. As previously discussed, carbonyl compounds possess **partial positive charges** and **partial negative charges** and, consequently, are capable of **delocalizing** adjacent charges through resonance. Scheme 4.9 illustrates this fact as applied to an ester. However, if we consider a protonated carbonyl compound, the resulting positive charge residing on the carbonyl oxygen is delocalized to the associated carbon atom (Scheme 4.10). The net result renders the carbon atom highly susceptible to reaction with **nucleophiles** (Scheme 4.11).

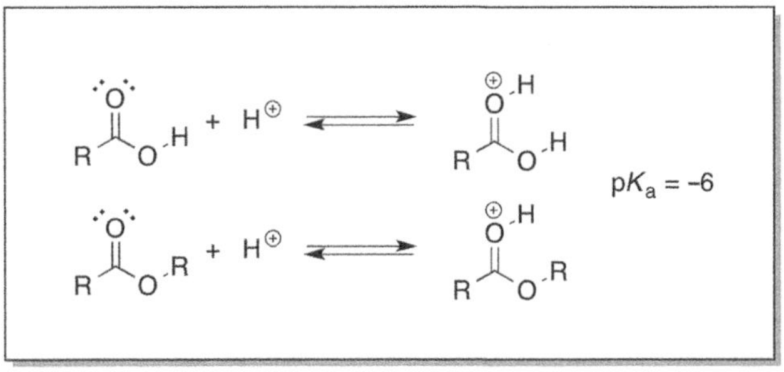

Scheme 4.7   *Carboxylic acids and esters can be protonated.*

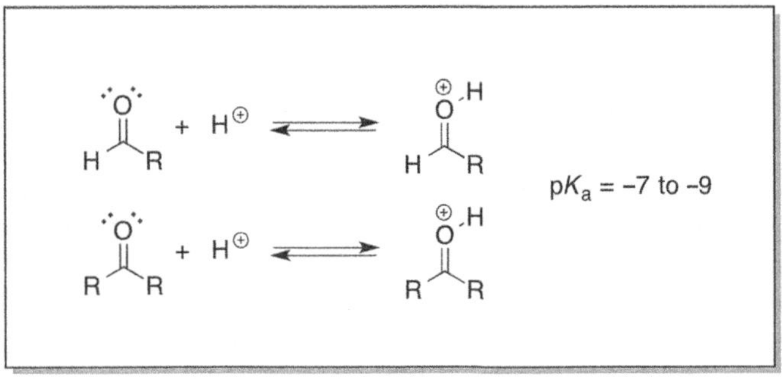

Scheme 4.8   *Aldehydes and ketones can be protonated.*

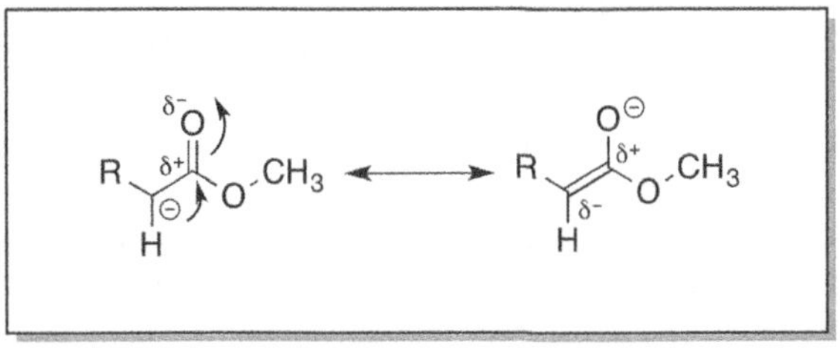

Scheme 4.9   *Carbonyl-based functional groups delocalize charges through resonance.*

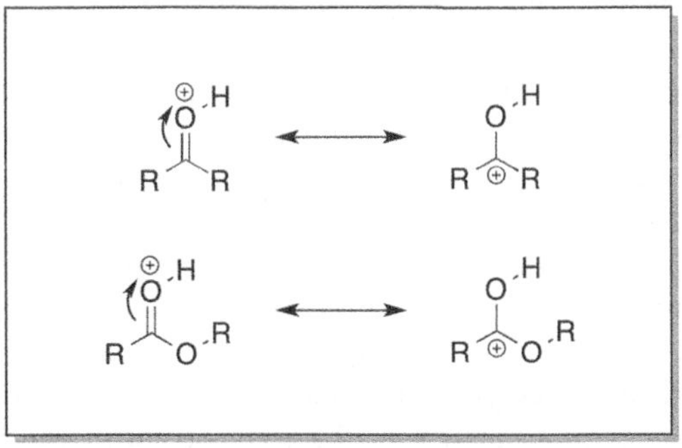

Scheme 4.10   *Protonated carbonyl-based functional groups delocalize their positive charges.*

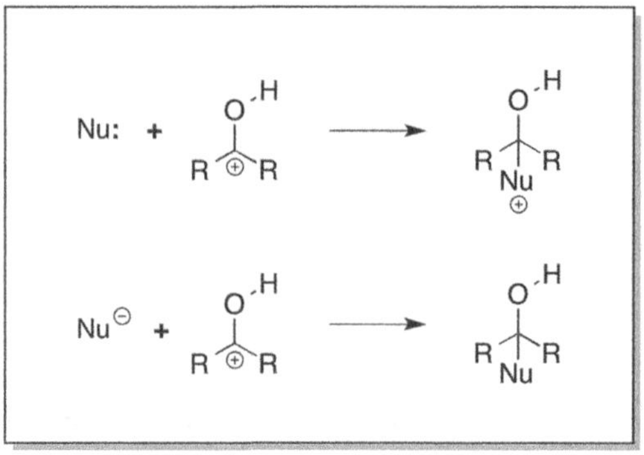

Scheme 4.11 *Protonated carbonyl-based functional groups are susceptible to reaction with nucleophiles.*

4.2 WHAT ARE NUCLEOPHILES?

As alluded to in Section 1.3 and in Section 4.1, heterolytic reactions generally involve species known as **nucleophiles** and complementary species known as **electrophiles**. By definition, a **nucleophile** is a compound that has an affinity for a positive charge. By analogy, an **electrophile** is a compound that has an affinity for a negative charge. **Nucleophiles** generally present themselves as either neutral species bearing available lone pairs of electrons or **anions** (negatively charged ions). When a **nucleophile** is an **anion**, the anion is generally the **conjugate base** of an acid. Figure 4.2 lists common **conjugate bases** used as **nucleophiles** along with their starting acids and associated pK_a values.

In considering the information presented in Figure 4.2, it is important to become familiar with the general trends that influence the degree of **nucleophilicity** associated with the **conjugate bases** of various acids. From our discussions of acids, we know that as the pK_a increases, the acidity decreases. Furthermore, as the acidity decreases, the basicity associated with **conjugate bases** increases. Since, by definition, bases attract protons and since protons are, by definition, positively charged, we can translate this relationship to infer that bases exhibit affinities for positive charges. Since **nucleophiles** are defined as substances that have affinities for positive charges, we can understand the statement from the previous paragraph equating nucleophiles with **conjugate bases** of acids. Taking this discussion to the next level, the weaker the acid (higher pK_a), the stronger the **conjugate base**. Furthermore, the stronger the **conjugate base**, the stronger the **nucleophile**.

While the general trend relating **basicity** and **nucleophilicity** stands, we cannot simply rely on the pK_a values listed in Figure 4.2 as a guide for these trends. In fact, with respect to overall **nucleophilicity**, there are relevant factors other than basicity. Among these are **polarizability** of the nucleophilic atom, **electronegativity** of the nucleophilic atom, **steric factors**, and **solvent effects**. For our purposes, **solvent effects** will be discussed in the context of **polarizability**.

Electronegativity, discussed in Chapter 1, is a measure of an atom's affinity for electrons. Thus, as **electronegativity** increases, affinity for electrons increases. Furthermore, as affinity for electrons increases, so does acidity. This is reflected in the decreasing pK_a values moving from methane to methylamine to methanol to hydrofluoric acid (Fig. 4.3). In this sequence, the trend relating increasing **basicity** of **conjugate bases** to increasing **nucleophilicity** holds true. Furthermore, this relationship holds true for each row in the **periodic table of elements** moving from left to right.

Nucleophile/Conjugate base	Acid	pK_a
$F^{\ominus}$ Fluoride Anion	HF Hydrofluoric Acid	3.18
$Cl^{\ominus}$ Chloride Anion	HCl Hydrochloric Acid	−2.2
$Br^{\ominus}$ Bromide Anion	HBr Hydrobromic Acid	−4.7
$I^{\ominus}$ Iodide Anion	HI Hydroiodic Acid	−10
$N \equiv C^{\ominus}$ Cyanide Anion	H−CN Hydrocyanic Acid	9.3
$^{\ominus}N = \overset{\oplus}{N} = N^{\ominus}$ Azide Anion	$H-N_3$ Hydrazoic Acid	4.6
$H_3C-O^{\ominus}$ Methoxide Anion	H_3C-OH Methyl Alcohol	15
$H_3C-\overset{\ominus}{N}H$ Methylamide Anion	H_3C-NH_2 Methylamine	35
$N \equiv C-\overset{\ominus}{C}H_2$ Acetonitrile Anion	$N \equiv C-CH_3$ Acetonitrile	25
$H_3C \overset{O}{\diagup} \underset{\ominus}{CH_2}$ Acetone Anion	$H_3C \overset{O}{\diagup} CH_3$ Acetone	20
$H_3C-O \overset{O \quad O}{\diagdown \diagup} O-CH_3$ with $\ominus$ and H Dimethyl Malonate Anion	$H_3C-O \overset{O \quad O}{\diagdown \diagup} O-CH_3$ with H H Dimethyl Malonate	13
$H_3C^{\ominus}$ Methyl Anion	CH_4 Methane	50–75

Figure 4.2 *Representative nucleophiles and their corresponding acid forms.*

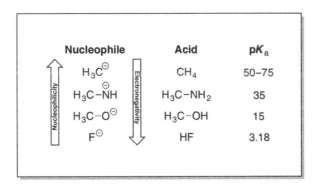

Nucleophile	Acid	pK_a
$H_3C^{\ominus}$	CH_4	50–75
$H_3C-\overset{\ominus}{N}H$	H_3C-NH_2	35
$H_3C-O^{\ominus}$	H_3C-OH	15
$F^{\ominus}$	HF	3.18

Figure 4.3 *Relationship between nucleophilicity, electronegativity, and basicity as illustrated using first row elements.*

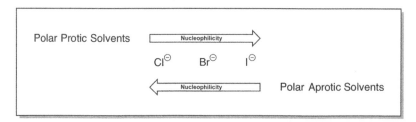

Figure 4.4   *The order of increasing nucleophilicity of halide ions is influenced by polarizing influences such as solvent effects.*

Polarizability refers to the ability of an atom to become polarized in the presence of external influences such as **solvent effects**. In general, **polarizability** increases as **electronegativity** decreases. Another way of looking at this relationship involves atomic size. Essentially, the larger an atom, the more diffuse its outer shell of electrons. As this electron shell becomes more diffuse, it also becomes more susceptible to polarizing influences. Furthermore, these polarizing influences can dramatically impact the order of **nucleophilicity** among atoms represented in any given column of the **periodic table of elements**. In fact, **polarizability** can override the relationship between **nucleophilicity** and **pK_a**. This effect is illustrated using the relative **nucleophilicities** of Cl⁻, Br⁻, and I⁻. If we refer to the **pK_a** values listed in Figure 4.2, we would expect the order of **nucleophilicity** among these halide ions to be Cl⁻ > Br⁻ > I⁻. This is, in fact, the case in the presence of **polar aprotic solvents** (solvents not possessing a dissociable proton) such as dimethylformamide. However, in the presence of **polar protic solvents** (solvents possessing a dissociable proton) such as water or alcohols, the order of **nucleophilicity** is I⁻ > Br⁻ > Cl⁻. This effect, shown in Figure 4.4, illustrates that relative **nucleophilicities** are not absolute.

Another factor influencing **nucleophilicity** and related to **polarizability** is the **hardness** or **softness** of the nucleophilic base. Specifically, a **hard base** is high in **electronegativity** and low in **polarizability**. Alternatively, a **soft base** is low in **electronegativity** and high in **polarizability**. Using these definitions, F⁻ is considered a **hard base** because it is high in **electronegativity**, is small in size, and holds its electrons very tightly. On the other hand, I⁻ is considered a **soft base** because its large size causes it to hold its electrons loosely and renders it highly **polarizable**.

The relationship between **hard bases** and **soft bases** now relates back to **solvent effects**. If we consider a **hard base** in a polar solvent, we find that the concentrated electron density associated with a **hard base** attracts a tight shell of solvent surrounding it. This solvent shell blocks the **hard base** from reacting as a **nucleophile**. On the other hand, this **solvent effect** is minimized for a **soft base** due to its large size and diffuse electron concentration. The absence of a solvent shell around a soft base enhances its ability to react as a **nucleophile**. This effect is illustrated in Figure 4.5. Thus, while **solvent effects** can influence the order of **nucleophilicity** observed for halide ions, the general rule of thumb is that **nucleophilicity** increases as we move from the top to the bottom of any given column in the **periodic table elements**.

Of the factors influencing **nucleophilicity**, **steric effects** have, perhaps, the greatest influence. **Steric effects** occur when groups attached to a **nucleophilic** atom affect the reactivity of the **nucleophile** beyond the expected reactivity based on **electronic effects** alone. In fact, it is this effect that can differentiate between a reactive **nucleophile** and a

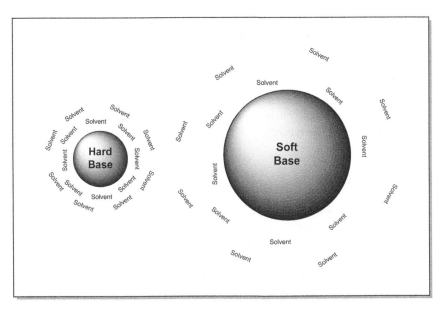

Figure 4.5 *Solvent shells surround hard bases more closely, making them less reactive nucleophiles compared with soft bases.*

Nucleophile/ Conjugate base	Acid	pK_a
$H_3C-O^\ominus$	$H_3C-O^{\,}H$	15
$H_3C\diagdown_{H\ H}O^\ominus$	$H_3C\diagdown_{H\ H}O^{\,}H$	15–16
$H_3C\diagdown_{H_3C\ H}O^\ominus$	$H_3C\diagdown_{H_3C\ H}O^{\,}H$	16–17
$H_3C\diagdown_{H_3C\ CH_3}O^\ominus$	$H_3C\diagdown_{H_3C\ CH_3}O^{\,}H$	18–19

Nucleophilicity ↑

Figure 4.6 *Steric effects can override the influence of pK_a values on nucleophilicity.*

true base. Consider the alkoxide ions illustrated in Figure 4.6. If we take into account only the **pK$_a$** values listed, we may expect that **nucleophilicity** would increase as we move from methoxide to ethoxide to isopropoxide to *tert*-butoxide. However, if we look at these **conjugate bases** not as a function of their corresponding **pK$_a$** values but as a function of their overall **structure**, we find a very different story. Specifically, a methyl group is small compared to an ethyl group. Furthermore, an ethyl group is small compared to an isopropyl group. Finally, an isopropyl group is small compared to a *tert*-butyl group. Therefore, if we consider that increasing **molecular volume** around a **nucleophilic** atom results in decreasing **nucleophilicity**, it is easy to reconcile the trend illustrated in Figure 4.6. Furthermore, when a **nucleophilic** atom becomes sterically blocked from acting as a **nucleophile**, the characteristics of this atom as a base become manifested. In fact, *tert*-butoxide anions are commonly used as bases in organic chemistry due to the diminished **nucleophilicity** surrounding the oxygen. Finally, it is important to remember that this effect is generally applicable to all **nucleophiles** regardless of the atoms involved.

4.3 LEAVING GROUPS

No discussion of **nucleophiles** would be complete without addressing **leaving groups**. As mentioned in Chapter 1 and illustrated in Scheme 4.12, **leaving groups** are important components in **nucleophilic** reactions because they represent molecular fragments that detach from parent molecules during the courses of the reactions. Specifically, a **nucleophile**, Nu, approaches an organic molecule and displaces the **leaving group**, L. This type of reaction raises the question of what relationship links **nucleophiles** to **leaving groups**. The answer to this question begins with pK_a values.

As previously discussed, relative acidities are important with respect to our ability to predict which proton will be removed first. Furthermore, relative acidity brings our attention to the relative stability of **conjugate bases**. Specifically, the more stable the **conjugate base** is, the higher will be its **acidity** and, in general, the lower will be its **nucleophilicity**. This trend, as discussed in the previous section, holds true when comparing **nucleophiles** from the same row in the **periodic table of elements**. Keeping this trend in mind, we can argue that since a **leaving group** is essentially the opposite of a **nucleophile**, the trend regarding **nucleophilicity** should roughly reverse when considering trends regarding the efficiency of **leaving groups**. This is, in fact, the case, and, as we will see as discussions progress through this book, the acid forms of **leaving groups** will generally exhibit higher pK_a values than the acid forms of competing **nucleophiles**.

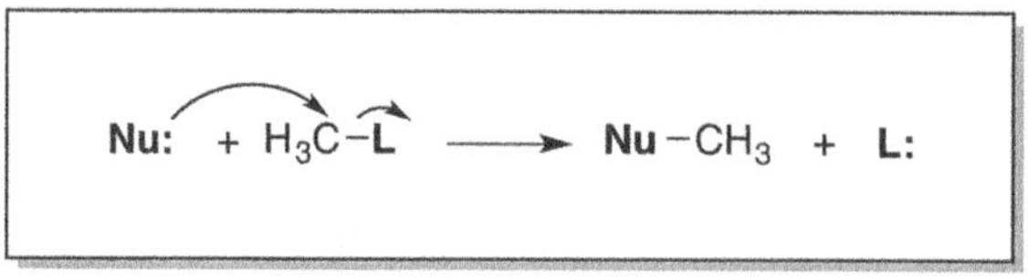

$$Nu: \; + \; H_3C-L \longrightarrow Nu-CH_3 \; + \; L:$$

Scheme 4.12 *Example of a nucleophilic reaction.*

4.4 SUMMARY

In this chapter, the concepts of organic bases and basicity were presented. These discussions were expanded to define nucleophiles and nucleophilicity. Trends associated with conjugate bases of acids and nucleophilicity were presented and translated to define the concept of leaving groups. As discussions continue, all of these concepts will play important roles in the various organic reaction mechanistic types presented in the following chapters.

PROBLEMS

1. In each case, circle the better nucleophile. Explain your answers.

a. H_3C-OH H_3C-NH_2

b. $H_3C-O^{\ominus}$ $H_3C-\overset{\ominus}{NH}$

c. $H_3C-O^{\ominus}$ H_3C-NH_2

d. H_3C-OH $H_3C-\overset{\ominus}{NH}$

e.   Cl$^{\ominus}$ I$^{\ominus}$

f.   N≡C$^{\ominus}$ HC≡C$^{\ominus}$

g.   H$_3$C$^{\ominus}$ N≡C–$\overset{\ominus}{C}$H$_2$

h.

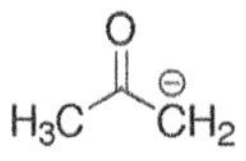

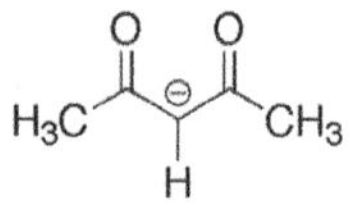

2. Nucleophiles often participate in nucleophilic substitution reactions. The general form of these reactions may be represented by the following equation where Nu_1^- and Nu_2^- are nucleophiles.

a. $Nu_1^{\ominus}$ +

b.   Explain what type of relationship between Nu_1^- and Nu_2^- is necessary in order for this reaction to be favored.

c.   What does this say about the relative basicities of Nu_1^- and Nu_2^- ?

d.   Which nucleophile has the larger pK_a?

e.   What generalization can be concluded about the relationship between bases and nucleophiles?

3. How can pK_a values be used to describe basicity?

4. As electron-donating and electron-withdrawing substituents will affect the acidity of organic molecules, so will they affect the basicity. How will the following substituents affect (raise, lower, or no change) the pK_a of aniline (aminobenzene)? Explain using partial charges to illustrate inductive effects. Remember, *o* refers to *ortho* positions, *m* refers to *meta* positions, and *p* refers to the *para* position. **In addressing these problems, assume that the acidity of the amine is influenced solely by the partial charges induced by additional ring substituents.**

a.   *o*-NO$_2$

b.   *p*-NO$_2$

c.   *m*-NO$_2$

d.   *p*-NH$_2$

e.   *m*-CH$_3$

f.   *p*-CH$_3$

g.   *p*-OCH$_3$

h.   *p*-Cl

i.   *m*-Cl

5. Arrange the following groups of molecules in order of increasing basicity. Explain your results using partial charges and inductive effects.

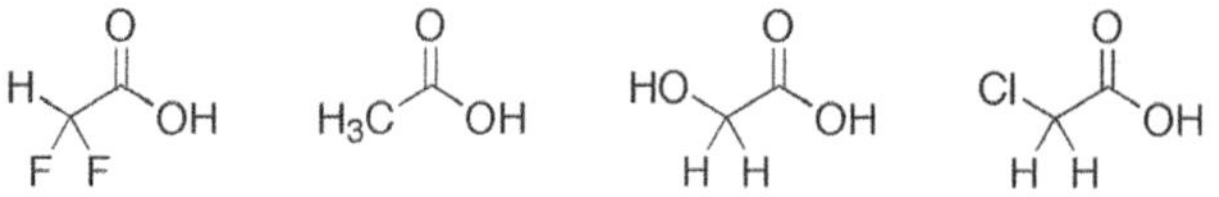

6. Predict the order of protonation of the basic sites on the following molecules. Back up your answers with pK_as.

a.

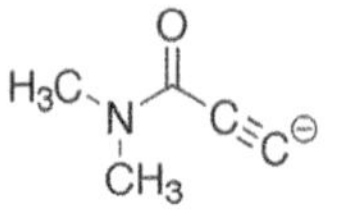

b.

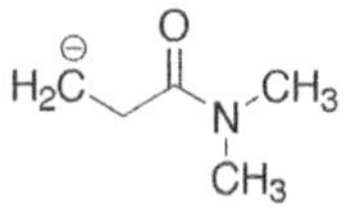

c.

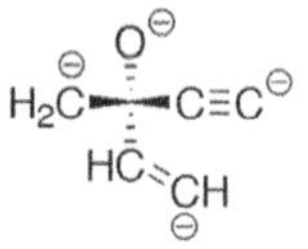

d.

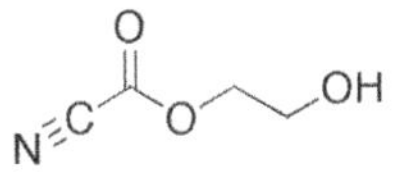

7. Of the protons attached to the heteroatoms, which proton is the least acidic? Explain your answer.

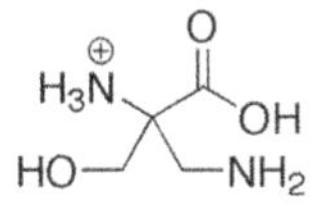

8. Separate the following group of bases into a group of hard bases and a group of soft bases. Rationalize your answers based on electronegativity and polarizability.

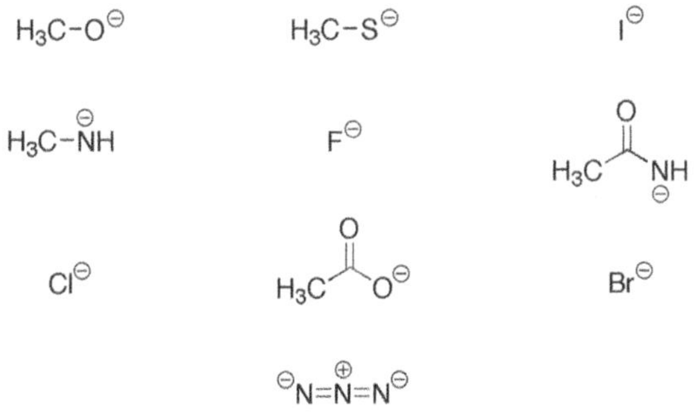

9. Arrange the following structures in the order of increasing nucleophilicity.

a.

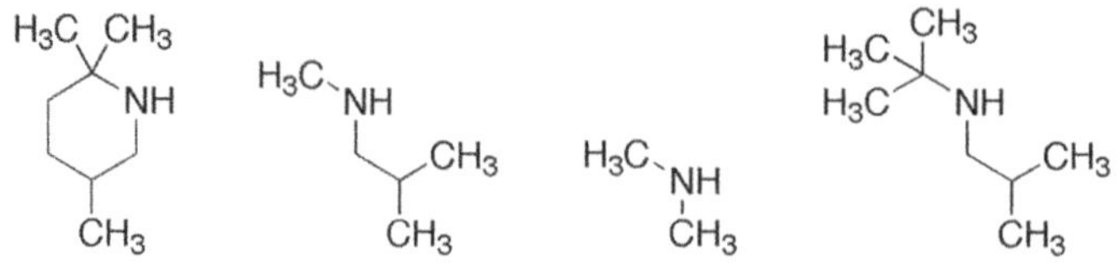

b.

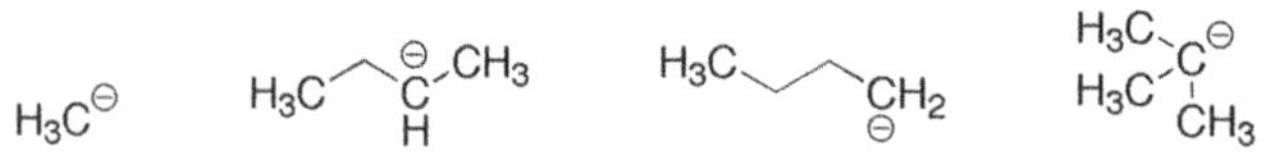

10. For the following pairs of structures, circle the better leaving group.

a. $Cl^{\ominus}$ $I^{\ominus}$

b. $H_3C-O^{\ominus}$ $H_3C-\overset{\ominus}{N}H$

c. $CH_3CH_2-O^{\ominus}$ $CF_3CH_2-O^{\ominus}$

d. $H_3C-S^{\ominus}$ $H_3C-O^{\ominus}$

e. $F^{\ominus}$ $Br^{\ominus}$

f. $H_3C-\overset{\ominus}{N}H$ $H_3C-S^{\ominus}$

g. $Br^{\ominus}$ $H_3C-O^{\ominus}$

h. 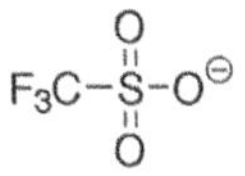 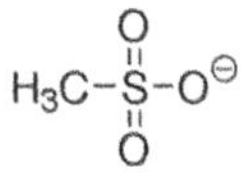 $F_3C-\overset{\overset{O}{\|}}{\underset{\underset{O}{\|}}{S}}-O^{\ominus}$ $H_3C-\overset{\overset{O}{\|}}{\underset{\underset{O}{\|}}{S}}-O^{\ominus}$

Chapter 5

S_N2 *Substitution Reactions*

As alluded to in previous chapters, the study of organic chemistry requires an understanding of the mechanistic types that drive reactions. While the detailed mechanisms associated with some complex reactions may lie beyond the scope of an introductory organic chemistry course, the fundamental components are easily recognized and applied to the reactions contained within generally presented curricula. As stated in earlier discussions, this book presents the concept of **arrow-pushing** with a focus on heterolytic reaction mechanisms. However, it is important to remember that the lessons presented herein are applicable to organic chemistry regardless of the mechanistic type.

While the fundamental mechanistic components of organic chemistry can be combined to describe complex mechanisms associated with complex reactions, the individual mechanistic components fall into a relatively small and well-defined group of four. These are S_N1, S_N2, **E1**, and **E2** reactions. In this chapter, the fundamentals associated with S_N2 reactions are presented.

5.1 WHAT IS AN S_N2 REACTION?

Among the mechanistic types relevant to organic chemistry, the S_N2 reaction mechanism is the simplest. In progressing from starting materials to products, these reactions generally consist of a **nucleophile** displacing a **leaving group**. Specifically, as illustrated in Scheme 5.1, consider a molecule where L^- is a leaving group. As shown, a nucleophile, Nu^-, can be introduced with displacement of the leaving group, thus generating a new molecule.

While discussions of **stereochemistry** are left to the organic chemistry textbooks adopted for introductory classes, it is important to recognize the **stereochemical** implications of S_N2

Arrow-Pushing in Organic Chemistry: An Easy Approach to Understanding Reaction Mechanisms,
Second Edition. Daniel E. Levy.
© 2017 John Wiley & Sons, Inc. Published 2017 by John Wiley & Sons, Inc.

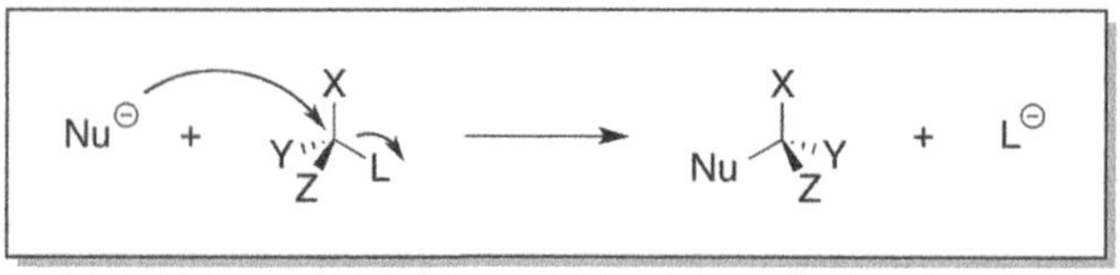

Scheme 5.1   *Representation of an S$_N$2 reaction.*

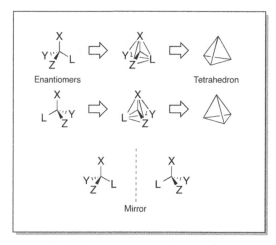

Figure 5.1   *Enantiomers are mirror images, not superimposable and dependent upon the tetrahedral arrangement of carbon atom substituents.*

reactions. In this respect, molecular bonds are drawn with either a straight line, a wedge, or a hashed wedge. Bonds drawn with straight lines are understood to lie in the same plane as the two-dimensional page on which they are drawn. Bonds drawn with wedges are understood to project above the plane of the page on which they are drawn. Finally, bonds drawn with hashed wedges are understood to project below the plane of the page on which they are drawn.

Recognizing that the substituents residing on a tetrasubstituted carbon atom are spherically spaced equidistant from one another, if all substituents are connected with lines, a **tetrahedron** is formed. Furthermore, as shown in Figure 5.1, if all four substituents are unique, they can be arranged in two configurations where the two molecules are mirror images and not superimposable. Because these two molecules are identical in composition but not in configuration in three-dimensional space, they are referred to as **stereoisomers**. Furthermore, when two molecules differ only by the spatial arrangement of their substituents rendering them mirror images of each other, these molecules are called **enantiomers**.

The aforementioned discussion of **stereochemistry** is important to the context of **S$_N$2** reactions because, as illustrated in Scheme 5.1, when a **nucleophile** displaces a **leaving group**, the **configuration** of substituents X, Y, and Z with respect to L becomes **inverted**. Thus, the **configuration** of X, Y, and Z with respect to Nu is opposite to their **configuration** with respect to L. As shown in Scheme 5.2, this **inversion** of **configuration** is mechanistically explained through the simultaneous formation of the Nu–carbon bond and cleavage of the L–carbon bond. Elongated hashed lines are used to illustrate the partial

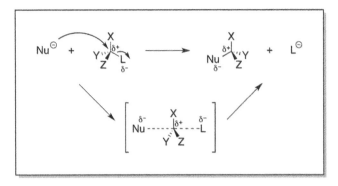

Scheme 5.2 *Mechanistic explanation of S$_N$2 reactions.*

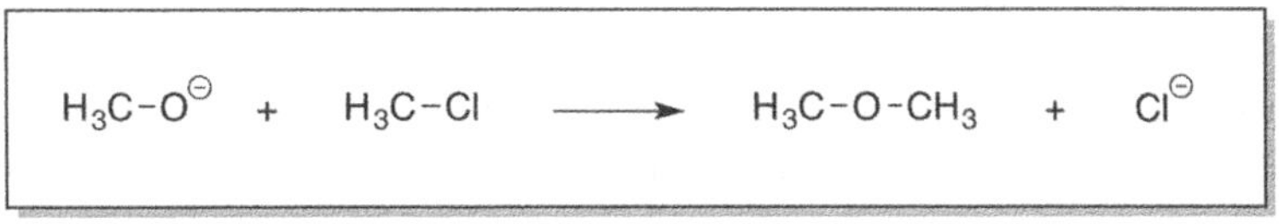

$$H_3C-O^\ominus \quad + \quad H_3C-Cl \quad \longrightarrow \quad H_3C-O-CH_3 \quad + \quad Cl^\ominus$$

Scheme 5.3 *S$_N$2 reactions proceed when incoming nucleophiles are more nucleophilic than outgoing leaving groups.*

formation or cleavage of molecular bonds. Electronically, this mechanism is explained with the placement of **partial positive** and **partial negative** charges.

Looking at Scheme 5.2, we recognize that an **S$_N$2** reaction proceeds with the substitution of a leaving group with a **n**ucleophile leading to the **S$_N$** designation. Because this mechanism proceeds with the initial approach of two species, it is referred to as a bimolecular reaction. The involvement of **2** species enhances the mechanistic designation to **S$_N$2**.

5.2 WHAT ARE LEAVING GROUPS?

The concept of **leaving groups** was introduced in Chapter 4 and presented in the context of **nucleophiles** and acid–base chemistry. However, with respect to **S$_N$2** reactions, there are additional perspectives, relating **nucleophiles** to **leaving groups**, deserving attention. For example, in the context of **nucleophilic substitution** reactions, it is reasonable to conclude that **leaving groups** are **nucleophiles**. In fact, this is generally true and one requirement for a **nucleophilic substitution** reaction to proceed is that the incoming **nucleophile** be a better **nucleophile** than the **leaving group**.

In consideration of the relationship between **nucleophiles** and **leaving groups**, recall that the **pK_a** value for methanol is 15–16 and the **pK_a** value for hydrochloric acid is −2.2. Because methanol is less acidic than hydrochloric acid, methoxide ions are expected to be better **nucleophiles** than chloride ions. This is, in fact, the case and the **S$_N$2** reaction illustrated in Scheme 5.3 will proceed to completion. Furthermore, as illustrated in Scheme 5.4, the reverse reaction will not proceed because relative to chloride ions, methoxide ions are poor **leaving groups**.

In summary, **S$_N$2** reactions are defined by the principles surrounding organic **acids** and their **conjugate bases** as discussed in Chapters 3 and 4. Specifically, as discussed in Chapter 4, relative **nucleophilicities** can be estimated based on **pK_a** values. Finally, relative

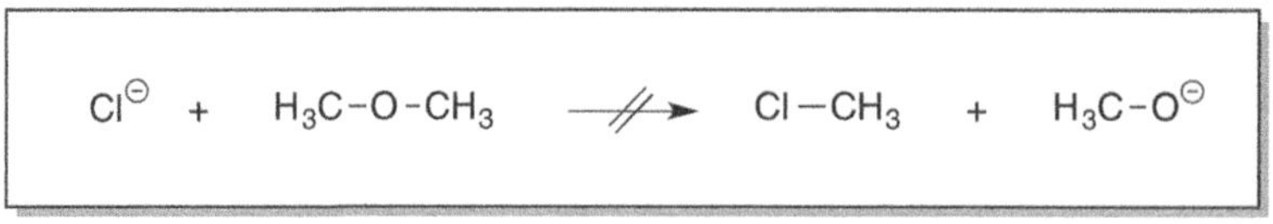

$$Cl^{\ominus} \;\;+\;\; H_3C-O-CH_3 \;\;\xrightarrow{\;\;/\!\!/\;\;}\;\; Cl-CH_3 \;\;+\;\; H_3C-O^{\ominus}$$

Scheme 5.4   *S$_N$2 reactions do not proceed when incoming nucleophiles are less nucleophilic than outgoing leaving groups.*

pK_a values can be used to predict whether a reaction will proceed and what component of the starting material will be the **leaving group**.

5.3   WHERE CAN S$_N$2 REACTIONS OCCUR?

With an understanding of what **S$_N$2** reactions are, it is now important to understand where, on a given molecule, such reactions will take place. In fact, the answer to this question can be found by recognizing **electronegativity** trends. As previously discussed, an atom's **electronegativity** relates to how strongly it holds onto its electrons. This translates into greater **partial negative charges** residing on more **electronegative** atoms and smaller **partial negative charges** residing on less **electronegative** atoms. Regarding trends, as previously discussed, relative **electronegativities** of atoms are readily identified using the **periodic table of elements**. For example, moving across each row of the **periodic table**, we find that **electronegativity** increases among the atoms, thus indicating that oxygen is more **electronegative** than nitrogen and that fluorine is more **electronegative** than oxygen.

As discussed in previous chapters, **pK_a** values are directly related to **electronegativities**. This is clearly reflected in the nitrogen–oxygen–fluorine trend discussed earlier considering that amine **pK_a** values are approximately 35, alcohol **pK_a** values are approximately 15–19, and the **pK_a** value for hydrofluoric acid is approximately 3. The increase in **acidity** is related to increased **electronegativity** because more **electronegative** atoms are more stable in their anionic form. Thus, fluorine can stabilize a negative charge better than oxygen, which, in turn, can stabilize a negative charge better than nitrogen.

Since **nucleophiles**, by definition, are attracted to positively charged centers, we must consider how such centers become attractive to **nucleophiles**. This effect relates to the **electronegativity** of the atoms attached to the center. Since an **electronegative** atom retains a **partial negative charge**, the **electronegative** atom pulls electron density from the atom on which it resides (usually carbon) and leaving it with a **partial positive charge**. This effect is illustrated in Figure 5.2 where an **electronegative** chlorine atom is attached to carbon and inducing a **partial positive charge**.

When a bond joins an atom bearing a **partial positive charge** to an atom bearing a **partial negative charge**, the bond is said to be **polarized**. Much in the same way a magnet possesses a positive pole and a negative pole, the respective ends of a **polarized bond** are positively and negatively charged. When referring to **polarity** and **polarized bonds**, the direction of **polarity** is, by convention, from **positive** to **negative**. As shown in Figure 5.3, this is commonly illustrated using a special arrow with a + at the positive end and the tip pointing in the direction of the negative end.

Because **nucleophiles** are attracted to sites of **positive** or **partial positive charges**, understanding the direction of **polarity** associated with a given bond serves three purposes. First, the site to which a given **nucleophile** is attracted is readily identified. Second, the

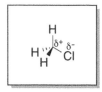

Figure 5.2 *Chloromethane bears a partial negative charge on the electronegative chlorine atom and a partial positive charge on the carbon atom.*

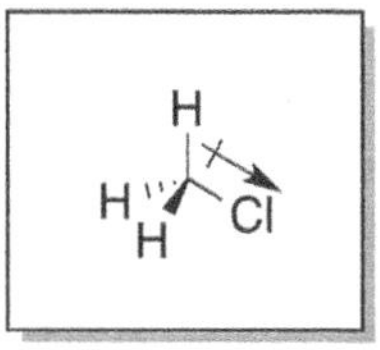

Figure 5.3 *The carbon–chlorine bond in chloromethane is polarized.*

spatial direction or trajectory of the reaction is readily identified. Lastly, the **leaving group** is readily identified. These are all graphically summarized in Scheme 5.5 using a different rendering of the **S$_N$2** reaction illustrated in Scheme 5.3.

In Chapter 3, the relationship between **nucleophiles** and **bases** as influenced by **steric bulk** was addressed. What was not addressed is the complementary issue surrounding the accessibility of **electrophilic** (positively charged or partially positively charged) sites to **nucleophiles**. In fact, in the same way that **nucleophilicity** decreases with increasing **steric bulk** around the **nucleophilic** atom, the ability of a **nucleophile** to react with an **electrophile** also decreases with increasing **steric bulk** around the site of potential **S$_N$2** reactions. This effect is illustrated in Scheme 5.6 using the reaction introduced in Scheme 5.3. As illustrated, successive introduction of methyl groups adjacent to the **S$_N$2** reaction site results in **decreased reaction rates**.

Thus, when identifying sites where **S$_N$2** reactions can occur, the following criteria must be met. First, **S$_N$2** reactions occur at **tetrahedral** carbon atoms. Second, **S$_N$2** reactions occur at molecular sites bearing the greatest degree of **positive charge**. Lastly, **S$_N$2** reactions occur at sites that are **sterically accessible** to the incoming **nucleophile**.

5.4 S$_N$2′ REACTIONS

In Section 5.3, detailed discussions were presented illustrating conditions and criteria relevant to the initiation of **S$_N$2** reactions. Furthermore, in previous chapters, the concept of **resonance** was introduced. When considering mechanistic organic chemistry, **resonance** is frequently recognized as contributing to the outcome of chemical reactions. In the following paragraphs, these contributions relative to **S$_N$2** reactions are presented.

Resonance, as introduced in Chapter 3, explains stability of **anions** and rationalizes trends in **pK_a** values. However, **resonance** can also be used to rationalize the stability of **cations** (positively charged ions). As shown in Scheme 5.7, the stability of the cycloheptatriene cation is explained by its resonance forms. There is, of course, another reason for the stability of the cycloheptatriene cation that relates to the principles of **aromaticity** and will not be discussed in detail in this book.

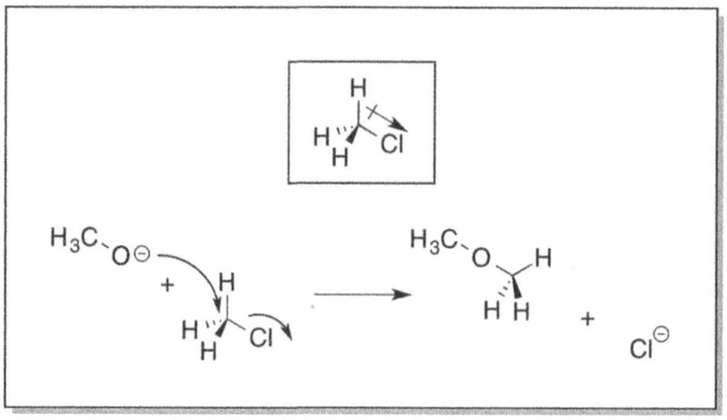

Scheme 5.5   *Understanding the direction of bond polarity allows identification of reaction site, trajectory of nucleophile, and identification of the leaving group.*

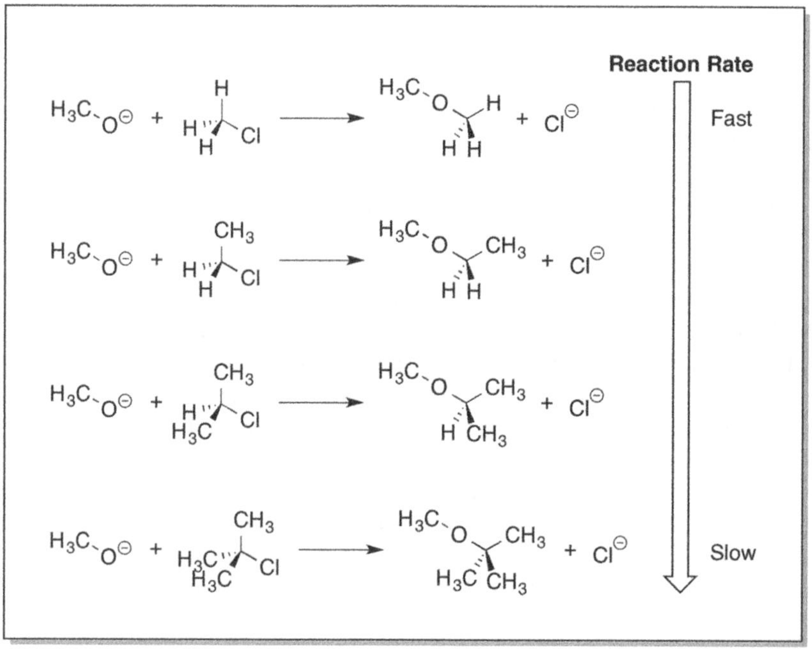

Scheme 5.6   *Steric bulk slows down reaction rates for S$_N$2 reactions.*

As positively and negatively charged ions can be stabilized through resonance forms, so can species bearing partial positive charges and partial negative charges. As previously discussed, an electronegative atom attached to a carbon atom will induce a partial positive charge onto the carbon atom. As illustrated in Figure 5.4, when substituents possessing bond unsaturation are also attached to the partially positive carbon, the partial positive and partial negative charges are extended into the unsaturated system.

When **partial positive charges** are **delocalized** through **unsaturated** bonds, the result, as illustrated earlier, is the presence of multiple sites to which **nucleophiles** will be attracted. This principle is illustrated in Scheme 5.8 and explained through **arrow-pushing**. As shown, when a **nucleophile** reacts with the terminal carbon atom, electrons from the double bond shift to displace the chloride ion. Since this is a **bimolecular nucleophilic**

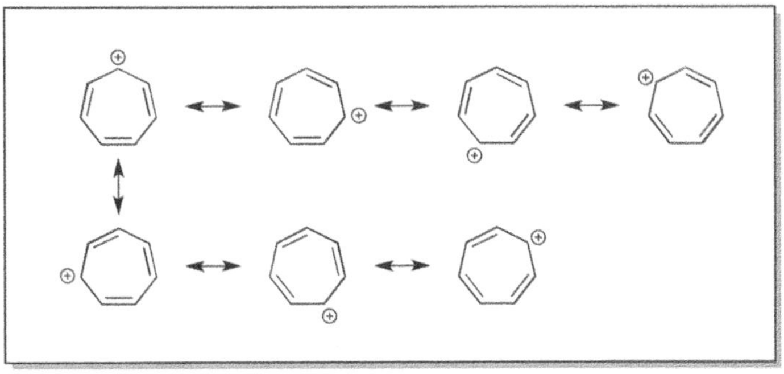

Scheme 5.7 *Resonance forms can be used to rationalize the stability of cations adjacent to sites of bond unsaturation.*

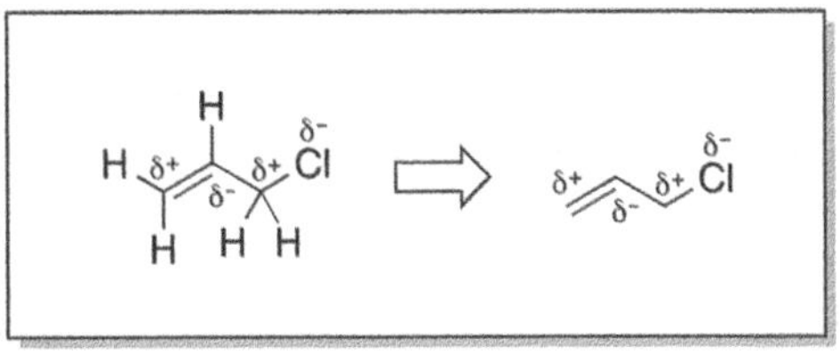

Figure 5.4 *Partial charges can be delocalized through unsaturated bonds.*

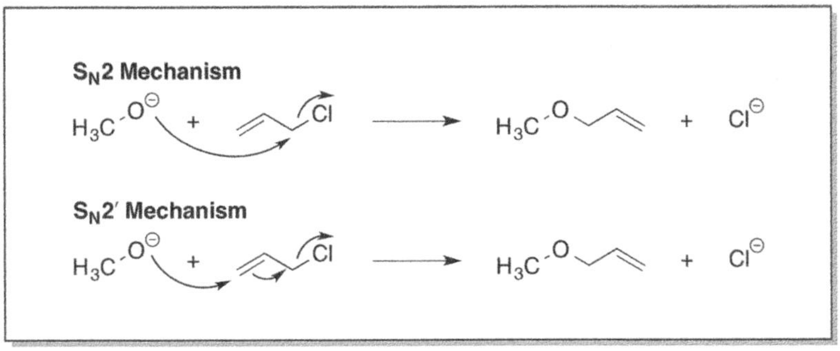

Scheme 5.8 *Comparison of S_N2 and S_N2' reactions as explained with arrow-pushing.*

substitution, the mechanism type falls within the definition of S_N2 reactions. However, since this reaction occurs through a double bond, or an extended **conjugated** system, it is designated S_N2'. In the case of the example shown in Scheme 5.8, the product is formed through a combination of S_N2 and S_N2' mechanisms.

In the case of the reaction illustrated in Scheme 5.8, the product is not dependent upon which site the **nucleophile** is drawn to or which mechanism the reaction proceeds through. However, when the double bond possesses an additional substituent, product mixtures can form as illustrated in Scheme 5.9. In general, when predicting the outcome of reactions where both S_N2 and S_N2' reactions are possible, the major product will be dependent upon the **steric** constraints around the various reactive (or partially positively charged) sites. In cases where the **nucleophile** has comparable accessibility to **electrophilic** sites, product mixtures are usually the result.

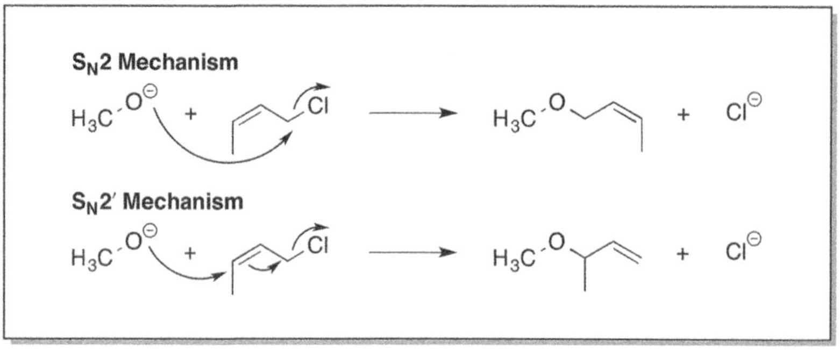

Scheme 5.9 *Competing S_N2 and S_N2' reaction mechanisms can lead to product mixtures.*

5.5 SUMMARY

In this chapter, S_N2 reaction mechanisms were defined and presented in the context of **nucleophiles** displacing **leaving groups** at **electrophilic** centers. Furthermore, the conditions required for S_N2 reactions to proceed were discussed as well as factors that influence the progression of such reactions. In this context, discussions of S_N2 reactions were extended into the related S_N2' reaction mechanisms.

As the principles of this chapter, by nature, build upon those presented in previous chapters, the same will be for topics discussed in the remainder of this book. The study of organic chemistry is a progressive task with many overlapping principles. As should be apparent from topics discussed thus far, many of these principles reduce to the acid–base chemistry presented in the earliest chapters. This is also the case for the remaining topics presented herein.

PROBLEMS

1. In many S_N2 reactions, the nucleophile is generated by deprotonation of an organic acid. For each molecule, chose the base best suited to completely remove the labeled proton. (Consider pK_a values and recognize that, in some cases, dianions should be considered.) Explain your answers.

a. 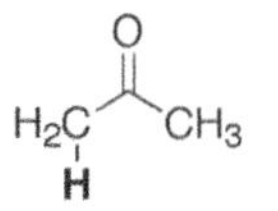 H_2C—$\overset{O}{\overset{\|}{C}}$—$CH_3$ **NaOCH$_3$: (CH$_3$)$_2$NLi : CH$_3$Li**
            $\underset{H}{|}$

b. 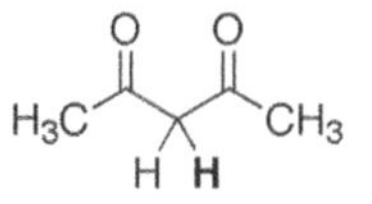 H_3C—$\overset{O}{\overset{\|}{C}}$—$\underset{H}{\overset{|}{C}}$—$\overset{O}{\overset{\|}{C}}$—$CH_3$ **NaOCH$_3$: (CH$_3$)$_2$NLi : CH$_3$Li**

c. 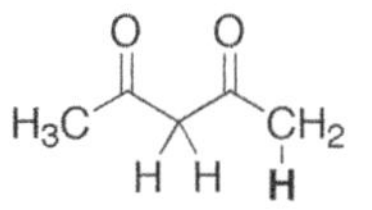 H_3C—$\overset{O}{\overset{\|}{C}}$—$\underset{H}{\overset{|}{C}}$—$\overset{O}{\overset{\|}{C}}$—$CH_2$ **NaOCH$_3$: (CH$_3$)$_2$NLi : CH$_3$Li**
                $\underset{H}{|}$

d. 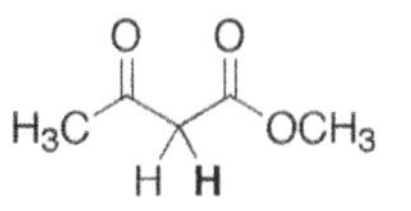 H_3C—$\overset{O}{\overset{\|}{C}}$—$\underset{H}{\overset{|}{C}}$—$\overset{O}{\overset{\|}{C}}$—$OCH_3$ **NaOH : NaOCH$_3$: NaOCH$_2$CH$_3$**

2. In predicting the course of S$_N$2 reactions, it is important to recognize groups most likely to act as nucleophiles. For each molecule, label the most nucleophilic site.

a.

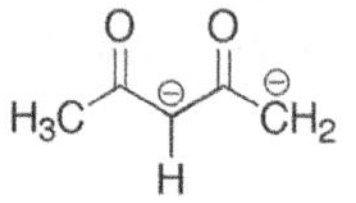

b.

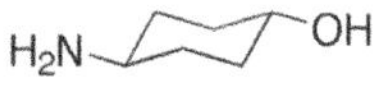

c.

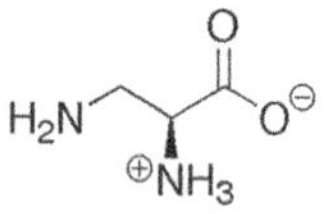

d. 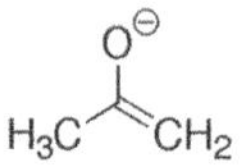 (Hint: consider resonance)

3. For each molecule, show the partial charges, bond polarity, and where a nucleophile is most likely to react.

a.

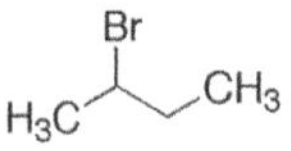

b.

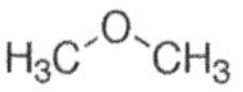

c.

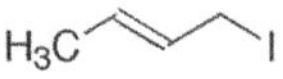

d.

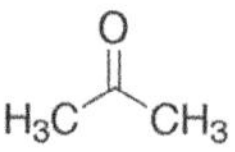

4. For each molecule, identify the leaving group assuming that H_3C^- is the nucleophile.

a.

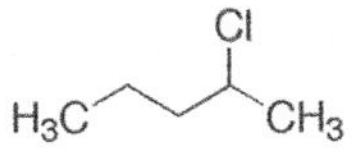

b.

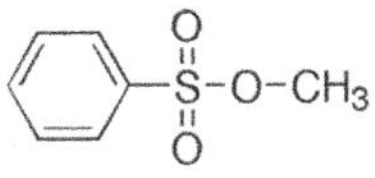

c.

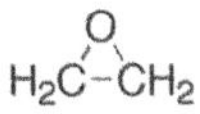

5. For each molecule, label the most likely leaving group. Explain your answers.

a.    Br⌒⌒OCH₃

b.  $(H_3C)_3\overset{\oplus}{N}$—⟨⟩—$\overset{\oplus}{O}(CH_3)_2$

c.    Br⌒⌒Cl

d.

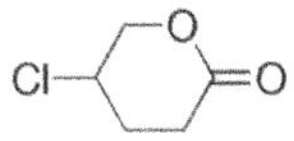

6. Detailed discussions focused on stereochemistry are not within the scope of this book. However, considering the products of typical S_N2 reactions, in addition to the transition state shown in Scheme 5.2, one may deduce the stereochemical course of this type of reaction. Predict the product of the following reaction and show the correct stereochemistry.

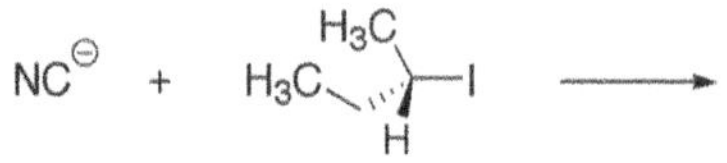

7. Predict the products of the following reactions by pushing arrows.

a. 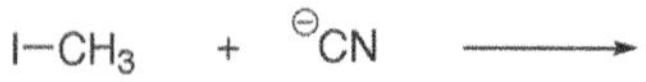 $I-CH_3$ + $^{\ominus}CN$ ⟶

b.

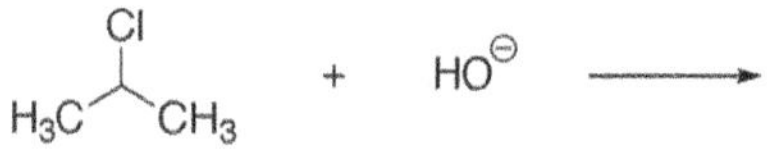

c.

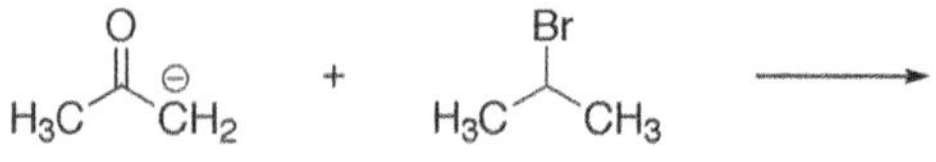

d. 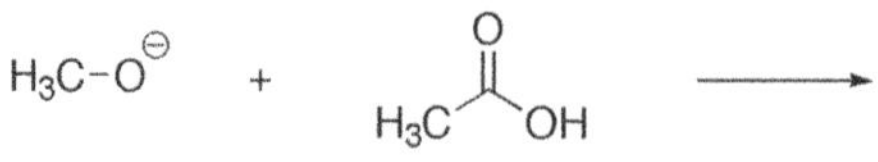 $H_3C-O^{\ominus}$ +

e. 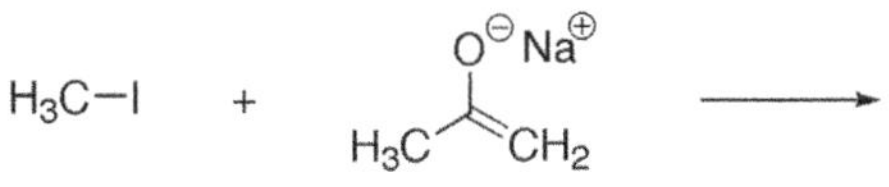 $H_3C{-}I$ + $\longrightarrow$

f. $K^{\oplus}F^{\ominus}$ 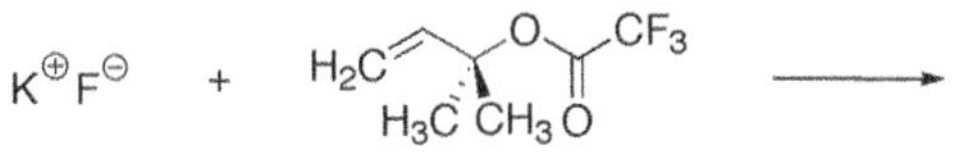   + $\longrightarrow$

g. 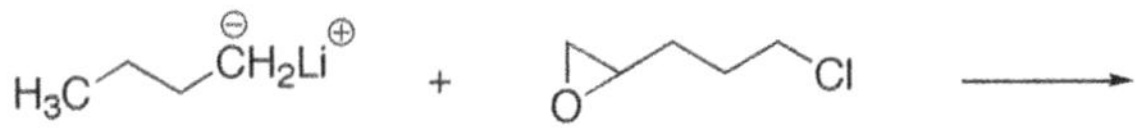   + $\longrightarrow$

h. 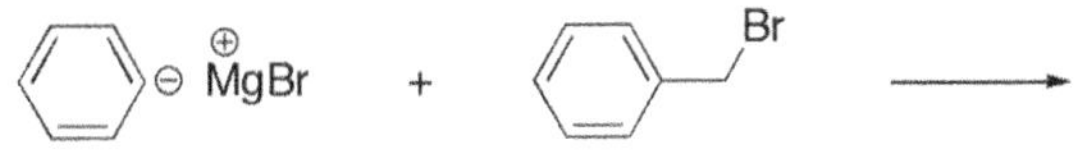   + $\longrightarrow$

i. 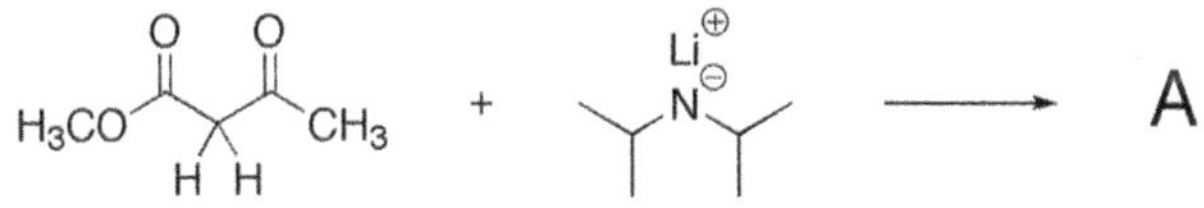   + $\longrightarrow$ A

j.   A   +   ⟶

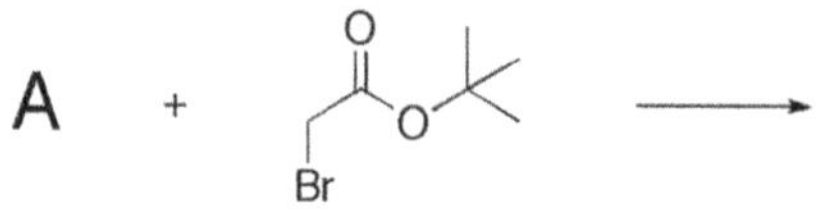

k. $\xrightarrow{\text{NaOH}}$ B   ⟶   C

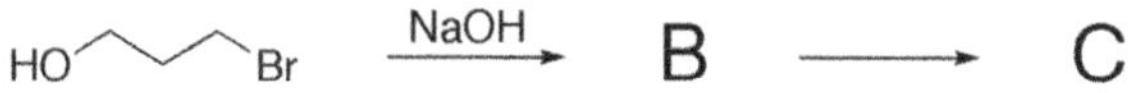

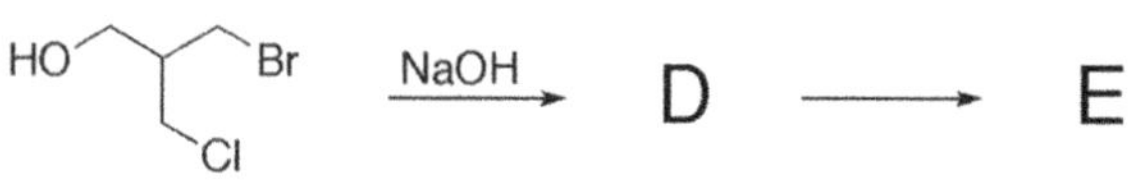

l. $\xrightarrow{\text{NaOH}}$ D   ⟶   E

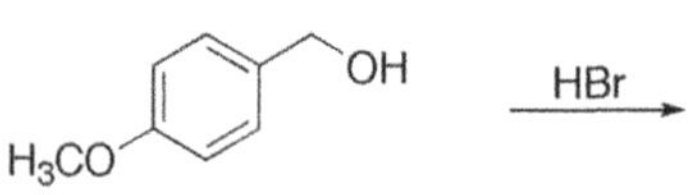

m. $\xrightarrow{\text{HBr}}$

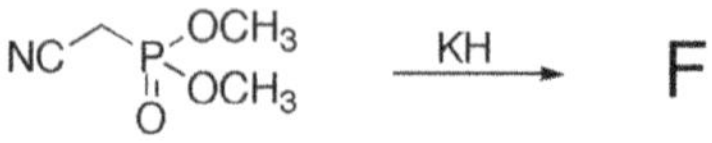

n. $\xrightarrow{\text{KH}}$ F

o.   F   +   ⟶

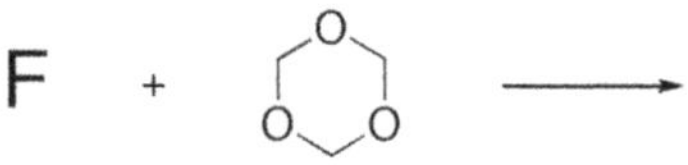

8. Addition reactions and conjugate addition reactions, to be discussed in Chapter 8, are related to S_N2 and S_N2' reactions, respectively. We can make these comparisons if we recognize that the carbonyl double bond contains a leaving group. Specifically, if a nucleophile adds to the carbon of a carbonyl, the carbonyl double bond becomes a carbon–oxygen single bond with a negative charge residing on the oxygen. Additionally, the trigonal planar geometry of the carbonyl carbon is converted to tetrahedral geometry. With these points in mind, predict the products of the following reactions and explain your answers. For Problem 8b, the nucleophile is a methyl anion associated with the illustrated cuprate.

a.

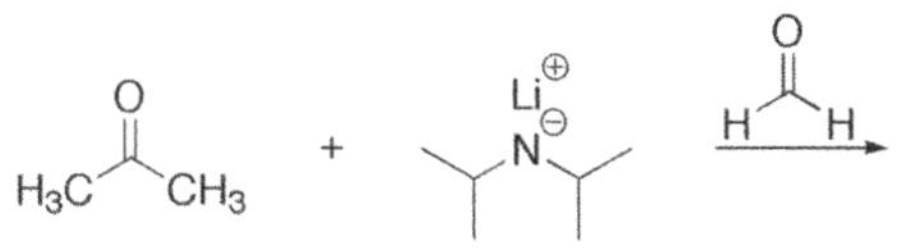

b.

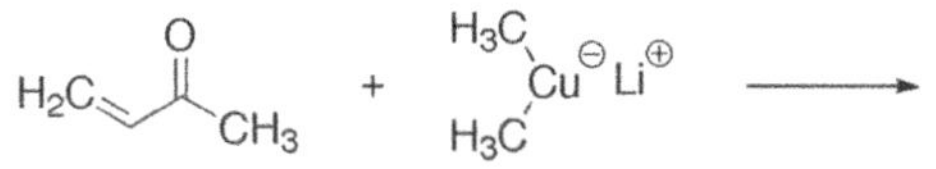

9. Propose a reasonable mechanism for each of the following reactions. Explain your answers by pushing arrows.

a.

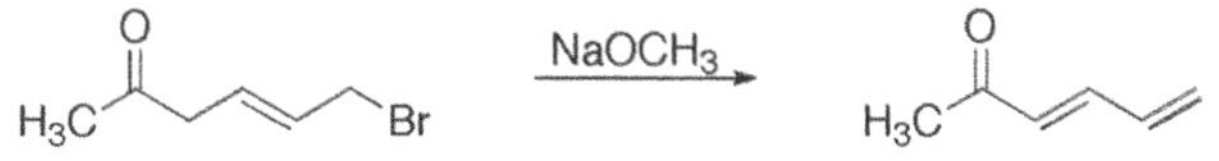

b.

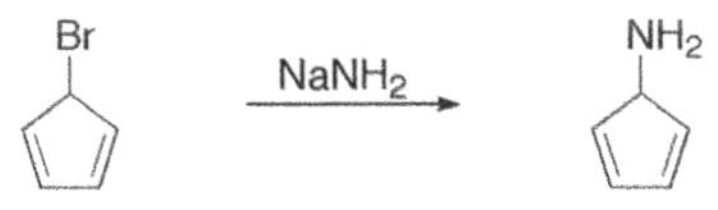

c.    HO ⌒⌒⌒ Br   —NaH→   [image: vinyl tetrahydrofuran]

d.    [image: vinyl tetrahydrofuran]   —NaNH$_2$→   HO ⌒⌒⌒ NH$_2$

10.   α,β-Unsaturated carbonyls are readily formed from the corresponding β-hydroxy ketones. Explain the product of the following reaction.

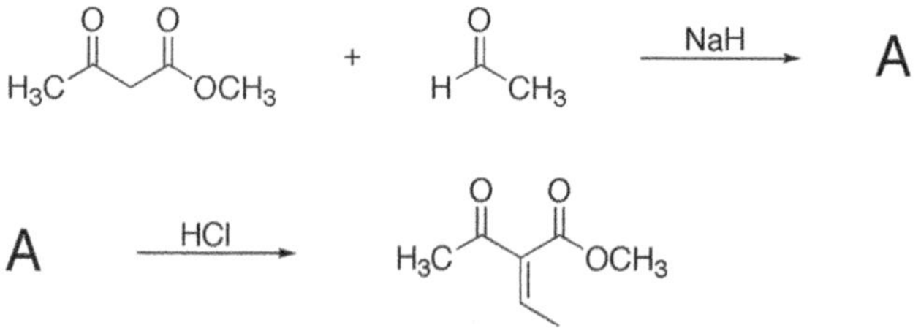

Chapter 6

S_N1 Substitution Reactions

In Chapter 5, S_N2 reactions were defined and presented in context of the various conditions necessary for such reactions to take place. However, as mentioned in the introductory comments of Chapter 5, there are additional fundamental mechanistic types relevant to organic chemistry of which understanding is essential in order to advance in this subject. This chapter focuses on a study of the mechanism of S_N1 reactions. While conditions required for S_N1 reactions to proceed are quite different from those essential for S_N2 reactions, the products of S_N1 reactions, in many cases, resemble those derived from S_N2 mechanisms. Additionally, unlike S_N2 reactions, S_N1 reaction mechanisms sometimes result in unwanted or, in some planned cases, preferred side reactions.

6.1   WHAT IS AN S_N1 REACTION?

As discussed in Chapter 5, an S_N2 reaction proceeds with the substitution of a leaving group with a **nucleophile** leading to the S_N designation. Because this mechanism proceeds with the initial approach of two species, it is referred to as a bimolecular reaction. The involvement of **2** species enhances the mechanistic designation to S_N2. Extrapolating from this definition, an S_N1 reaction also proceeds with the substitution of a leaving group with a **nucleophile** leading to the S_N designation. However, unlike S_N2 reactions, S_N1 reactions proceed through initial dissociation of the **leaving group** from the starting material prior to participation of the **nucleophile** as shown in Scheme 6.1. This occurs because of differences in reactants and reaction conditions as compared to those relevant to S_N2 processes. Once the leaving group has dissociated, the resulting **carbocation** (a carbon atom possessing a positive charge) is free to react with a **nucleophile** as shown in Scheme 6.2. Because the initial step in this reaction involves a single molecule dissociating from its leaving group, initial stage of this

Arrow-Pushing in Organic Chemistry: An Easy Approach to Understanding Reaction Mechanisms,
Second Edition. Daniel E. Levy.
© 2017 John Wiley & Sons, Inc. Published 2017 by John Wiley & Sons, Inc.

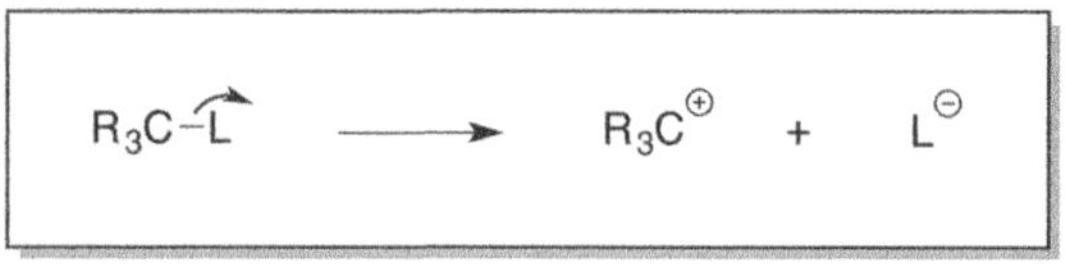

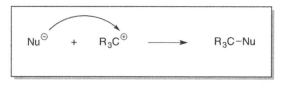

Scheme 6.1 *The initial phase of an S$_N$1 reaction involves dissociation of a leaving group from the starting molecule generating a carbocation.*

Scheme 6.2 *The second phase of an S$_N$1 reaction involves reaction of a carbocation with a nucleophile generating a new product.*

process is considered a unimolecular reaction. The involvement of only **1** species in the initial phase of the reaction enhances the mechanistic designation to **S$_N$1**.

Because, with **S$_N$1** reactions, a reactive **carbocation** is formed before incorporation of a **nucleophile**, other products may form in addition to the simple substituted materials anticipated. These additional products arise from the specific properties of **carbocations**. The properties of **carbocations** and their related mechanistic outcomes are presented in the following sections.

6.2 HOW ARE S$_N$1 REACTIONS INITIATED?

In order for an **S$_N$1** reaction to proceed, initial formation of a **carbocation** is required. A primary method for the formation of **carbocations** occurs during **solvolysis** reactions. **Solvolysis** reactions, illustrated in Scheme 6.3, involve the reaction of an organic molecule with the surrounding **solvent**. Furthermore, these reactions generally proceed through initial separation of a **carbocation** from its **leaving group** followed by reaction of the **carbocation** with a surrounding **solvent** molecule forming a new compound. The example shown in Scheme 6.3 illustrates these steps for the reaction of *tert*-butylbromide with methanol forming methyl *tert*-butylether (MTBE).

As shown in Scheme 6.4, the previously described **solvolysis** reaction can be explained using **arrow-pushing**. Specifically, initial separation of the bromide **leaving group** from the *tert*-butyl **cation** proceeds with electrons residing on the bromide **anion**. Subsequent reaction of the *tert*-butyl **cation** with **lone pairs** of **electrons** donated by the **solvent** (methanol) molecules results in the formation of a new carbon–oxygen bond. Dissociation of hydrogen from the resulting **oxonium** (oxygen cation) ion liberates the product (MTBE) and hydrobromic acid. As a direct reference to the definition of **S$_N$1** reactions, it is important to recognize that the first step (the rate-limiting step) involves only *tert*-butylbromide, thus rendering this step **unimolecular**.

In general, **solvolysis** reactions occur under circumstances where a molecule possessing an exceptionally good **leaving group** is dissolved in a polar **solvent**. Under these conditions, the **polarity** of the **solvent** renders formation of the **carbocation** more favorable by selectively **solvating** either the **carbocation**, its accompanying **anion**, or both. Once the **carbocation** is separated from its **anion**, it may undergo typical **S$_N$1** reactions as discussed

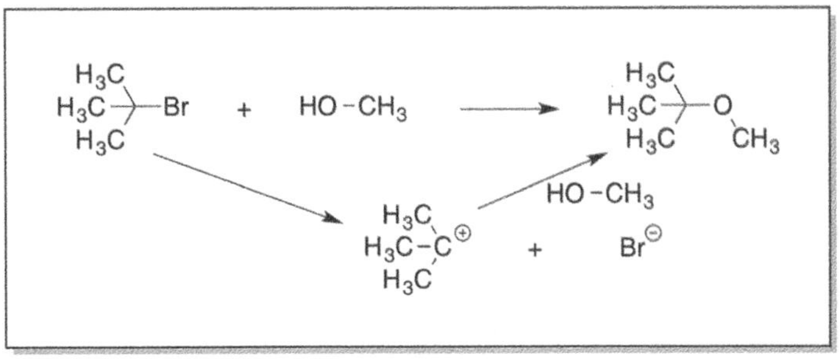

Scheme 6.3 *Solvolysis of* tert-*butylbromide in methanol produces MTBE via an S$_N$1 mechanism.*

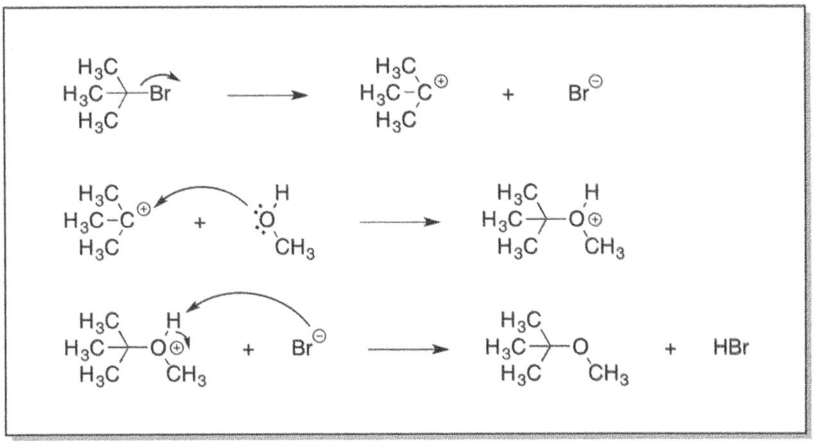

Scheme 6.4 *Explanation of the solvolysis of* tert-*butylbromide in methanol using arrow-pushing.*

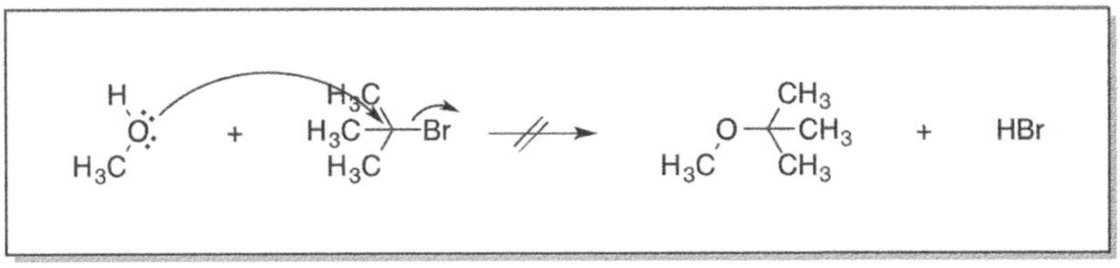

Scheme 6.5 *Methanol will not react with* tert-*butylbromide via an S$_N$2 mechanism.*

in the following paragraphs. As depicted in Scheme 6.5, the reaction illustrated in Scheme 6.4 will not proceed by an **S$_N$2** mechanism because of the **steric bulk** of the starting *tert*-butylbromide. Additional discussions surrounding the influence of **steric factors** were presented in Chapters 4 and 5.

6.3 THE CARBOCATION

As defined in the previous sections of this chapter, **carbocations** are positively charged carbon ions. However, simply defining this unique species of cations without exploring its associated properties does little to promote understanding of **S$_N$1** reactions and the related side reactions observed for this mechanistic type. Therefore, this section focuses on the nature, stability, and reactivity of **carbocations** as explained using **arrow-pushing**. While the alluded to side reactions include both **elimination** reactions and **rearrangements**, only **rearrangements** are presented in this chapter. Discussions focused on **eliminations** are found beginning in Chapter 7.

6.3.1 Molecular Structure and Orbitals

Before delving into more details regarding the reactive nature and stability of **carbocations**, it is important to understand the structure of these species. Recall that S_N2 reactions occur at carbon atoms bearing four substituents. Furthermore, recall that **electrophilic** carbon centers participating in S_N2 reactions are **tetrahedral** in geometry with all bond angles measuring approximately 109.5°—the tetrahedral bond angle. This equal spacing, illustrated in Figure 6.1, is only possible if the natures of all four bonds connecting the central carbon atom to its four substituents are identical.

Since an understanding of **orbital** theory is critical to understanding organic reaction mechanisms, **review of the relevant material presented in primary organic chemistry textbooks is essential**. For the purposes of this discussion, recall that ground state first row elements (including C, N, and O) all possess one *s*-orbital and three *p*-orbitals. Figure 6.2 illustrates the shapes of *s*- and *p*-orbitals.

If we consider methane (CH_4), we find that not only does the central carbon atom possess four hydrogen substituents, but also these four hydrogens are equally spaced in a **tetrahedral** arrangement with equal bond lengths. As *s*-orbitals and *p*-orbitals are spatially different, this level of structural equality cannot be explained through bonding with one *s*-orbital and three *p*-orbitals. Instead, this equality is explained by combining the single *s*-orbital with the three *p*-orbitals forming four equal *sp³* hybrid orbitals. Figure 6.3 illustrates the various **hybrid orbitals** that are involved in most chemical bonds found in organic chemistry.

Expanding upon Figure 6.3, an *sp* hybrid orbital is made up of one part *s*-orbital and one part *p*-orbital. Furthermore, an *sp²* hybrid orbital is made up of one part *s*-orbital and two parts *p*-orbital. Finally, an *sp³* hybrid orbital is made up of one part *s*-orbital and three parts *p*-orbital. In cases such as *sp* and *sp²* hybridization where only a subset of the three *p*-orbitals are used in forming hybrid orbitals, the **unhybridized *p*-orbitals** are utilized in the formation of double and triple bonds.

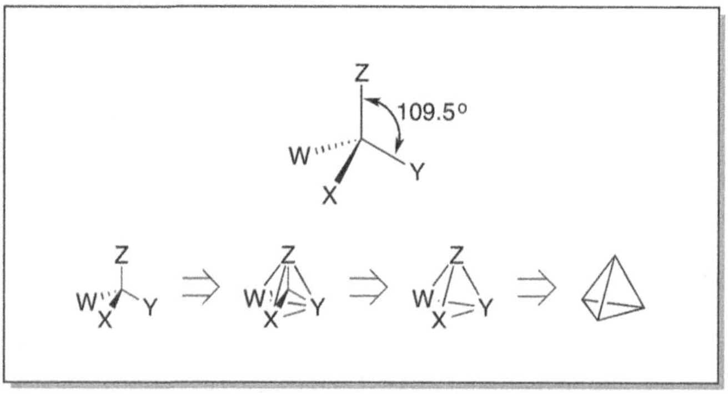

Figure 6.1 *Fully substituted carbon atoms present substituents in tetrahedral arrangements.*

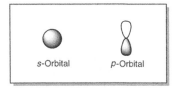

Figure 6.2 s-Orbitals are spherical and p-orbitals are shaped like hourglasses.

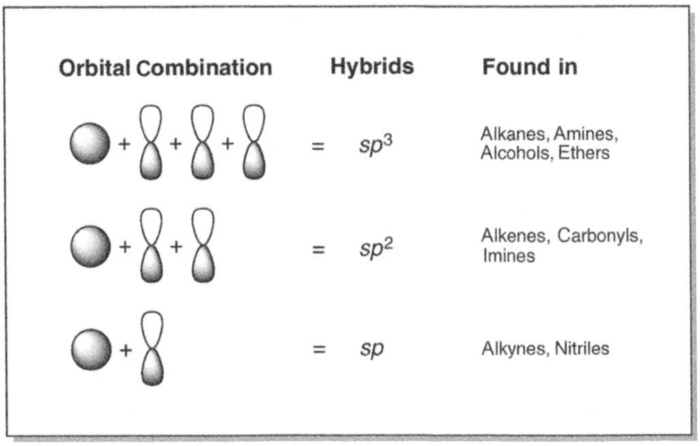

Orbital Combination	Hybrids	Found in
⬤ + ∞ + ∞ + ∞ =	sp^3	Alkanes, Amines, Alcohols, Ethers
⬤ + ∞ + ∞ =	sp^2	Alkenes, Carbonyls, Imines
⬤ + ∞ =	sp	Alkynes, Nitriles

Figure 6.3 Hybrid orbitals result from combinations of s- and p-orbitals.

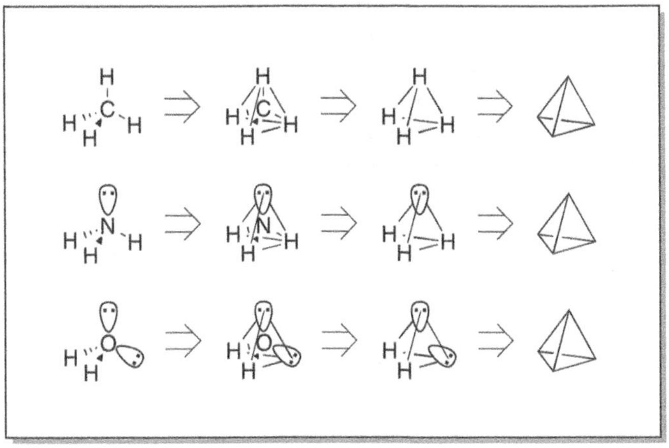

Figure 6.4 Like substituents, lone pairs influence molecular geometry.

While the present discussions focus on **orbital hybridization** relative to bonds between atoms, it is important to recognize that non-bonding electron pairs (**lone pairs**) also partici- pate in **orbital hybridization**. Thus, as illustrated in Figure 6.4 and relating to *sp³* **hybrid- ized** centers, for the purposes of determining **orbital hybridization**, **lone pairs** can be treated as bonds between a central atom and nothing.

As alluded to in Figure 6.3, *sp³* **hybridization** occurs when a central atom possesses a total of four substituents comprised of any combination of atoms and **lone pairs**.

Furthermore, *sp²* **hybridization** occurs when a central atom possesses a total of three substituents comprised of any combination of atoms and **lone pairs**. Finally, *sp* **hybridization** occurs when a central atom possesses a total of two substituents comprised of any combination of atoms and **lone pairs**. While thus far attention has been focused on the **tetrahedral** nature of *sp³* **hybridized** atoms, exploring the geometric consequences of *sp²* and *sp* **hybridized** atoms reveals very different spatial relationships between substituents. Specifically, as shown in Figure 6.5, the three substituents of an *sp²* **hybridized** atom adopt a **trigonal planar** relationship with bond angles of 120° and all substituents residing in the same plane. Furthermore, the two substituents of an *sp* **hybridized** atom adopt a **linear** relationship with bond angles of 180°.

Having addressed the geometric consequences of **orbital hybridization**, the previous discussions can now be related back to **carbocations**. Recalling the rules relating the number of substituents to specific **orbital hybrids**, we recognize that a **carbocation** possesses a maximum of three substituents and is thus rendered as no more than *sp²* **hybridized**. Furthermore, the **carbocation** positive charge resides in an unoccupied *p*-orbital. The **trigonal planar** structure of an *sp²* **hybridized carbocation** is illustrated in Figure 6.6 and enhanced with the placement of a *p*-orbital at the cationic center.

Having established the three-dimensional structures of **carbocations** as **planar**, we can now study the **stereochemical progression** of S_N1 reactions as compared to S_N2 reactions. As shown in Scheme 6.6, the **stereochemical** course of an S_N2 reaction is well defined because **nucleophilic displacement** of a **leaving group** proceeds with **inversion** of **stereochemistry**. Thus, the stereochemical outcome is defined by the stereochemistry of the starting material. As for S_N1 reactions, since the step required for initiation of these reactions involves formation of a **planar** species, incoming **nucleophiles** have equal accessibility to both sides of the reactive **carbocation**. As shown in Scheme 6.7, this results in complete elimination of **stereochemical control** over these reactions. Thus, where S_N2 reactions on **stereochemically pure** starting materials proceed with generation of a single **stereoisomer**, S_N1 reactions proceed with complete loss of **stereochemical identity** even when the starting material is **stereochemically pure**. Specifically, an S_N2 reaction on a chiral starting material yields **one chiral product**, and an S_N1 reaction on a chiral starting material yields a **racemic mixture of two stereoisomers**.

Figure 6.5 *Different orbital hybridizations results in different molecular geometries.*

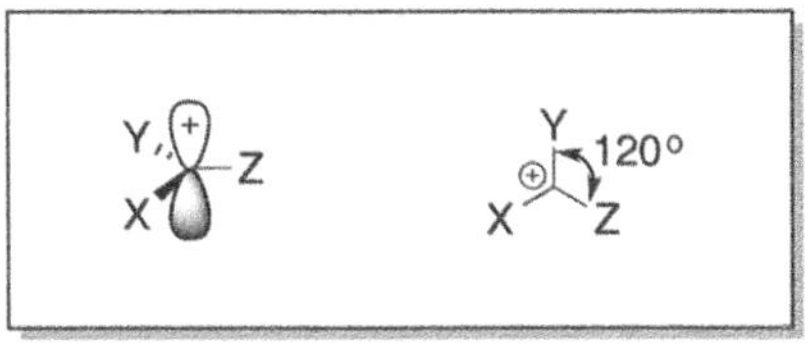

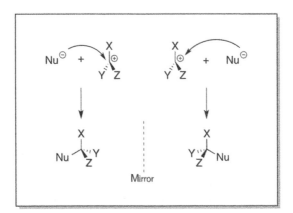

Figure 6.6 sp² *Hybridized carbocations possess trigonal planar geometries.*

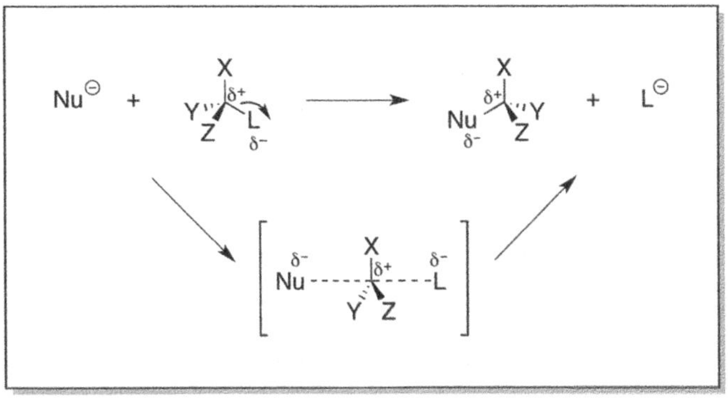

Scheme 6.6 *The stereochemical courses of S$_N$2 reactions are defined by the stereochemical configuration of the starting materials—one product is formed.*

Scheme 6.7 *The stereochemical identities of starting materials subjected to S$_N$1 reactions are lost due to the planarity of reactive carbocations—two products are formed.*

6.3.2 Stability of Carbocations

As alluded to at the beginning of this section, **carbocations** generated during **S$_N$1** mechanisms are subject to **side reactions** that include **eliminations** and **rearrangements**. Considering the possibility of these **side reactions**, one must consider the **stability** of **carbocationic** species. To clarify, if **carbocations** were inherently stable, they would not be readily subject to additional transformations. Having already addressed the structure of **carbocations**, attention can now be focused on the factors influencing stability.

In studying **carbocations**, it is important to recognize that **tertiary carbocations** are more stable than **secondary carbocations**. Furthermore, **secondary carbocations** are

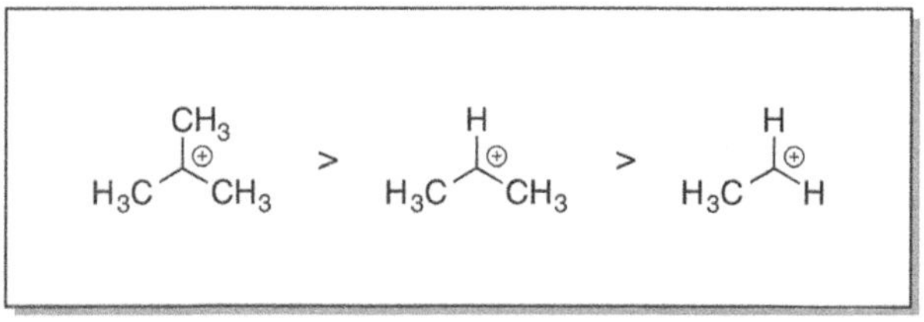

Figure 6.7 *Tertiary carbocations are more stable than secondary carbocations, and secondary carbocations are more stable than primary carbocations.*

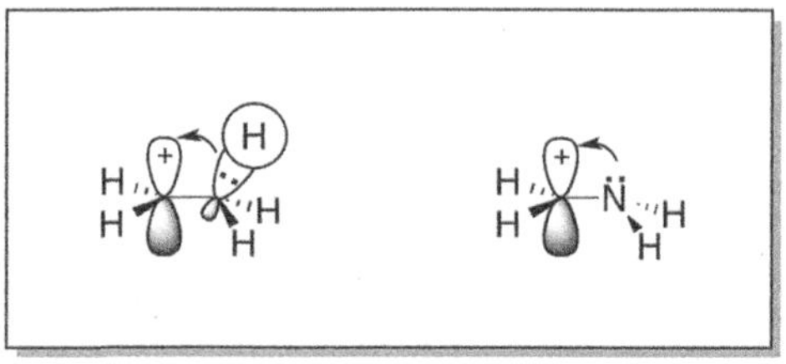

Figure 6.8 *Hydrogen atom s-orbitals can donate electron density to adjacent cationic centers as can heteroatoms bearing lone electron pairs.*

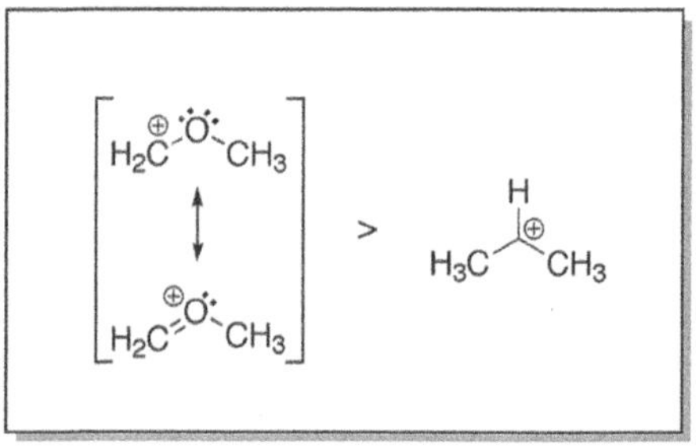

Figure 6.9 *Heteroatoms stabilize carbocations better than hyperconjugation effects.*

more stable than **primary carbocations**. This relationship, shown in Figure 6.7, results from an effect known as **hyperconjugation**. Specifically, **hyperconjugation**, illustrated in Figure 6.8, defines the ability of a hydrogen atom to donate electron density from its **s-orbital** to sites of neighboring electron deficiency. This effect is similar to the stabilization of **carbocations** bearing heteroatoms with lone electron pairs. Thus, the greater the number of carbon–hydrogen bonds located adjacent to a positive charge, the greater the stability of the cation.

As **hyperconjugation** can be related to **cationic** stabilization by neighboring **lone pairs**, relationships between these types of effects must be noted. As shown in Figure 6.9, heteroatom-induced stabilization is a stronger effect than **hyperconjugation**.

With the understanding that **hyperconjugation** and heteroatoms both stabilize **cations** through **resonance effects**, the influence of full **conjugation** to sites of unsaturation deserves mention. As shown in Figure 6.10, direct conjugation is generally a stronger effect than **hyperconjugation**. This effect is illustrated with an **allylic carbocation** compared to a **secondary carbocation**. However, if we consider a **tertiary carbocation**, as shown in

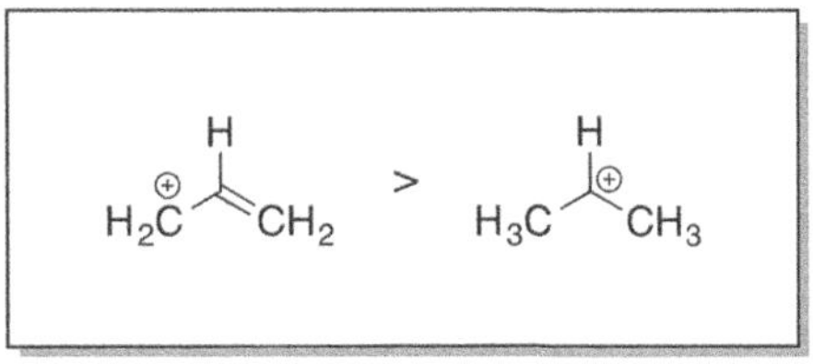

Figure 6.10 *Allylic carbocations are more stable than secondary carbocations.*

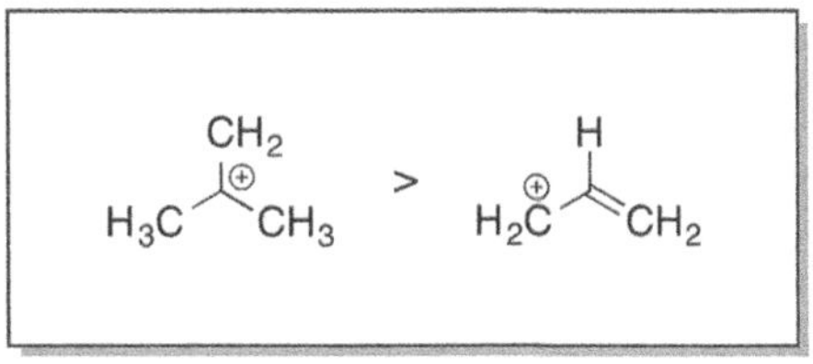

Figure 6.11 *Tertiary carbocations are more stable than allylic carbocations.*

Figure 6.11, this trend is reversed, thus emphasizing that while **resonance** stabilization is good, it is not as good as the stabilization obtained by having three alkyl groups associated with the cation.

6.4 CARBOCATION REARRANGEMENTS

Having addressed the **structure** and **stability** of **carbocations**, discussions will now be centered on the specific **side reactions** to which **carbocations** are susceptible. Specifically, this section focuses on rearrangements of carbocations known as **hydride shifts** and **alkyl shifts**.

6.4.1 1,2-Hydride Shifts

Recalling the role played by **hyperconjugation** in the **stabilization** of **carbocations**, a more detailed examination of this phenomenon is warranted. Looking back at Figure 6.6, we note that **carbocations** are **planar** with an unoccupied *p*-orbital extending both above and below the plane of the ion. Furthermore, looking back at Figure 6.8, the electrons in a carbon–hydrogen bond adjacent to a **carbocation** can **conjugate** toward the **positive charge** residing in the vacant *p*-orbital. This donation of electron density can only occur if the carbon–hydrogen bond is aligned with the vacant *p*-orbital as shown in Figure 6.12 using several perspective views. Specifically, the carbon–hydrogen bond must lie in the same plane as the vacant *p*-orbital.

When the **alignment** of a carbon–hydrogen bond with a vacant *p*-orbital takes place allowing for **hyperconjugation**, a "pseudo double bond" develops. As illustrated in Figure 6.13, this can be envisioned as a double bond with a closely associated hydrogen ion.

If, as shown in Figure 6.13, **hyperconjugation** results in the formation of species possessing both double bond character and associated hydrogen ions, **equilibrium-controlled migration** of the associated hydrogen ion can be expected. This transformation, shown in

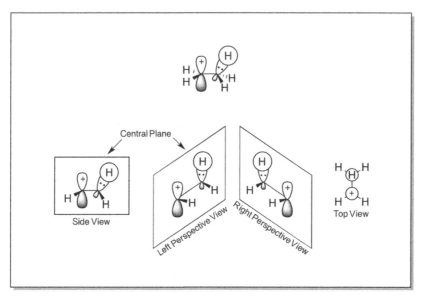

Figure 6.12   *Hyperconjugation occurs when a carbon–hydrogen bond lies in the same plane as a carbocation's vacant p-orbital.*

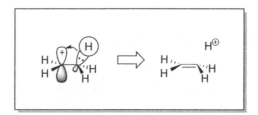

Figure 6.13   *Hyperconjugation can be viewed as formation of a "pseudo-double bond."*

Scheme 6.8, is known as a **1,2-hydride shift** and results in the migration of a proton from carbon 1 to carbon 2.

While the example illustrated in Scheme 6.8 shows **equilibrium** between two chemically identical **carbocations**, there are factors influencing the direction of these transformations when applied to more complex systems. If we consider Scheme 6.9, we notice that the positive charge migrates exclusively to the tertiary center reflecting the increased stability of **tertiary carbocations** over **primary carbocations**. In general, where **1,2-hydride** shifts are possible, rearrangement of less stable **carbocations** to more stable **carbocations** is expected.

6.4.2   1,2-Alkyl Shifts

Moving from discussion of **1,2-hydride shifts** to **1,2-alkyl shifts**, it is important to remember that **hydride shifts** occur much more readily than the corresponding **alkyl shifts**. In fact, as a general rule, **alkyl shifts** will not occur unless a **hydride shift** cannot take place.

Among the most famous examples of a reaction involving a **1,2-alkyl shift** is the **pinacol rearrangement**. This reaction, shown in Scheme 6.10, results in the conversion of a **vicinal diol** to a **ketone**.

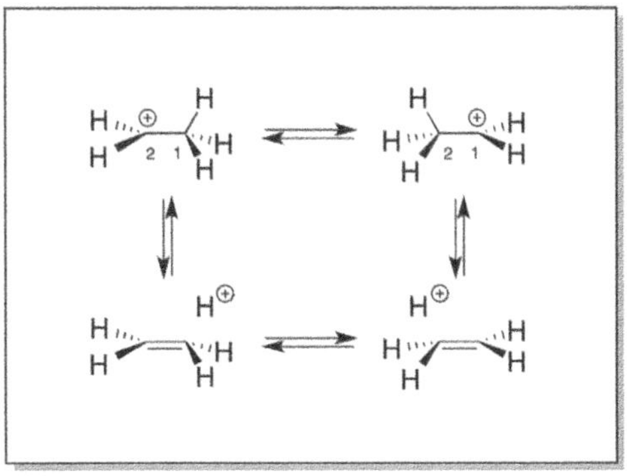

Scheme 6.8 *Hyperconjugation leads to migration of hydrogen atoms through a 1,2-hydride shift.*

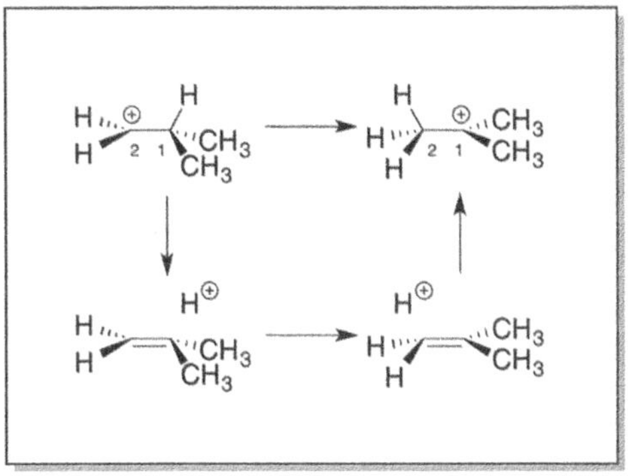

Scheme 6.9 *Rearrangements via 1,2-hydride shifts generate more stable carbocations from less stable carbocations.*

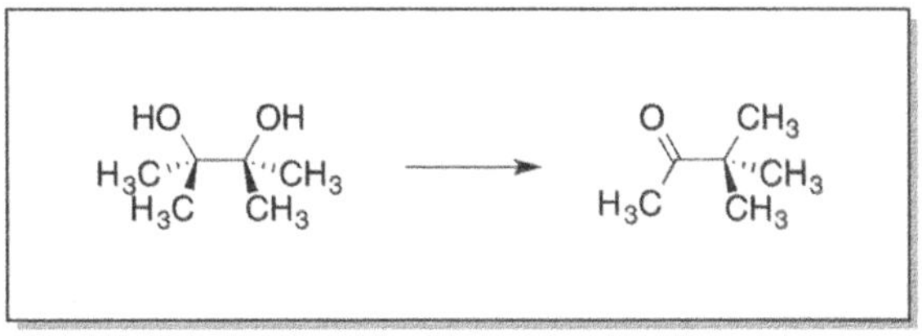

Scheme 6.10 *The pinacol rearrangement.*

Mechanistically, the **pinacol rearrangement** is explained by initial **carbocation** formation through **solvolysis**. This step, illustrated in Scheme 6.11, involves protonation of an alcohol followed by water leaving and generating a cation. In looking at this cation, one may imagine that a **1,2-hydride shift** is possible. However, the only sources of hydrogens for such a shift are the methyl groups adjacent to the cationic center. If a hydride migrates from one of these methyl groups, as illustrated in Scheme 6.12, the result would be generation of a **primary carbocation**. As previously discussed, since **primary carbocations** are less stable than **tertiary carbocations**, this migration will not occur.

While the hydride shift illustrated in Scheme 6.12 cannot occur as a part of the **pinacol rearrangement**, the intermediate **carbocation** is subject to **alkyl migrations**. As shown in Scheme 6.13, a **1,2-alkyl shift** results in the transfer of the cation from a **tertiary center** to

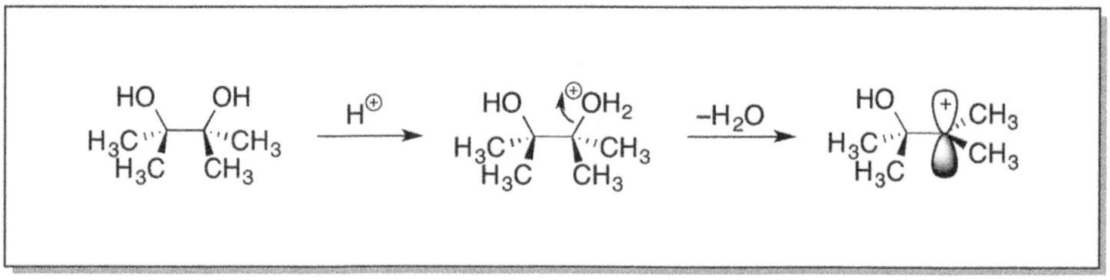

Scheme 6.11 *The pinacol rearrangement proceeds through solvolysis-mediated cation formation.*

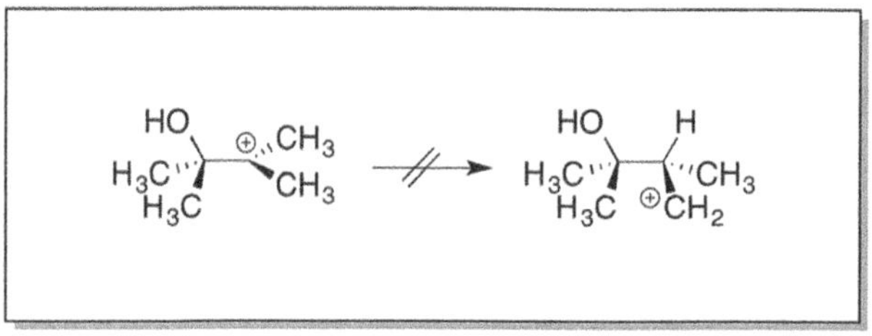

Scheme 6.12 *1,2-Hydride shifts will not occur when the product cation is less stable than the starting cation.*

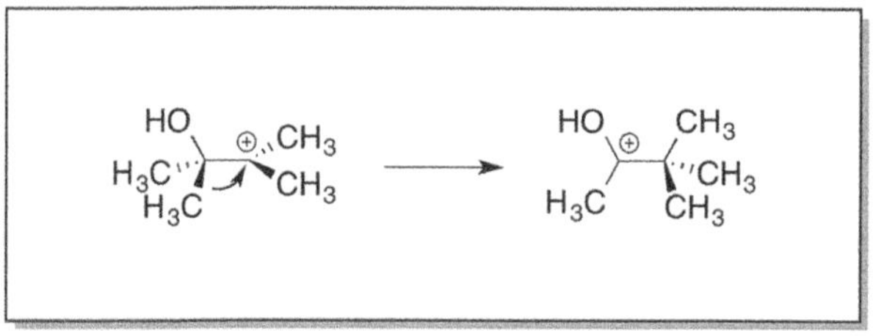

Scheme 6.13 *Alkyl migrations occur when the resulting carbocation is more stable than the starting carbocation.*

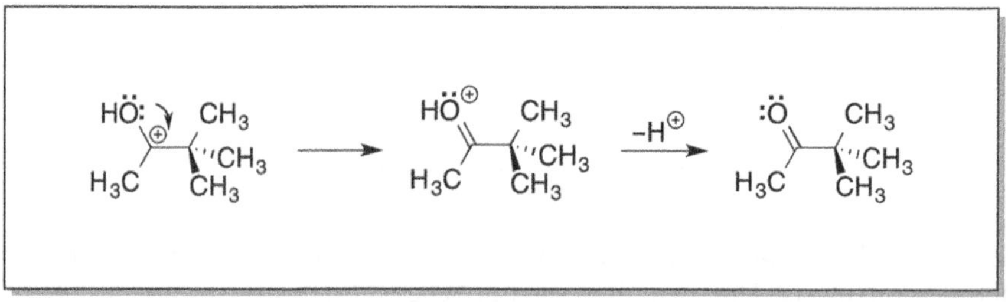

Scheme 6.14 *Conclusion of the pinacol rearrangement involves migration of the positive charge to the adjacent oxygen atom followed by deprotonation.*

a center adjacent to a **heteroatom**. As the oxygen **heteroatom** possesses **lone electron pairs**, these **lone pairs** serve to stabilize the cation. Thus the illustrated **1,2-alkyl shift** transforms a **carbocation** into a more stable **carbocation**.

Mechanistic conclusion of the **pinacol rearrangement** is illustrated in Scheme 6.14 and involves initial donation of an oxygen **lone pair** to the cation, thus migrating the charge to the oxygen atom. The resulting oxygen cation then releases a proton liberating the illustrated neutral **ketone**.

As the mechanistic steps discussed for the **pinacol rearrangement** have been illustrated using **arrow-pushing**, it is important to recognize that in all cases, the arrows have been drawn pushing **electrons** toward **positive charges**. This point has been previously discussed and will continue to be emphasized.

6.4.3 Preventing Side Reactions

Because of **1,2-hydride and alkyl shifts**, it is possible to obtain **multiple products** from S_N1 reactions. Thus, to induce one product to predominate, we must find a way to stabilize the **carbocation**. This is done by using highly **polar solvents** such as **acetic acid, dimethylformamide**, and **dimethyl sulfoxide**. In using this strategy, the lifetime of a carbocation can be extended allowing the most stable product more time to form. As a result, **control** over formation of desired products in reasonable yields from S_N1 reactions can be achieved.

6.5 SUMMARY

In this chapter, S_N1 reactions were introduced, compared to S_N2 reactions, and discussed mechanistically. Through these discussions, the involvement of atomic **orbitals**, and their various **hybrid** combinations, was addressed. Furthermore, complicating **side reactions** such as **hydride and alkyl migrations** were presented. As discussions move into more advanced mechanistic types, it is important to maintain awareness of the involvement and orientation of **orbitals**, the **steric** environment at reactive centers, and the overall reactivity of **nucleophiles** and **electrophilic centers**.

PROBLEMS

1. For the following molecules, state the hybridization (*sp*, *sp²*, *sp³*) of the orbitals associated with the highlighted bond. Also, state the geometry of the bound atomic centers (linear, bent, trigonal planar, tetrahedral).

    a. 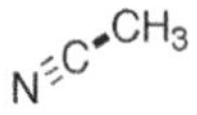 N≡C•CH₃

    b. N≡C•C(H)=CH₂

    c. N≡C•C≡CH

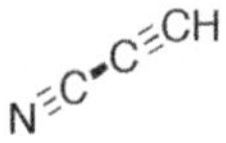

    d. 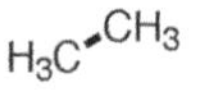 H₃C•CH₃

    e. 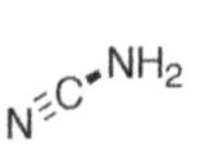 N≡C•NH₂

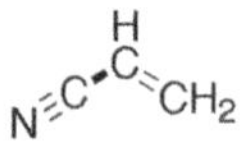

f. 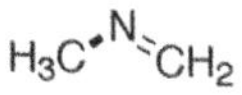 $H_3C\text{-}N\text{=}CH_2$

g. $N\text{≡}C\text{-}OH$

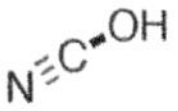

h. $H_2C\text{=}C\text{=}O$ answer for both double bonds

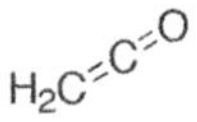

i. 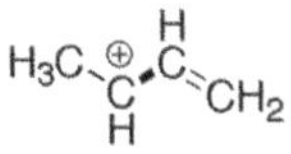 $H_3C\text{-}\overset{\oplus}{C}\text{-}C\text{=}CH_2$ (with H above and H below the second carbon)

j. 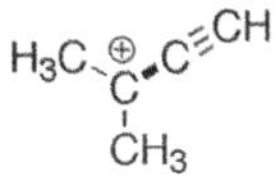 $H_3C\text{-}\overset{\oplus}{C}\text{-}C\text{≡}CH$ (with CH_3 below the central carbon)

2. Predict all of the products of the following reactions.

a.

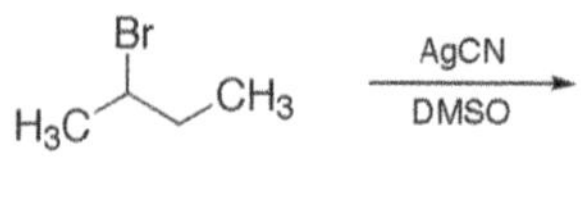

$$\text{AgCN} \over \text{DMSO}$$

b.

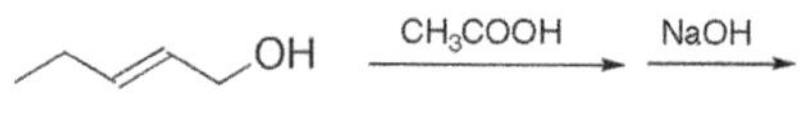

$$\text{CH}_3\text{COOH} \qquad \text{NaOH}$$

c.

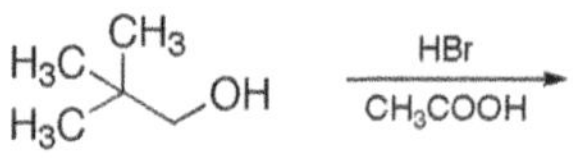

$$\text{HBr} \over \text{CH}_3\text{COOH}$$

d.

$$\text{HBr} \over \text{CH}_3\text{COOH}$$

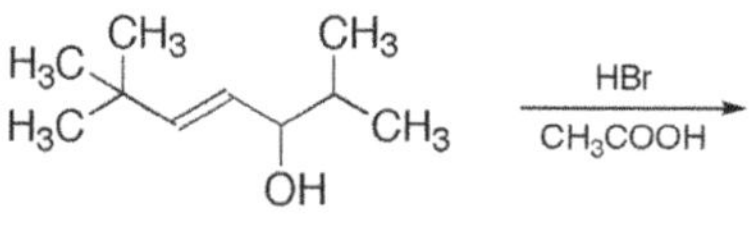

3. For each of the following reactions, determine which will proceed via an S_N1 or an S_N2 mechanism. In cases where both may be applicable, list appropriate reaction conditions (e.g., solvents, reagents) that would favor S_N1 over S_N2 and vice versa. Explain your answers.

a.

b.

c.

4. In studying 1,2 alkyl and hydride shifts, we explored the observation that shifts will not occur unless the newly formed carbocation is more stable than the starting carbocation. Additionally, as illustrated in Figure 6.12, these shifts were explained using hyperconjugation, thus requiring that the orbital containing the positive charge and the bond containing the shifting group lie within the same plane. This is necessary in order to allow sufficient orbital overlap for the shift to take place.

In addition to 1,2-shifts, which occur between adjacent bonds, other shifts are possible where the migrating group apparently moves across space. As with 1,2-shifts, these additional shifts can only occur when the positively charged empty p-orbital lies within the same plane as the bond containing the migrating group, thus allowing sufficient orbital overlap. With this in mind, explain the following 1,5-hydride shift. (Hint: Consider different structural conformations. You may want to use models.) Asterisk (*) marks enrichment with ^{13}C.

Chapter 7

Elimination Reactions

Until now, discussions have focused only on how **carbanions** and **carbocations** behave under conditions favorable for **nucleophilic substitutions**. However, these species may undergo other types of reactions in which **unsaturation** is introduced into the molecule. Such reactions are called **elimination reactions** and should be considered whenever charged species are of importance to the mechanistic progression of a molecular transformation. In previous chapters, S_N1 and S_N2 reactions were discussed. In this chapter, the corresponding **E1** and **E2** elimination mechanisms are presented.

7.1 E1 ELIMINATIONS

Having addressed the chemistry of **carbocations** and associated S_N1 reaction mechanisms, it is appropriate to begin discussions of **elimination reactions** with the related **E1** mechanism. As addressed in Chapter 6, **carbocations** generated from **solvolysis** reactions can undergo various types of **rearrangements** that include hydride and alkyl shifts. Furthermore, these shifts were rationalized when the empty ***p*-orbital** associated with the positive charge is aligned in the same plane with the migrating group. Figure 7.1 reiterates the concept of **hyperconjugation** necessary for these shifts to occur. Furthermore, Figure 7.2 reiterates that **hyperconjugation** can be viewed as introducing **double bond character** to a **carbocation**. Carrying this rationale one step further, if the **double bond character** in a given **carbocation** becomes stabilized through full **dissociation** of a proton, the result, illustrated in Scheme 7.1, is formation of a full double bond through an **E1** elimination mechanism.

As previously alluded, **E1** reactions are integrally related to S_N1 reactions by virtue of the **carbocations** common to both mechanisms. Thus, reiterating the **solvolysis** reaction leading to the conversion of *tert*-butyl bromide to MTBE illustrated in Scheme 7.2, we

Arrow-Pushing in Organic Chemistry: An Easy Approach to Understanding Reaction Mechanisms,
Second Edition. Daniel E. Levy.
© 2017 John Wiley & Sons, Inc. Published 2017 by John Wiley & Sons, Inc.

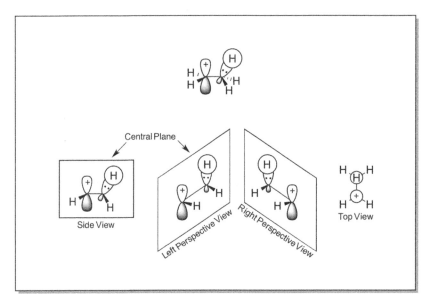

Figure 7.1 *Hyperconjugation occurs when a carbon–hydrogen bond lies in the same plane as a carbocation's vacant p-orbital.*

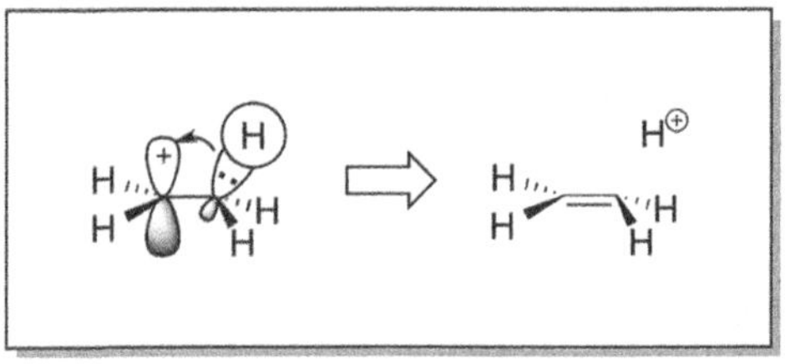

Figure 7.2 *Hyperconjugation can be viewed as formation of a "pseudo double bond."*

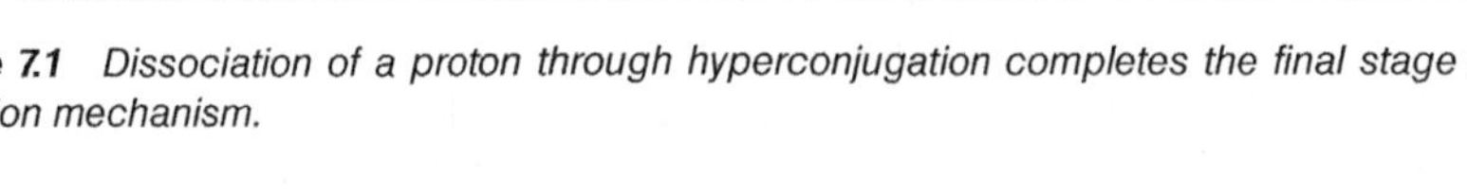

Scheme 7.1 *Dissociation of a proton through hyperconjugation completes the final stage of an E1 elimination mechanism.*

understand how the formation of isobutylene occurs. Formation of isobutylene only occurs through the **E1** process and comprises approximately 20% of the reaction mixture.

As can be deduced from discussions presented earlier and in Chapter 6, it is very important to recognize that when designing reactions involving **carbocations**, both **migration reactions** and **elimination reactions** can complicate the outcome of intended S_N1 transformations. An example illustrating the potential formation of multiple side products is shown in Scheme 7.3 with the **solvolysis** of 2-bromo-2,3-dimethylpentane in methanol.

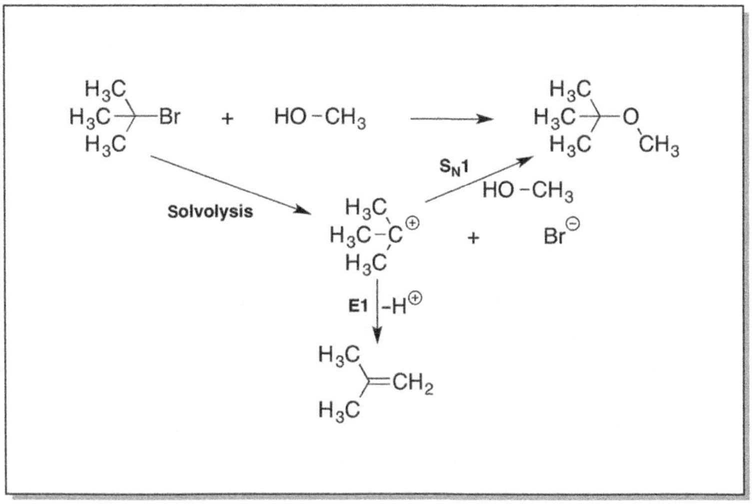

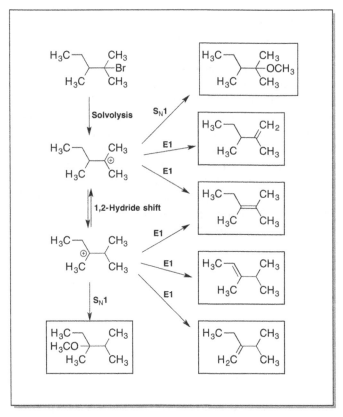

Scheme 7.2 *E1 mechanisms explain additional products observed during S$_N$1 reactions.*

Scheme 7.3 *Solvolysis of 2-bromo-2,3-dimethylpentane in methanol leads to formation of up to six different products via multiple mechanistic pathways.*

Returning to Scheme 7.1, we recognize that an **E1** reaction proceeds with the **E**limination of a leaving group leading to the **E** designation. Because this mechanism proceeds with the initial dissociation of a single starting material forming a carbocation, this process is considered a unimolecular reaction. The involvement of only **1** species in the initial phase of the reaction enhances the mechanistic designation to **E1**.

7.2   E1cB ELIMINATIONS

As discussed in Section 7.1 and in Chapter 6, **carbocations** are integrally involved in the **E1** and **S$_N$1** reaction mechanisms. This focus on intermediate carbocation species is not to imply that **carbanion** intermediates cannot be involved in such reactions. In fact, carbanions are very useful intermediates leading to the formation of double bonds.

Referring to Chapter 3, an **acid** is a molecule that liberates **hydrogen ions**. Furthermore, Appendix 1 lists pK_a values for various types of functional groups including **neutral functional groups**. It is important to recognize that protons adjacent to certain neutral functional groups can be removed under appropriate conditions. Thus, molecules containing these **neutral functional groups** may be considered **weak acids**.

As discussed in Chapter 4, when an acid, HA, reacts with a base, B$^-$ or B, an **anionic component**, A$^-$, is formed from the acid (Scheme 7.4). This anionic component is said to be the **conjugate base** of the starting acid, HA. It is important to understand that A$^-$ may be any organic molecule from which a hydrogen atom has been removed. This is independent of the **molecular structure** and/or associated **functional groups**. Recognizing that HA may be an organic molecule, the reader is strongly encouraged to become fully acquainted with the relative acidities summarized in Appendix 1.

In Chapter 3, the concept or **resonance** was introduced as the **delocalization of charge** across multiple atoms in a given molecule. This concept was further developed to rationalize the **relative acidities** of organic molecules. When an acid, HA, is deprotonated and the resulting **conjugate base**, A$^-$, is **stabilized** through **resonance**, the **stabilized carbanion** becomes an available nucleophile. If there is a leaving group adjacent to a stabilized carbanion **negative charge**, through **elimination** of the **leaving group**, a **double bond** is formed. This is illustrated in Scheme 7.5 using **2-iodomethyl dimethyl malonate** as an example. As shown using **arrow-pushing**, **sodium hydride** (base) initially reacts to form **hydrogen gas** and a **malonate anion** (the **conjugate base**). Again, using **arrow-pushing**, the **resonance stabilization** of the **malonate conjugate base** is illustrated. As shown in Scheme 7.6, the malonate anion then undergoes **β-elimination** to displace the iodide forming the illustrated **double bond**.

$$B + HA \; \rightleftharpoons \; BH^{\oplus} + A^{\ominus}$$

$$B^{\ominus} + HA \; \rightleftharpoons \; BH + A^{\ominus}$$

Scheme 7.4 *General representation of bases (B or B⁻) reacting with acids (HA) forming conjugate bases (A⁻).*

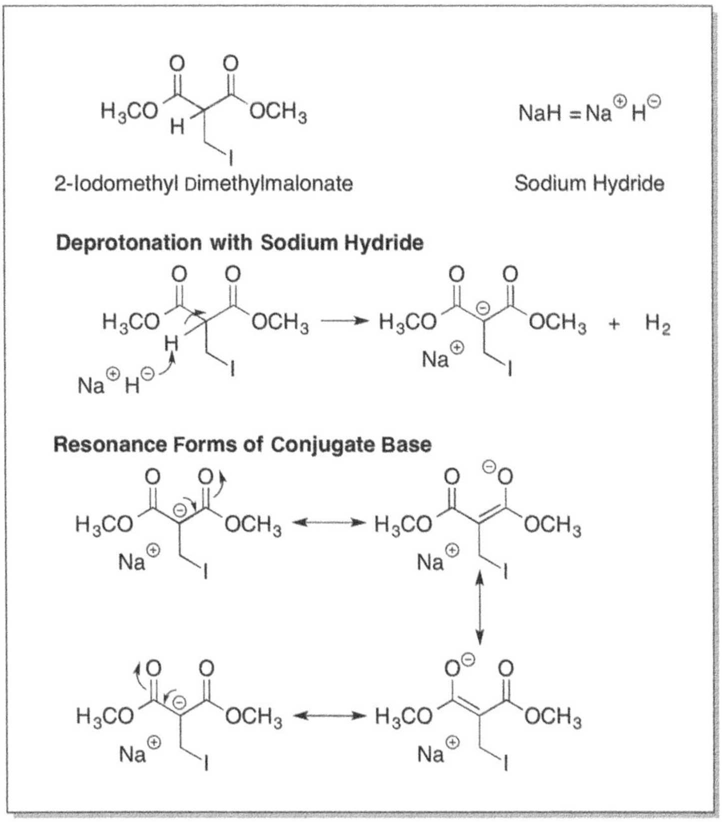

Scheme 7.5   *Formation of the conjugate base and associated resonance structure resulting from the reaction of 2-iodomethyl dimethylmalonate with sodium hydride.*

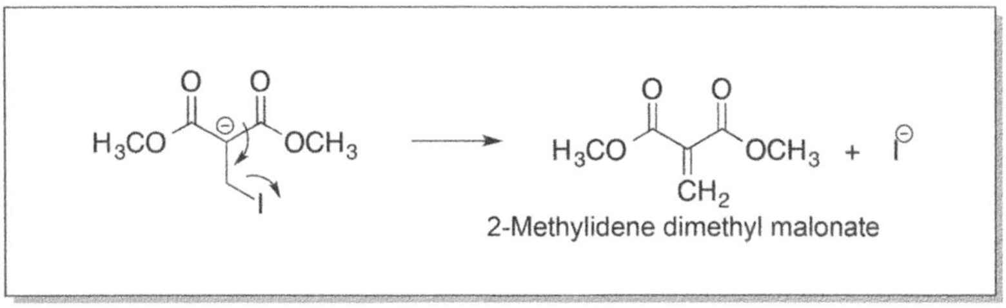

Scheme 7.6   *β-Elimination of the iodide completes the E1cB mechanism converting the 2-iodomethyl dimethylmalonate anion to 2-methylidene dimethyl malonate.*

Considering the reaction illustrated in Schemes 7.5 and 7.6, one can understand that attempts to replace the iodide of **2-iodomethyl dimethyl malonate** through an S_N2 reaction will be very difficult (Scheme 7.7). This is an example of a common general consideration chemists must deal with when designing strategies for the synthesis of novel molecules. Without an understanding of the various **reaction mechanism subtypes** and how they relate to one another, synthetic strategies may fail.

In studying Schemes 7.5 and 7.6, we recognize that an **E1cB** reaction proceeds through initial extraction of a proton by a base leading to formation of a conjugate base. Elimination

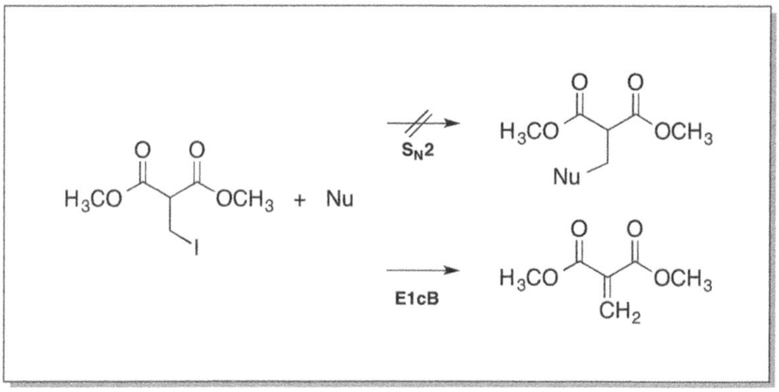

Scheme 7.7 *Reaction of 2-iodomethyl dimethyl malonate with a nucleophile results in predominant formation of the E1cB elimination product.*

of a leaving group justifies the **E** designation. Because this mechanism proceeds with the initial dissociation of a single starting material forming a carbanion, this process is considered a unimolecular reaction. The involvement of only **1** species in the initial phase of the reaction enhances the mechanistic designation to **E1**. Furthermore, because the initially formed carbanion is a **c**onjugate **b**ase, the designation for this reaction becomes **E1cB**.

7.3   E2 ELIMINATIONS

To this point, considerable time has been spent discussing **acids**, **bases**, **nucleophiles**, and **leaving groups**. These were ultimately all presented in the context of S_N2 reactions. Like the complicating side reactions associated with **carbocations** formed during S_N1 reactions, depending upon the nature of substituents adjacent to acidic protons, S_N2 reaction conditions can induce similar complications. For example, consider a molecule with an acidic proton and a leaving group, **L**, on the carbon adjacent to the acidic proton. Consider, also, that **nucleophiles** are **bases**. As shown in Scheme 7.8, an alternative to **nucleophilic displacement** of a **leaving group** is found in the simultaneous **extraction** of a hydrogen atom and **elimination** of a **leaving group** resulting in formation of an **olefin**. In general,

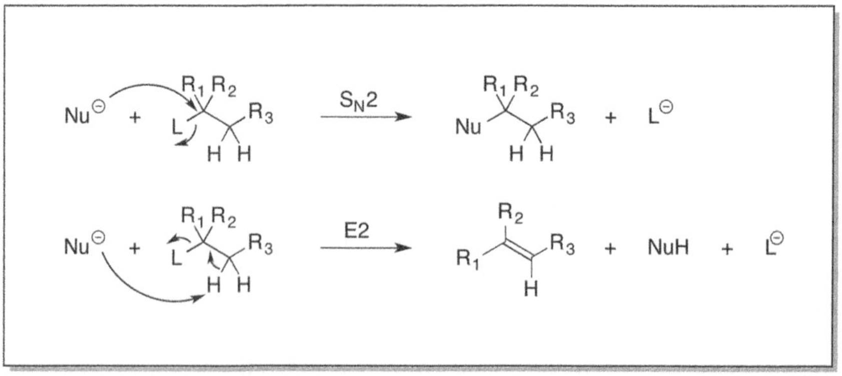

Scheme 7.8 *S_N2 substitution reactions can occur in competition with E2 elimination reactions.*

more **basic nucleophiles** will favor the **E2** pathway, while stronger **nucleophiles** will favor the S_N2 pathway.

In studying Scheme 7.8, we recognize that an **E2** reaction proceeds through initial extraction of a proton by a base or nucleophile leading to **e**limination of a leaving group justifying the **E** designation. Because this mechanism proceeds through the interaction of two species (substrate and base/nucleophile), **E2** reactions are recognized as bimolecular. Thus, the involvement of **2** species in the initial phase of the reaction enhances the mechanistic designation to **E2**. Finally, it is important to note that while **E1** reactions proceed through cationic intermediates and **E1cB** reactions proceed through anionic intermediates, formation of a full formal negative charge is not essential for **E2** reactions to proceed.

7.4 HOW DO ELIMINATION REACTIONS WORK?

In addressing the mechanistic basis behind **elimination** reactions, we must refer back to discussions surrounding **carbocations** in the context of S_N1 reactions. Furthermore, consideration of **carbocation**-associated **hydride/alkyl shifts** and E1-related products is essential. Recall that **carbocations** are stabilized by phenomena such as **hyperconjugation**. Furthermore, recall that **hydride shifts**, **alkyl shifts**, and **E1 eliminations** are dependent upon the planar alignment of an empty *p*-orbital and an adjacent bond bearing either a migrating group or a dissociable hydrogen atom as illustrated in Figure 7.1.

The mechanistic basis behind the stability and reactivity of **carbocations**, regardless of the reaction outcome, depends on the alignment of an empty *p*-**orbital** and the **orbitals** comprising an adjacent bond. Specifically, if there are no **planar alignments** then **hyperconjugation**, **hydride/alkyl shifts**, or **eliminations** cannot occur. Perhaps there is no better illustration of this fact than a comparison of the **stability** of primary, secondary, and tertiary **carbocations**. As reiterated from Chapter 6, Figure 7.3 illustrates the order of stability from most stable to least stable. This trend in stability is directly related to the number of adjacent carbon–hydrogen bonds available for **hyperconjugation**.

Looking at the structures shown in Figure 7.3, we notice that the *tert*-butyl **carbocation** possesses nine carbon–hydrogen bonds adjacent to the cation, while the secondary **carbocation** possesses six and the primary **carbocation** possesses only three. This tabulation of bonds is relevant in that the more adjacent carbon–hydrogen bonds, the more opportunities there are for **hyperconjugation** to occur. In this discussion, the term "opportunities" is important because single bonds employing sp^3 **orbitals** are not rigid and can rotate around the bond axis as shown in Figure 7.4 in much the same way a wheel rotates on an axel. Thus, when empty *p*-**orbitals** and adjacent bonds are not in alignment, there can be no

Figure 7.3 *Tertiary carbocations are more stable than secondary carbocations, and secondary carbocations are more stable than primary carbocations.*

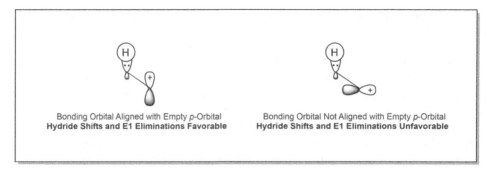

Figure 7.4   *When a carbon–hydrogen (or carbon alkyl) bond is aligned with an empty p-orbital, 1,2-hydride/alkyl shifts and E1 eliminations are favorable.*

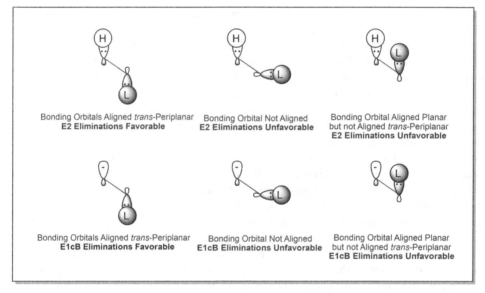

Figure 7.5   *When a carbon–hydrogen bond or a negatively charged orbital is aligned trans-periplanar with a carbon leaving group bond, E2 eliminations and E1cB eliminations are favorable.*

associated reaction, and the observed reactions are only possible due to the intermittent alignment of a system that is continually in motion.

As already discussed, **E1**, **E1cB**, and **E2 eliminations** differ, in part, by the electronic nature of the mechanism. Specifically, **E1 eliminations** depend on **cationic** intermediates, **E1cB eliminations** depend on **anionic** intermediates, and **E2 eliminations** are not dependent upon formation of a full and formal charge. These differences, however, do not eliminate the mechanistic similarities of these reactions as related to the necessary alignment of adjacent bonds between atoms. While, as shown in Figure 7.4, **E1 eliminations** require alignment of a carbon–hydrogen bond with an adjacent empty ***p*-orbital**, **E1cB** and **E2 eliminations**, as shown in Figure 7.5, require alignment of a carbon–hydrogen bond or a negatively charged orbital with an adjacent carbon leaving group bond. Furthermore, as shown in Figure 7.5, the relationship between these bonds is critical for **elimination**

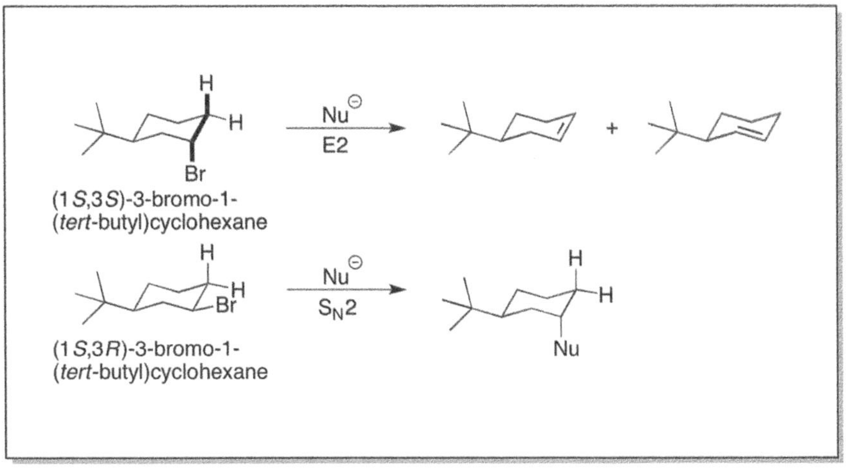

Scheme 7.9   *E2 eliminations depend upon the presence of* trans-periplanar *relationships.*

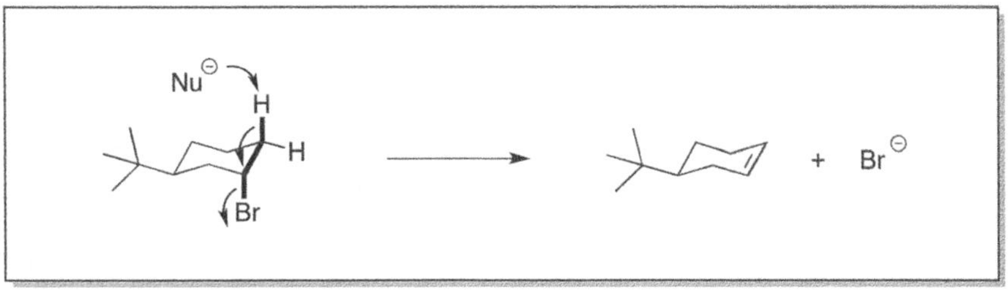

Scheme 7.10   *Mechanistic progression of E2 eliminations.*

to occur. Specifically, the relevant bonds must adopt a ***trans*** relationship within the same **plane**. This relationship is referred to as ***trans*-periplanar**.

A practical example demonstrating the importance of the ***trans*-periplanar** relationship between protons and leaving groups is illustrated in Scheme 7.9 and applied to **E2 elimination** reactions. As shown, when reacted with a nucleophile, **(1*S*,3*S*)-3-bromo-1-(*tert*-butyl)cyclohexane** is converted to the illustrated cyclohexene structures. However, the same reaction conditions applied to **(1*S*,3*R*)-3-bromo-1-(*tert*-butyl)cyclohexane** result in **S$_N$2 displacement** of the bromide. These observations are mechanistically explained based on the presence or absence of a ***trans*-periplanar** relationship between the bromide and an adjacent hydrogen atom. As illustrated in Scheme 7.9, one of two such ***trans*-periplanar** relationships is highlighted in the **(1*S*,3*S*)-3-bromo-1-(*tert*-butyl)cyclohexane** structure using boldface bonds. The absence of such a relationship between the bromide and adjacent hydrogen atoms in the **(1*S*,3*R*)-3-bromo-1-(*tert*-butyl)cyclohexane** structure is reflected in the absence of boldface bonds.

As illustrated in Scheme 7.10 using **arrow-pushing**, the **E2 elimination** from Scheme 7.9 proceeds through reaction of a nucleophile with a ***trans*-periplanar** hydrogen atom. This initial step is accompanied by the **concerted** shift of electrons from the carbon–hydrogen bond to the carbon bearing the bromide. Loss of the bromide completes formation of the illustrated cyclohexene. The fact that two products are illustrated in Scheme 7.9 is a reflection of the presence of two ***trans*-periplanar** carbon–hydrogen bonds relative to the bromide.

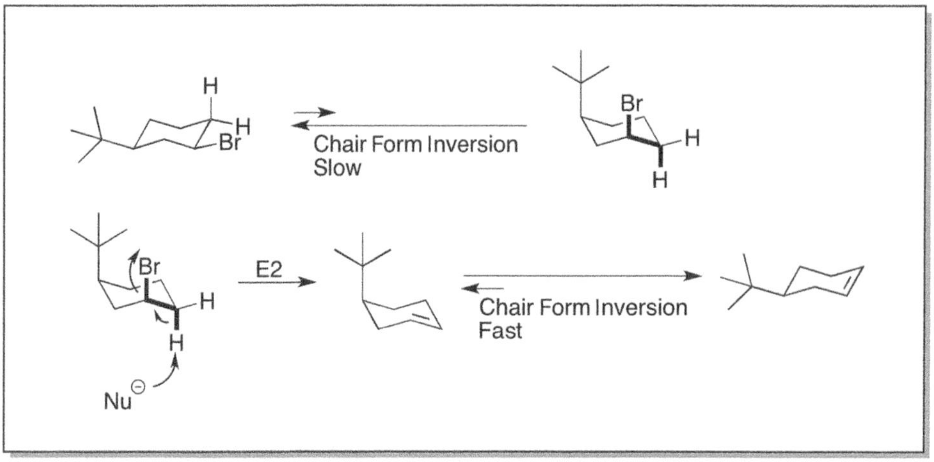

Scheme 7.11 *If* trans-*periplanar relationships can be established, E2 elimination products can form.*

In the case of **(1S,3R)-3-bromo-1-(*tert*-butyl)cyclohexane**, **E2 eliminations** are possible provided the illustrated **chair structure inverts** to place the *tert*-butyl and bromide groups in **axial** positions. As illustrated in Scheme 7.11, upon completion of this inversion, *trans*-**periplanar** relationships are established between the bromide and adjacent carbon–hydrogen bonds allowing for **E2 eliminations** to occur. Due to **steric factors** associated with the highly disfavored necessity to place the *tert*-**butyl group** in an **axial position**, this **chair form inversion** is very slow, thus minimizing the formation of **E2 elimination** products. Chair form inversion to the final structure is fast because **steric** issues are minimized when the *tert*-**butyl** group moves to an **equatorial** position.

7.5   E1cB ELIMINATIONS VERSUS E2 ELIMINATIONS

At first glance, the **E1cB elimination** mechanism (Scheme 7.6) may look like that corresponding to an **E2 elimination** (Scheme 7.10). Both reactions involve **bases/nucleophiles** and both require a *trans*-**periplanar** relationship in order for **elimination** to occur. In some cases, the only way to distinguish between these mechanistic pathways may be through analysis of the **reaction kinetics**. As **reaction kinetics** is a topic beyond the scope of this book, readers are referred to their general organic chemistry textbooks for more information. In general, if formation of an **anion** or **conjugate base** is predicted based upon the structure and pK_a of the starting material, the **E1cB** reaction pathway is more likely. However, if there is no structural feature or **functional group** capable of stabilizing an **anion** or a **conjugate base**, the **E2** mechanism can be predicted.

7.6   SUMMARY

In this chapter, **elimination** reactions were presented both independently and in association with their related **nucleophilic substitution** mechanisms. Furthermore, the processes by which molecules undergo **E1**, **E1cB**, and **E2 eliminations** were presented and explained using **bonding** and **non-bonding orbitals** and their required relationships to one another.

While much emphasis was placed on the **planar** relationships of orbitals required for both **elimination** reaction mechanisms, the special case of *trans*-**periplanar** geometries were described as necessary for efficient **E2 eliminations** to occur.

While *trans*-**periplanar** relationships are important to **E2 elimination** reactions, discussions focused on the **E1cB** mechanism were less specific in this area. Due to the **stabilized conjugate base** intermediate in **E1cB** reactions, bonds bearing leaving groups have time to rotate into the required *trans*-**periplanar** position. As a distinction between the **E1cB** and **E2** mechanisms, **E2 eliminations** proceed in a **concerted** manner while **E1cB eliminations** may proceed stepwise. As for the ability of **leaving groups** to move into *trans*-**periplanar** positions, it is important to remember that **rotation** around an acyclic **single bond**, as illustrated in Figures 7.4 and 7.5, occurs readily. Therefore, **elimination** reactions should not be removed from consideration if a molecule is drawn in a **conformation** that makes these reactions appear unfavorable. When looking at any type of **nucleophilic** reaction, initial identification of relevant *trans*-**periplanar** relationships will aid in the identification of potential side products and their respective mechanisms of formation.

PROBLEMS

1. E2 eliminations do not require acidic protons in order to proceed. Explain how this can occur.

2. When $CH_3OCH_2CH_2CH_2Br$ is treated with magnesium, we get the Grignard reagent $CH_3OCH_2CH_2CH_2MgBr$. However, when $CH_3OCH_2CH_2Br$ is treated with magnesium, the product isolated is $H_2C{-}CH_2$. Explain this result.

3. With an understanding of E1 mechanisms, one may realize that under S_N1 reaction conditions, multiple products may form. In addition to the products predicted in Chapter 6 for the following molecules, predict plausible elimination products.

a.

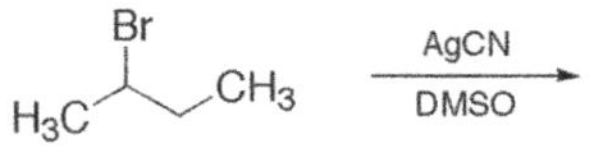

$$\underset{H_3C}{\overset{Br}{\bigwedge}}CH_3 \quad \xrightarrow[\text{DMSO}]{\text{AgCN}}$$

b.

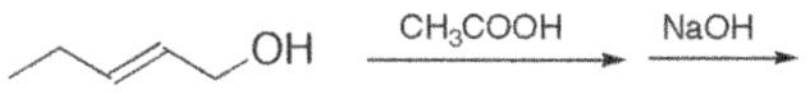

$$\diagdown\diagup\diagdown OH \quad \xrightarrow{CH_3COOH} \quad \xrightarrow{NaOH}$$

c.

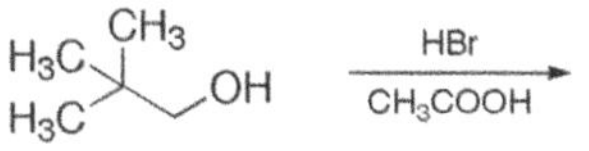

$$\underset{H_3C}{\overset{CH_3}{\underset{H_3C}{\bigvee}}}OH \quad \xrightarrow[CH_3COOH]{HBr}$$

d.

$$\underset{H_3C}{\overset{CH_3}{\underset{H_3C}{\bigvee}}}\diagup\underset{OH}{\overset{CH_3}{\bigvee}}CH_3 \quad \xrightarrow[CH_3COOH]{HBr}$$

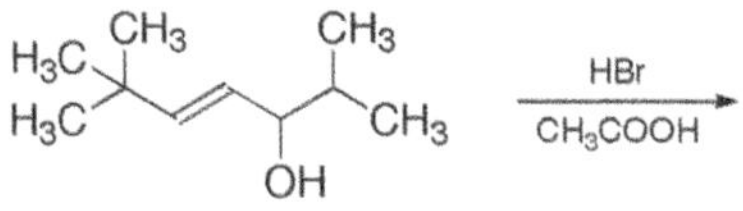

4. Presently, several different organic reaction mechanisms have been presented. Keeping all of these in mind, predict all of the possible products of the following reactions and list the mechanistic type or types leading to formation of these products.

a.

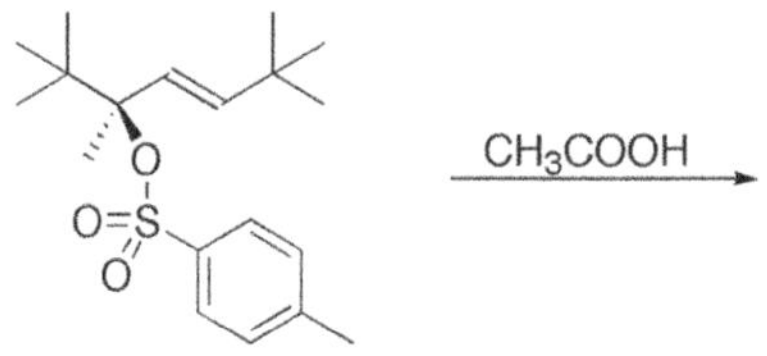

$$\xrightarrow{\text{CH}_3\text{COOH}}$$

b.

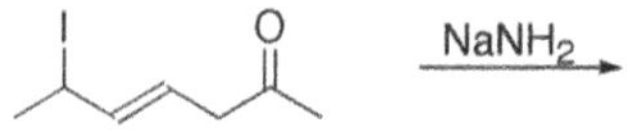

$$\xrightarrow{\text{NaNH}_2}$$

c.

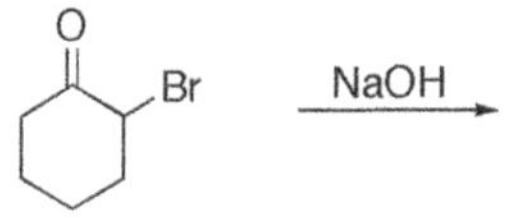

$$\xrightarrow{\text{NaOH}}$$

d.

$$\xrightarrow{\text{AgNO}_3}$$

e. 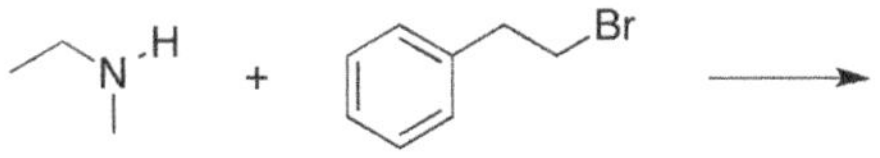   (ethylamine) + (phenethyl bromide with Br) ⟶

f. 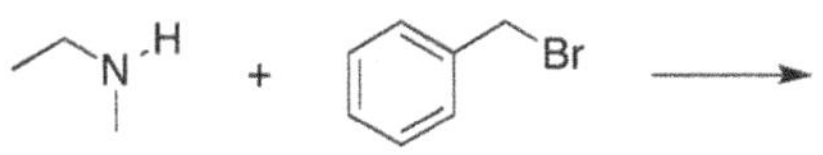   (ethylamine) + (benzyl bromide with Br) ⟶

5. As mentioned earlier, stereochemistry is not of great concern in the chapters of this book. However, certain mechanistic types will show specific stereochemical consequences when acting on chiral molecules. With this in mind, predict the product resulting from the E2 elimination of HBr when the shown isomer of 4-bromo-3-methyl-2-pentanone is treated with sodamide. Show all stereochemistry and explain your answer.

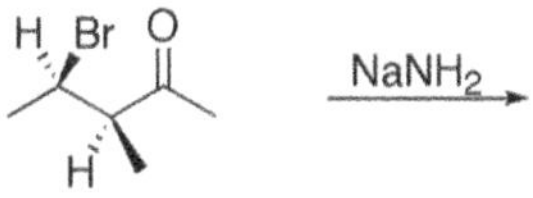

(structure: H, Br, O with NaNH₂ →)

$$\xrightarrow{\text{NaNH}_2}$$

6. Based on the answer to Problem 5, predict the product of the following reactions and show all stereochemistry.

a. 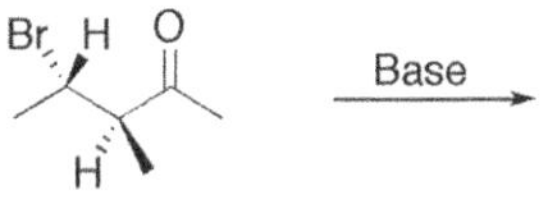   (structure: Br, H, O) $\xrightarrow{\text{Base}}$

b. 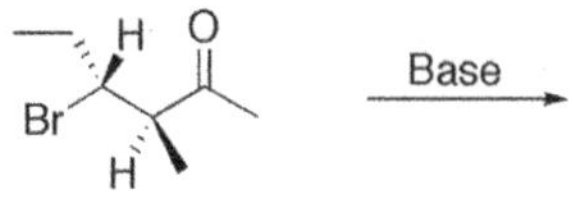 Base →

c. 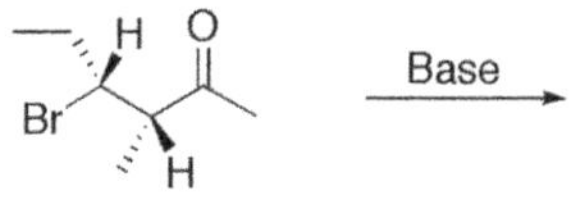 Base →

d. 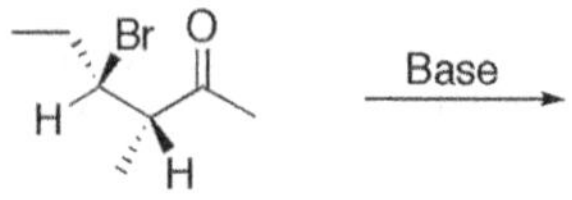 Base →

7. Explain the results of the following experiment.

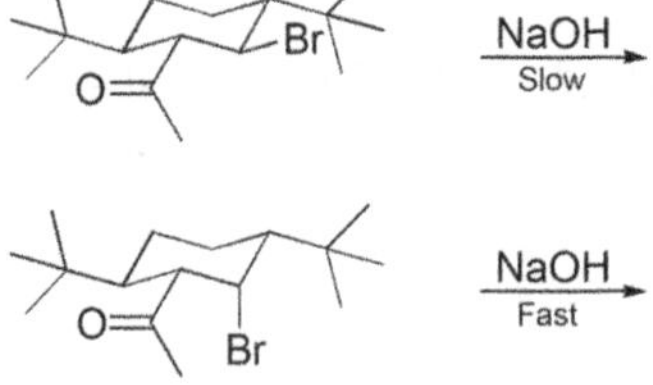

NaOH Slow →

NaOH Fast →

8. Reaction (I) proceeds through the E2 elimination mechanism and reaction (II) proceeds through the E1cB mechanism. Using arrow-pushing, explain these observations.

(I)

(II)

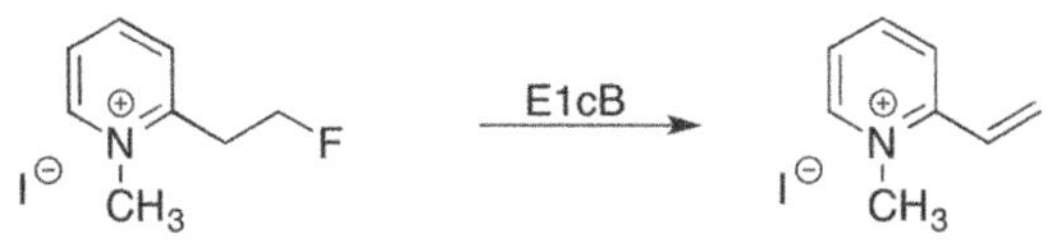

Chapter **8**

Addition Reactions

In Chapter 7, **elimination** reactions were presented. In the context of **elimination** reactions, the formation of **double bonds** was noted regardless of the **elimination mechanism** discussed. Continuing from the concept of using **elimination** reactions to form sites of **unsaturation**, one may reason that **addition** reactions can be used to remove sites of **unsaturation**. Thus, elaborating upon **addition** reactions, this chapter provides an introduction to relevant **mechanisms** applied to both carbon–carbon double bonds (**olefins**) and carbon–oxygen double bonds (**carbonyls**).

8.1 ADDITION OF HALOGENS TO DOUBLE BONDS

Throughout this book, the various mechanistic types driving reactions were shown to rely upon interactions between **charged species** such as **nucleophiles** and **electrophiles**. However, when looking at **ethylene**, the simplest of **olefins**, there are no **partial charges** (or **steric factors**) that distinguish one side of the **double bond** from the other. Due to its **symmetry**, there can be no pure **nucleophilic** or **electrophilic** sites. Furthermore, when looking at **bromine** in its natural form of Br_2, there are no interactions between the two atoms other than a single and **unpolarized** bond joining them. Nevertheless, when **ethylene** and **bromine** are brought together, the reaction illustrated in Scheme 8.1 occurs.

In order to explain this reaction, consider the fact that, due to the overlapping *p*-orbitals, **double bonds** are **electron rich**. This property allows **olefins**, under certain conditions, to act as **nucleophiles**. In the case of a **double bond** reacting with molecular **bromine**, the result is the formation of a **three-membered ring** containing a **positively charged bromine atom**. This three-membered ring is known as a bridged **bromonium ion**. Concurrent to

---

Arrow-Pushing in Organic Chemistry: An Easy Approach to Understanding Reaction Mechanisms, Second Edition. Daniel E. Levy.
© 2017 John Wiley & Sons, Inc. Published 2017 by John Wiley & Sons, Inc.

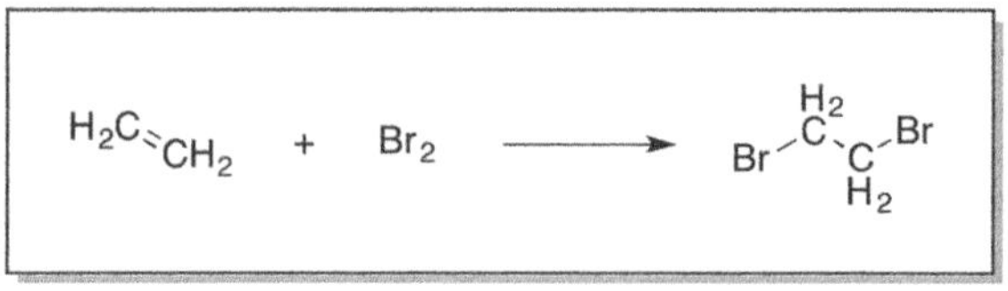

Scheme 8.1 *Addition of bromine to ethylene.*

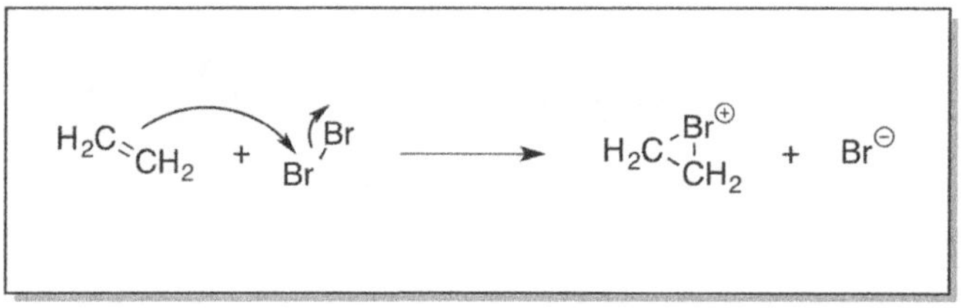

Scheme 8.2 *Molecular bromine reacts with double bonds generating a bromonium ion and a bromide anion.*

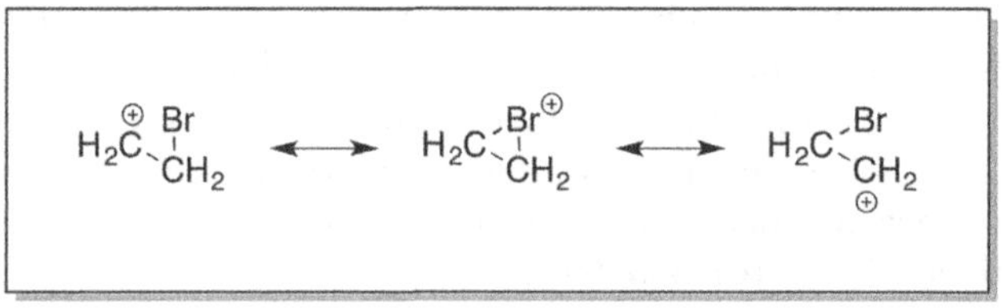

Scheme 8.3 *Bromonium ions possess electrophilic carbon atoms.*

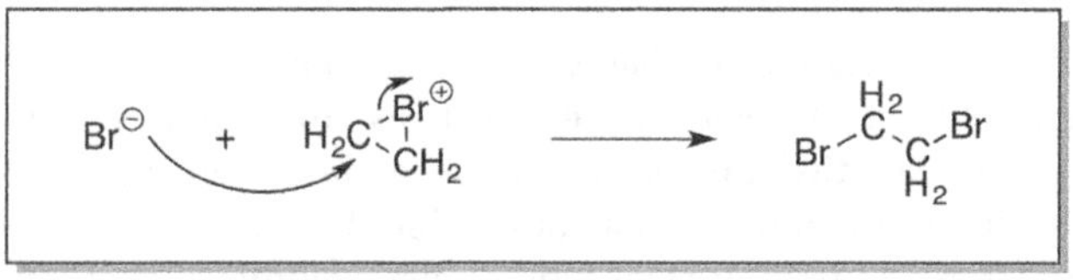

Scheme 8.4 *Nucleophilic reaction between a bromide anion and a bromonium ion generates 1,2-dibromoalkanes.*

formation of this species, a **bromide anion** is **displaced**. The initial reaction between bromine and ethylene is illustrated in Scheme 8.2 using **arrow-pushing**.

Once the **bromide anion** becomes liberated from its parent **molecular bromine**, it is free to act as a **nucleophile**. Due to the **positive charge** residing on the bridged **bromonium ion**, the adjacent carbon atoms now possess **partial positive charges**. This is due to the positively charged **bromine** pulling **electron density** from the carbon atoms. The **electrophilic** nature of the adjacent carbon atoms is illustrated in Scheme 8.3 using **resonance** structures. Because the **carbon atoms** are now **electrophilic**, they are susceptible to reaction with the **bromide anion** that has dissociated as shown in Scheme 8.2. As illustrated in Scheme 8.4, using **arrow-pushing**, this sequence of events leads to the formation of **1,2-dibromoethane**.

8.2 MARKOVNIKOV'S RULE

Diatomic halogen molecules such as **bromine** are not the only chemicals that can add across **double bonds**. In fact, any **protic acid**, under the proper conditions, can undergo such reactions. Specifically, as shown in Scheme 8.5, reaction of ethylene with an acid, HX, where X is OH, CN, or any halide, produces a substituted ethane.

Mechanistically, the addition of **acids** across **double bonds** is very similar to the reaction of **olefins** with **halogens**. In order to understand this, it is important to recognize the **electron-rich** character of **double bonds** described in Section 8.1. With this property of **olefins** in mind, one recognizes that **double bonds** can become **protonated** under acidic conditions. As illustrated in Scheme 8.6, **protonated olefins** are electronically very similar to the **bromonium ion** shown in Scheme 8.3 and, as such, can be described with **charge-delocalized resonance structures**. Furthermore, these **resonance structures** are identical to those conceptually presented in Chapters 6 and 7 during discussions of **hyperconjugation**. Recall that **hyperconjugation** is the effect leading to stabilization of **carbocations** (Chapter 6) as well as being the driving force being **1,2-hydride shifts** (Chapter 7). Bringing these concepts into the **addition** of **protic acids** to **olefins**, the step following **protonation** (illustrated in Scheme 8.7) is no different than the second step of an S_N1 substitution reaction.

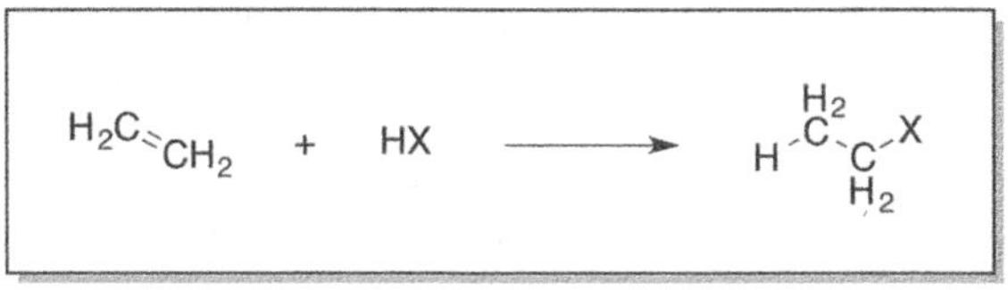

Scheme 8.5 *Protic acids can add across double bonds.*

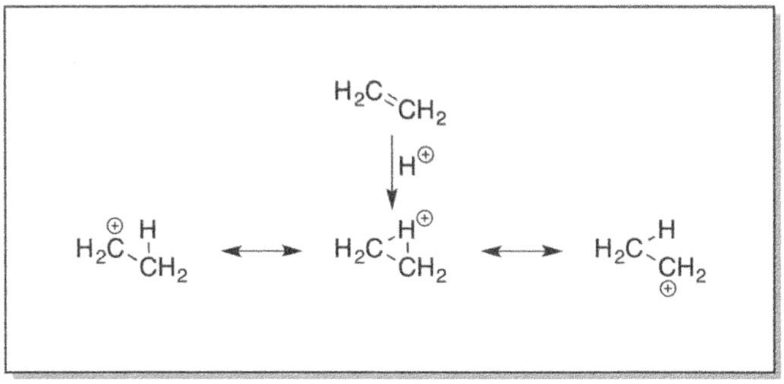

Scheme 8.6 *Double bonds can become protonated under acidic conditions.*

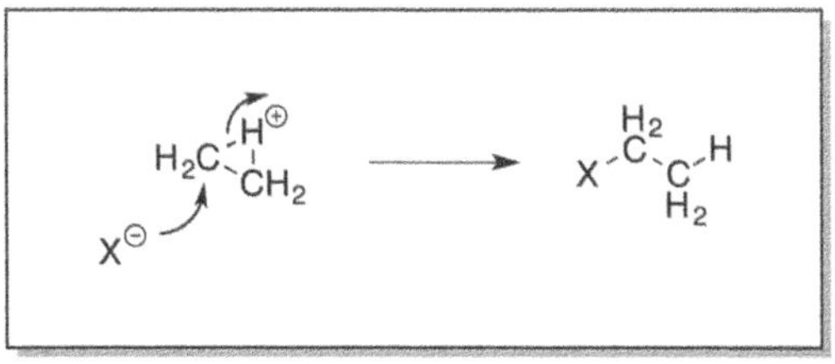

Scheme 8.7 *Nucleophiles add to protonated olefins.*

Unlike the **addition** of **halogens** across **double bonds, addition** of **acids** results in formation of **asymmetrical products**. Specifically, a **different group** is added to each side of the **double bond**. Thus, if this reaction is applied to **asymmetrical olefins** such as **propene**, multiple products might be expected to form as illustrated in Scheme 8.8. In fact, while a mixture of products is formed, there is an overwhelming presence of the **secondary substituted product** compared to that with substitution at a **primary position**. This preference of reaction products resulting from **addition** of **protic acids** across **double bonds** is governed by **Markovnikov's rule**.

In order to understand the mechanistic basis behind **Markovnikov's rule**, it is useful to refer to the mechanisms through which **acids** add across **double bonds**. Of particular relevance are the **resonance forms** of the **protonated olefins** illustrated in Scheme 8.6. Since, for **ethylene**, the two carbon atoms are both **primary**, there is no distinction between them. However, as illustrated in Scheme 8.9, in the case of **propene**, protonation of the olefin results in introduction of **cationic character** to both a **primary carbon atom** and a **secondary carbon atom**.

Referring to the discussions presented in Chapter 6 regarding the relative **stabilities** of **carbocations** (and **hyperconjugation**), we are reminded that **tertiary carbocations** are more stable than **secondary carbocations**, which, in turn, are more stable than **primary carbocations**. Since, as shown in Scheme 8.9, **protonation** of **propene** results in **cationic character** at both a **secondary carbon** and a **primary carbon**, a greater presence of **cationic character** on the **secondary site** is expected compared to the **primary**. This allows a **nucleophile** to add, preferentially, to the **secondary site** generating the reaction outcome presented in Scheme 8.8. Thus, in general, **Markovnikov's rule** states that **when an acid is added across a double bond, the conjugate base adds to the more substituted carbon atom**.

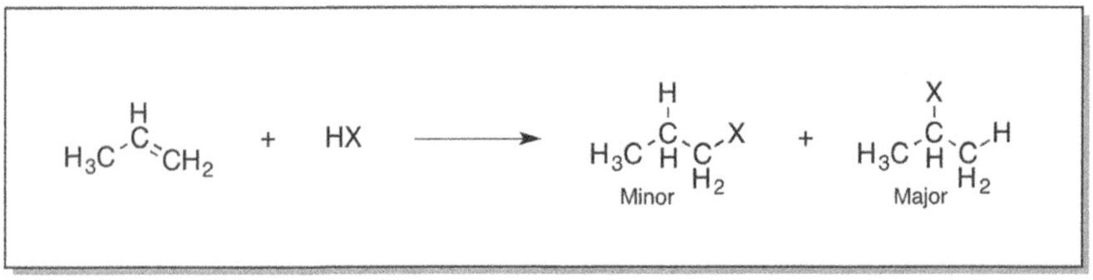

Scheme 8.8   *Multiple potential products are possible from addition of protic acids across double bonds.*

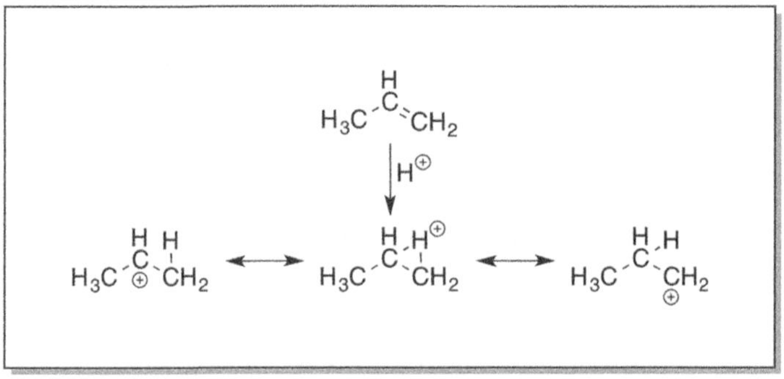

Scheme 8.9   *Protonation of propene introduces cationic character to both primary and secondary centers.*

8.3 ADDITIONS TO CARBONYLS

Olefins, in the absence of attached **polarizing groups**, generally react as described earlier with reactivity mediated through the **nucleophilicity** of the **double bond**. However, replacement of one of the **olefinic** carbon atoms with oxygen results in the formation of a polar **carbonyl** group. As shown in Figure 8.1, the **polarity** is described through placement of a **partial negative charge** on the oxygen and a **partial positive charge** on the carbon. Discussions describing the **polarity** of **carbonyls** (and other **functional groups**), based on the **electronegativities** of the various atoms involved, were presented in Chapter 1. **Addition reactions** involving **carbonyls** are discussed in the following paragraphs.

8.3.1 1,2-Additions

Because of the inherent **polarity** associated with **carbonyl** groups, **nucleophiles** are drawn to the **carbonyl** carbon atoms in much the same way that **nucleophiles** participate in S_N2 **reactions**. This mechanism, alluded to in several problems presented in previous chapters, is illustrated in Scheme 8.10 using **arrow-pushing**. As shown, a bonding pair of electrons joining the **carbonyl** oxygen atom to its associated carbon atom acts as the **leaving group**, placing a full **negative charge** on the oxygen atom. Generally, this type of reaction produces **alcohols** from **carbonyls**. Because of the trigonal planar geometry of a **carbonyl** group, there is no **stereochemical** preference associated with these addition reactions.

When considering reactions involving the **addition** of **nucleophiles** to **carbonyls**, it is important to understand that many **nucleophiles** can also serve as **leaving groups**. Therefore, in order to prevent the reverse reaction (**elimination** of the added **nucleophile**) illustrated in Scheme 8.11, **carbon-based nucleophiles** are generally utilized. Such **nucleophiles** include, but are not limited to, **Grignard reagents**, **alkyllithium reagents**, and

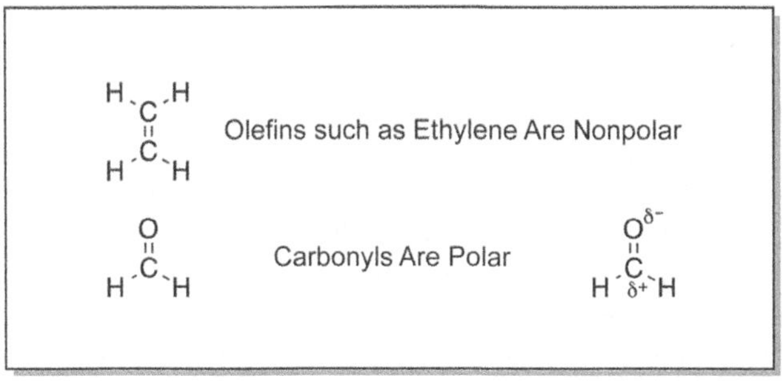

Figure 8.1 *While unsubstituted olefins are nonpolar, carbonyls are polar.*

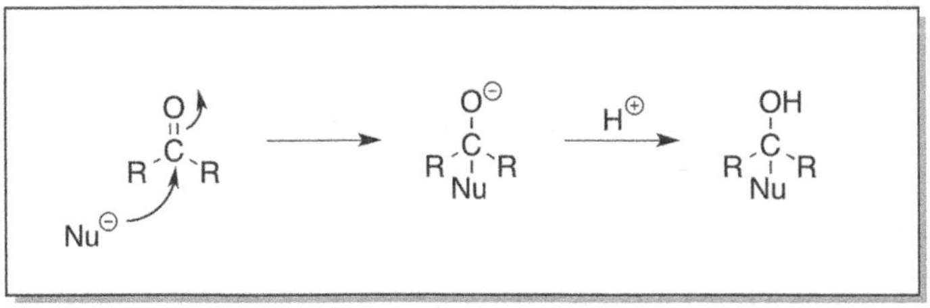

Scheme 8.10 *Nucleophiles can add to carbonyls to form alcohols.*

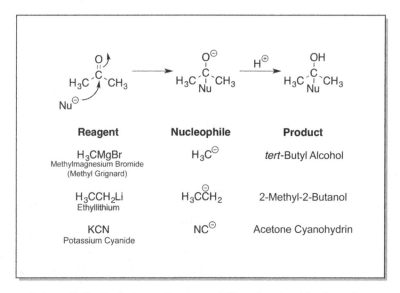

Scheme 8.11 *Addition of nucleophiles to carbonyls can be reversible.*

Reagent	Nucleophile	Product
H$_3$CMgBr Methylmagnesium Bromide (Methyl Grignard)	H$_3$C$^\ominus$	*tert*-Butyl Alcohol
H$_3$CCH$_2$Li Ethyllithium	H$_3$CCH$_2$$^\ominus$	2-Methyl-2-Butanol
KCN Potassium Cyanide	NC$^\ominus$	Acetone Cyanohydrin

Scheme 8.12 *Products resulting from addition of nucleophiles to acetone.*

potassium cyanide. In the case of **Grignard** and **alkyllithium reagents**, the result is the formation of **alcohols**. Using **potassium cyanide**, **cyanohydrins** are formed. These reagents and the products of their reactions with acetone are illustrated in Scheme 8.12.

Thus far, all examples related to the **addition** of **nucleophiles** to **carbonyls** involve **basic** (**anionic**) conditions. However, such conditions are not required. Recalling that a **carbonyl** oxygen atom possesses a **partial negative charge**, we recognize that under **acidic** conditions, it can be **protonated**. The **protonation** of **carbonyl** groups, illustrated in Scheme 8.13, was discussed in Chapter 4. Thus, as shown in Scheme 8.14 using acetone, treatment of **carbonyls** with **acids** such as HCN (**hydrocyanic acid**) provides another route for the formation of functional groups such as **cyanohydrins**.

If, as shown in Scheme 8.15, the atoms of a **carbonyl** are numbered with "1" representing the oxygen and "2" representing the **electrophilic carbonyl** carbon atom, we notice that addition of a **nucleophile** to the **carbonyl** results in the introduction of a new **substituent** at atom "2." Therefore, this type of **addition** is known as a **1,2-addition**.

8.3.2 1,4-Additions

The concept of **S$_N$2 reactions** was presented in Chapter 5. In the context of this discussion, the **S$_N$2 mechanism** was extended to **allylic** systems. These **allylic displacements**, because of their mechanistic similarities to **S$_N$2 reactions**, were designated **S$_N$2′ reactions**.

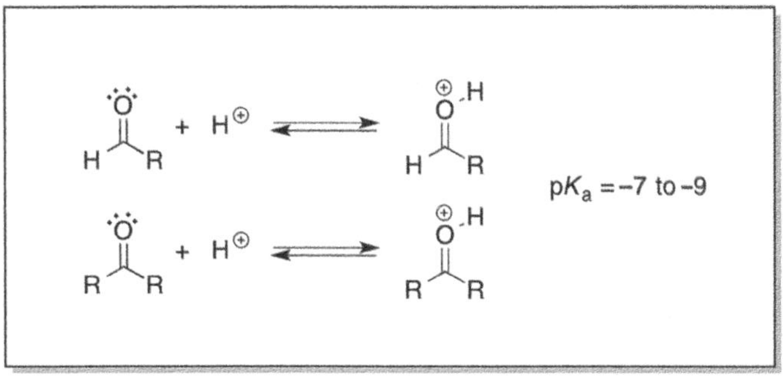

Scheme 8.13   *Carbonyls can become protonated.*

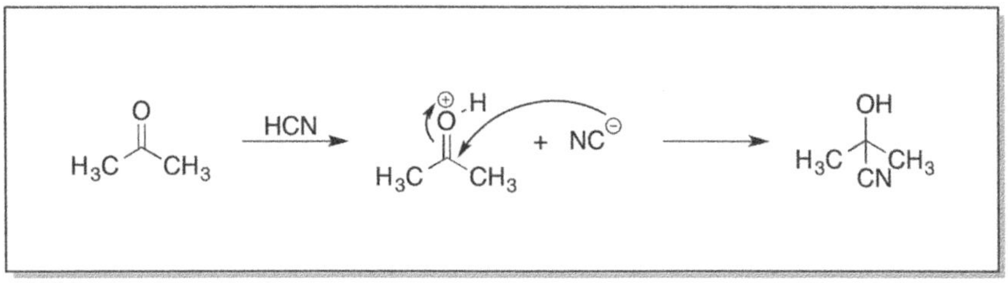

Scheme 8.14   *Addition of nucleophiles to carbonyls can occur under acidic conditions.*

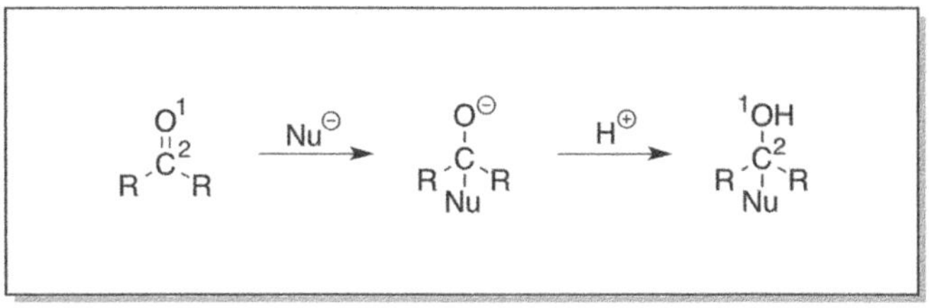

Scheme 8.15   *Addition of nucleophiles to simple carbonyls results in 1,2-additions.*

A representation of an S_N2' mechanism, compared to an S_N2 mechanism, is illustrated in Figure 8.2 using **arrow-pushing**.

In Section 8.3.1, the **addition** of **nucleophiles** to **carbonyls** was directly compared to S_N2 **reactions**. In recognition of these mechanistic similarities, one may anticipate that **nucleophiles** can similarly add to **α,β-unsaturated carbonyl** groups. Such additions are, in fact, common and, as such, are illustrated in Scheme 8.16 using **arrow-pushing**. As shown, the **nucleophile** initially **adds** to the **double bond** with **delocalization** of the negative charge into the carbonyl group generating an **enolate anion**. Once treated with acid, the **enolate anion** becomes **protonated** and forms an **enol**. **Enols**, being high-energy species, readily isomerize and regenerate the **carbonyl** functionality.

If, as shown in Scheme 8.17, the atoms of an **α,β-unsaturated carbonyl** are numbered with "1" representing the oxygen, "2" representing the **carbonyl** carbon atom, and "3" and "4" sequentially representing the adjacent two **olefinic** carbon atoms, we notice that addition of a **nucleophile** in the manner illustrated in Scheme 8.16 results in the introduction of a new **substituent** at atom "4." Therefore, this type of **addition** is known as a **1,4-addition**.

While **1,4-additions** to **unsaturated carbonyl** systems are common, it is important to recognize that the same **α,β-unsaturated** carbonyl systems are also subject to **1,2-additions**.

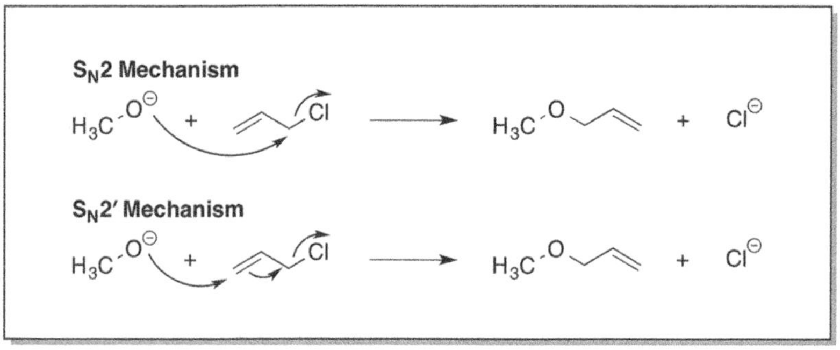

Figure 8.2 Comparison of S_N2 and S_N2' reactions as explained with arrow-pushing.

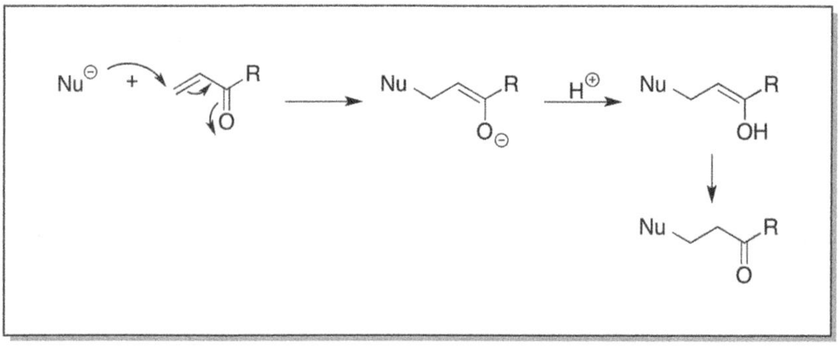

Scheme 8.16 Addition of nucleophiles to α,β-unsaturated carbonyl groups as explained using arrow-pushing.

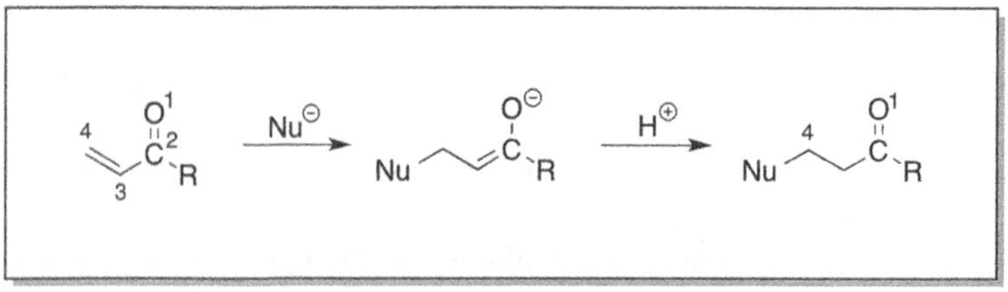

Scheme 8.17 Addition of nucleophiles to α,β-unsaturated carbonyls can result in 1,4-additions.

Fortunately, these two types of **additions** are highly dependent upon the form of the **nucleophiles** used. For example, simple **organometallic reagents** such as **alkyllithium reagents** and **Grignard reagents** tend to participate in **1,2-additions,** while **organocuprates** generally participate in **1,4-additions**. These trends, however, are not absolute, and the reader is referred to general organic chemistry textbooks for broader and more detailed treatments of these addition mechanisms.

In a final consideration regarding 1,2- and 1,4-addition reactions, **α,β-unsaturated** carbonyl systems can be sequentially subjected to both mechanisms. As illustrated in Scheme 8.18, if **methyl vinyl ketone** is treated first with **dimethyllithiocuprate** and then with **methylmagnesium bromide**, the resulting product is **2-methyl-2-pentanol**.

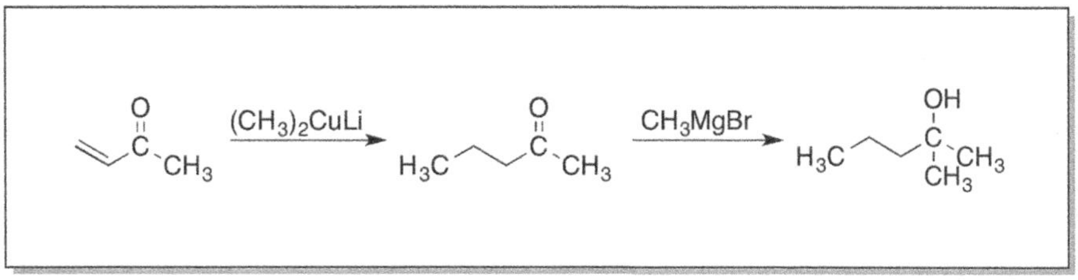

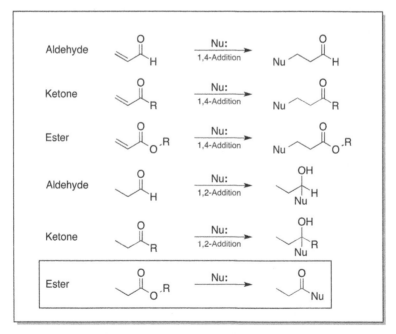

Scheme 8.18 *α,β-Unsaturated carbonyl systems can be sequentially subjected to 1,4-additions and 1,2-additions.*

8.3.3 Addition–Elimination Reactions

In our present discussions, 1,2- and 1,4-**additions** to **carbonyl** systems were introduced. However, these reactions were not presented in the context of specific **carbonyl-based functional groups**. Thus, the three types of **functional groups** generally used in **carbonyl addition reactions** are **aldehydes**, **ketones**, and **esters**.

With respect to all of the aforementioned functional groups, **1,4-additions** are generally applicable. However, of these three groups, only **aldehydes** and **ketones** are generally useful as substrates for **1,2-additions**. Figure 8.3 illustrates the products resulting from both 1,2- and 1,4-**additions** of **nucleophiles** to **aldehydes**, **ketones**, and **esters**. As shown, while the products of **1,4-additions** all result in retention of the **carbonyl** functionality, **1,2-additions** result in conversion of the respective **carbonyl** groups into **alcohols**. However, when an **ester** is involved, the illustrated product is a **ketone** and retains the **carbonyl** of the starting **ester**.

Figure 8.3 *Unlike most carbonyl-based functional groups, nonconjugated esters can react with nucleophiles and retain the carbonyl unit.*

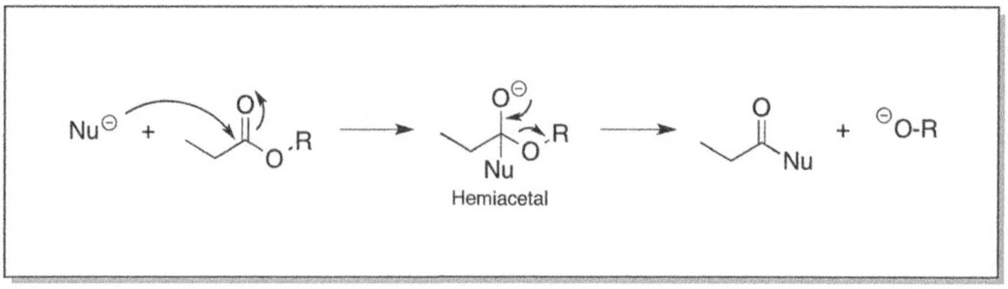

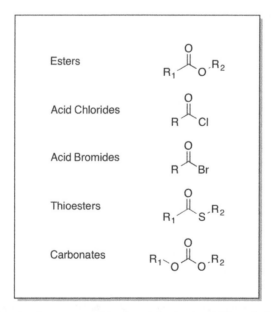

Scheme 8.19 *The addition–elimination mechanism illustrated with arrow-pushing.*

In examining the mechanism leading to the **nucleophile**-mediated conversion of an **ester** to a **ketone**, initial **addition** of a **nucleophile** to the **carbonyl** results in formation of a **hemiacetal** intermediate. Subsequent **collapse** of the **hemiacetal** intermediate liberates a **ketone** and an **alkoxide leaving group**. This mechanistic sequence, illustrated in Scheme 8.19 using **arrow-pushing**, is known as an **addition–elimination** and involves initial **addition** of a **nucleophile** to a **carbonyl** followed by **elimination** of an **alkoxide leaving group**. As a cautionary note, the conversion of **esters** to **ketones** can be difficult to control due to sequential reaction of the newly formed **ketones** with **nucleophiles** present in the reaction mixture.

Addition–elimination reactions are not exclusive to **esters**. In fact, these reactions can occur with any carbonyl-based system where the leaving group is a weaker **nucleophile** than that initially reacting. Such systems, illustrated in Figure 8.4, include, but are not limited to, **esters**, **acid halides**, **thioesters**, and **carbonates**. Finally, when predicting the products of potential **addition–elimination reactions**, guidance is readily obtained through consideration of the relative **pK_a values** of the respective **nucleophiles** and **leaving groups**.

Figure 8.4 *Functional groups capable of participating in addition–elimination reactions.*

8.4 SUMMARY

In this chapter, the principles presented in Chapter 5 (S_N2 **Reactions**) were extended into **olefinic-** and **carbonyl-based systems**. In exploring these areas, the electronic properties and **nucleophilic/electrophilic** nature of these groups were discussed. Finally, discussions of **nucleophilic additions** into these functionalities were extended into conjugated unsaturated systems leading to strategies for the incorporation of diverse modifications to relatively simple substrates. Specifically, this diversity of modifications becomes much more apparent when combining the principles presented in this chapter with those of Chapters 5 and 7. All of these principles will be useful when working through the problems of this chapter as well as advancing through introductory organic chemistry coursework.

PROBLEMS

1. Predict the products of the following reactions and then answer the following questions. Consider stereochemistry.

    I. 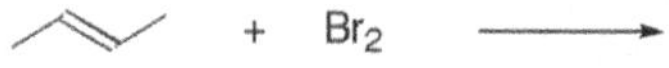   + Br₂ ⟶

    II.    + Br₂ ⟶

    III.    + Br₂ ⟶

    IV. 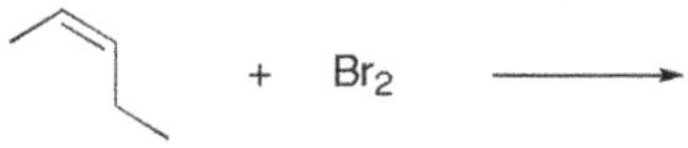   + Br₂ ⟶

    a.   Are the products of reactions I and II the same or are they different? Explain your answer.

    b.   How do you account for the products of reactions I and II?

    c.   Are the products of reactions III and IV the same or are they different? Explain your answer.

2. Predict all of the products of the following reactions.

a.    ⫻   +   HBr   ⟶

b.    ⟋⟍   +   HBr   ⟶

c.    ⫽⟍Υ   +   HBr   ⟶

3. Explain the results of the following reactions. Use arrow-pushing and specify mechanistic types.

a.

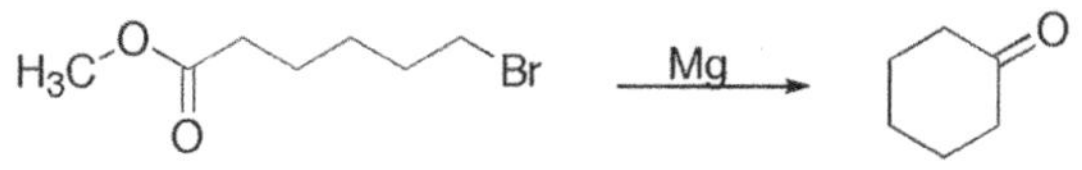

b.

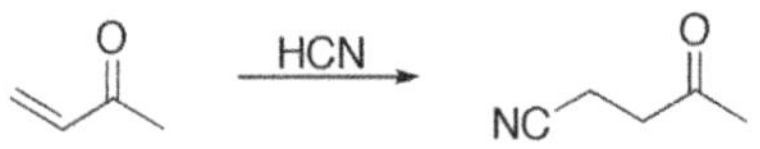

c.

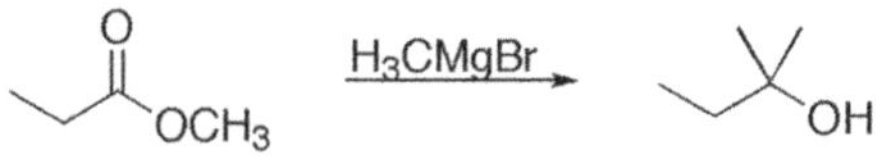

d.

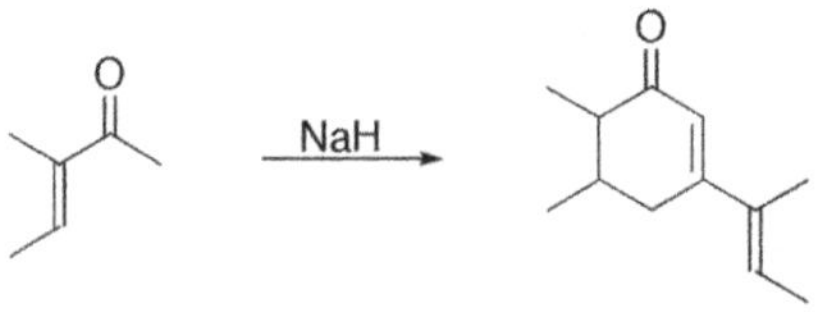

4. Explain the following reactions in mechanistic terms. Show arrow-pushing.

a. 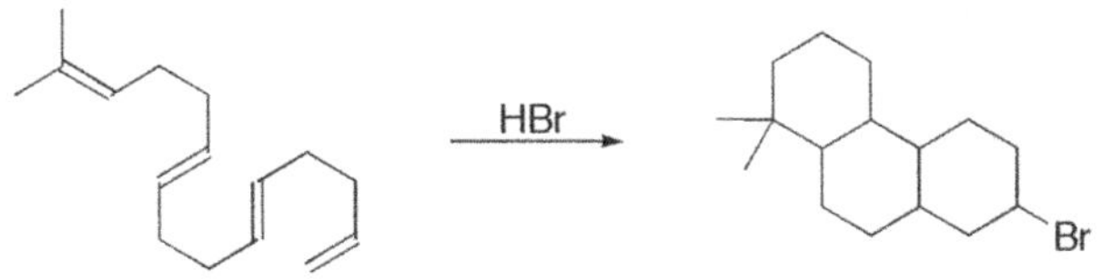 $\xrightarrow{\text{HBr}}$

b. 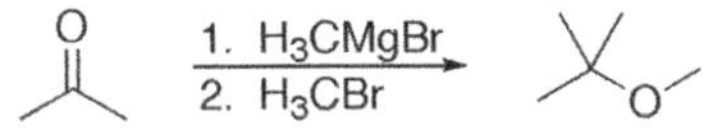 $\xrightarrow[\text{2. H}_3\text{CBr}]{\text{1. H}_3\text{CMgBr}}$

c. 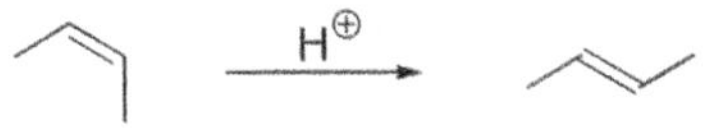 $\xrightarrow{\text{H}^{\oplus}}$

d. 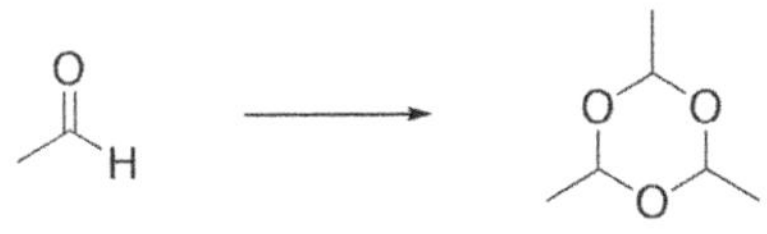 $\longrightarrow$

5. Explain the following products resulting from the reaction of amines with carbonyls. Use arrow-pushing and specify mechanistic types.

a. 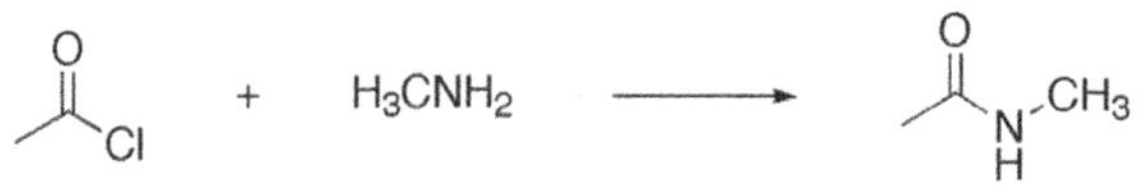 + H₃CNH₂ ⟶

b. 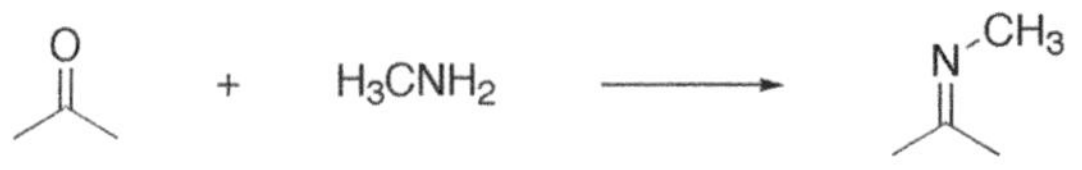 + H₃CNH₂ ⟶

c. 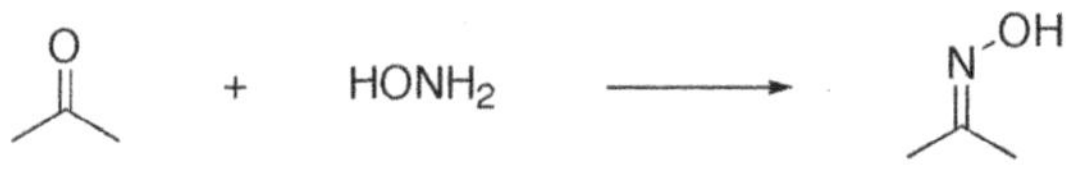 + HONH₂ ⟶

d. 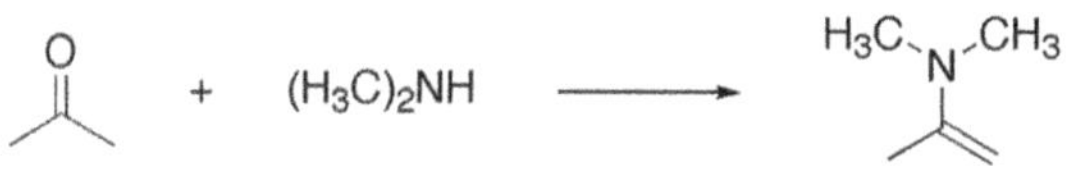 + (H₃C)₂NH ⟶

6. Provide mechanisms for the following reactions. Show arrow-pushing.

a.   +   HOCH$_3$ ⟶

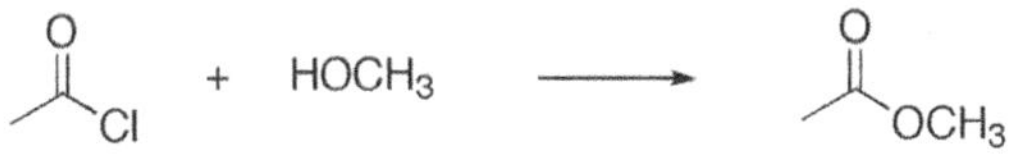

b.   +   H$_2$NCH$_3$ ⟶

c.   +   H$_2$NCH$_3$ ⟶

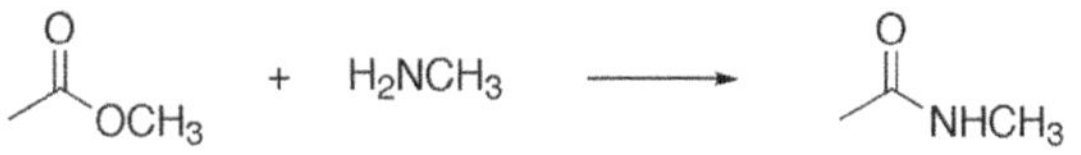

d.   +   H$_2$NCH$_3$ ⟶

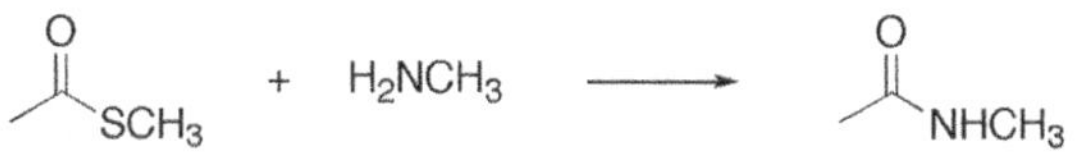

e.

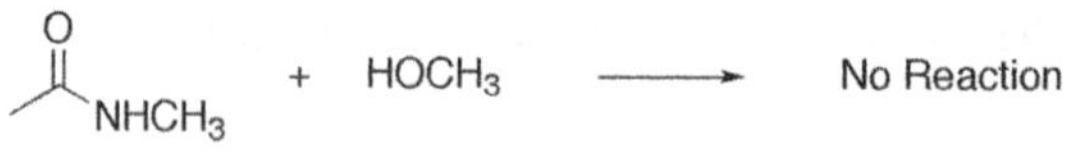

![structure: CH3-C(=O)-NHCH3]

+ HOCH₃ ⟶ No Reaction

f.

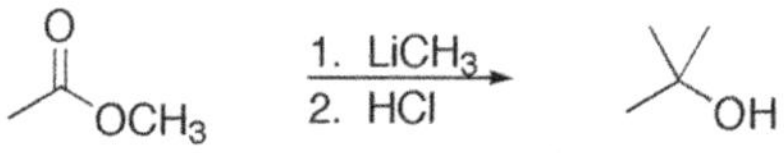

![structure: CH3-C(=O)-OCH3]

$$\xrightarrow[\text{2. HCl}]{\text{1. LiCH}_3}$$

![product: (CH3)3C-OH]

g.

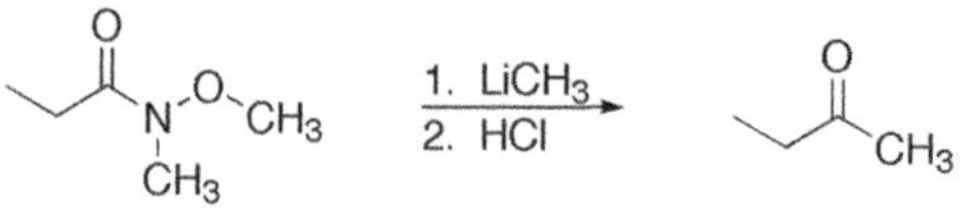

![structure: CH3CH2-C(=O)-N(CH3)-O-CH3]

$$\xrightarrow[\text{2. HCl}]{\text{1. LiCH}_3}$$

![product: CH3CH2-C(=O)-CH3]

h.

H₃C-C(=O)-Cl + H₃C-O-C₆H₅

$$\xrightarrow{\text{AlCl}_3}$$

![product: H3C-O-C6H4-C(=O)-CH3]

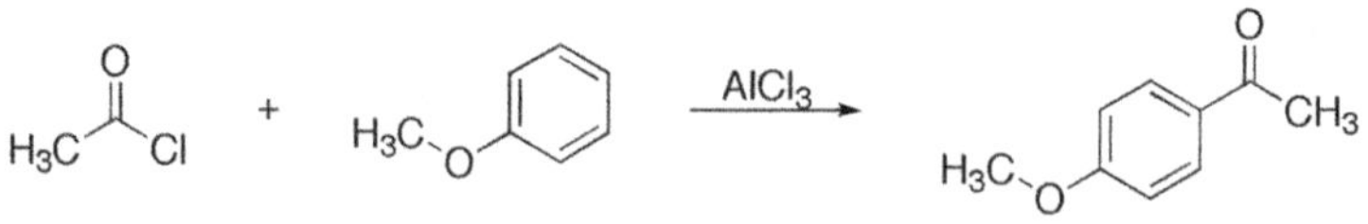

7. Explain the following amide-forming reactions using arrow-pushing. Specify the structures of **A**, **B**, and **C**, and show all relevant mechanistic steps.

a.

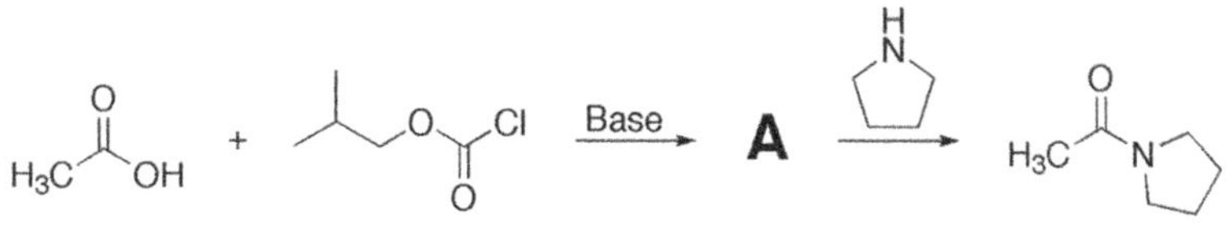

b.

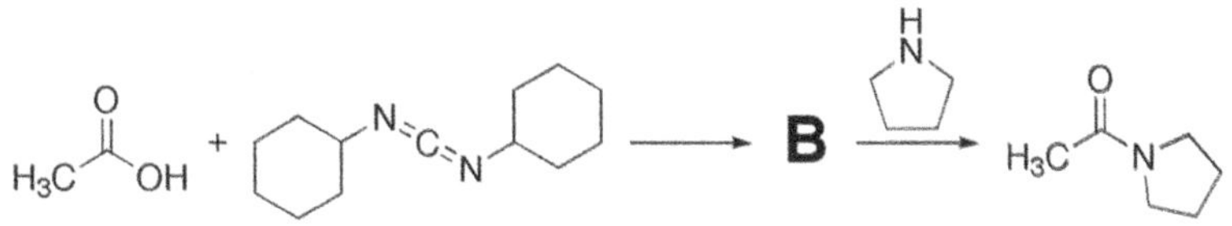

c.

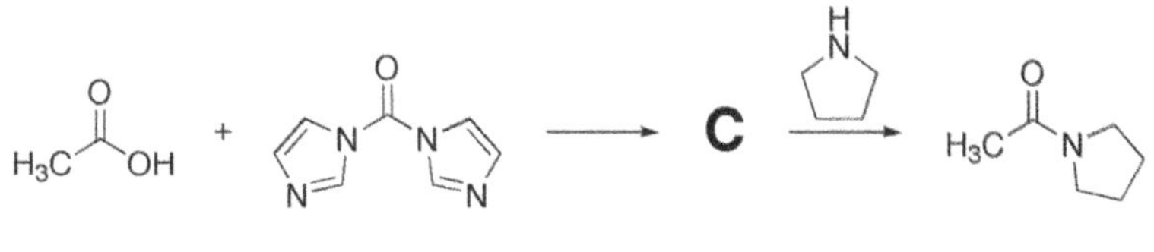

Chapter *9*

Carbenes

Chapters 3 through 8 present discussions of reactions and reaction mechanisms dependent upon positively charged species (**cations**) and negatively charged species (**anions**). As presented, **cations** result from removal of an **electron** from one **orbital** of a central atom or the loss of a **negatively charged leaving group**. Conversely, **anions** result from the addition of an **electron** to one **orbital** of a central atom or the removal of a **proton**. Typically these species form upon **heterolytic cleavage** of a given bond resulting in formation of both a component bearing a **non-bonded electron pair** (**anion**) and a component possessing an empty **orbital** (**cation**). In the case of **carbenes**, both a **nonbonded electron pair** and an empty **orbital** are present. This chapter introduces the concepts of **carbenes**, which contain both a **non-bonded electron pair** and an empty **orbital**. In addition, representative **carbene reactions** and associated mechanistic pathways are discussed.

9.1 WHAT ARE CARBENES?

A **carbene** is an intermediate species comprising a central carbon atom, a **non-bonded lone pair of electrons**, and an empty *p*-**orbital**, resulting in the central carbon being severely electron deficient. Furthermore, with both an empty *p*-**orbital** (generally found in **carbocations**) and a **non-bonded electron pair** (generally found in **carbanions**) present on the same carbon atom, **carbenes** are neutral species. As shown in Figure 9.1, **carbenes** have only six electrons associated with the central carbon atom. Because a **carbon atom** is at lowest energy with a full **octet** of **electrons**, **carbenes** are highly **electrophilic**.

Because of their high reactivity, **carbenes** are typically not stable for storage and must be generated as part of the reactions for which they are intended. Sections 9.2 and 9.3 will provide insight on **carbene formation** and useful **carbene reactions**.

Arrow-Pushing in Organic Chemistry: An Easy Approach to Understanding Reaction Mechanisms,
Second Edition. Daniel E. Levy.
© 2017 John Wiley & Sons, Inc. Published 2017 by John Wiley & Sons, Inc.

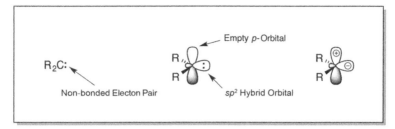

Figure 9.1 *Structural representations of carbenes using dot notation, inclusion of orbitals and representative illustration of neutralizing charges.*

9.2 HOW ARE CARBENES FORMED?

In general, **carbenes** are formed by reacting a **halogenated hydrocarbon** such as **chloroform** with a **strong base**. As shown in Scheme 9.1, when **chloroform** is treated with a base such as **potassium** *tert*-**butoxide**, **deprotonation** occurs. It is important to recognize that this initial **deprotonation** results in formation of the **trichloromethyl anion** and that the **anion** comprises a **non-bonded electron pair**.

Following **deprotonation**, the **trichloromethyl anion** undergoes a further transformation releasing a **chloride anion** and liberating **dichlorocarbene**. The process allowing formation of **dichlorocarbene** is called α-**elimination** and is compared to β-**elimination** in Scheme 9.2 using **arrow-pushing**. The concept of β-**elimination** was discussed in Chapter 7 (Scheme 7.6) as part of the **E1cB mechanistic pathway**. The term β-**elimination** is derived from the observation that the **leaving group** is associated with the carbon atom adjacent to the anion—the β-**position**. In the case of α-**elimination**, the **leaving group** is associated with the carbon atom bearing the anion—the α-**position**. It is α-**elimination** that typically results in **carbene formation**.

Referring to Scheme 9.2, it is important to recognize that there are other sources of **carbenes** aside from subjecting **halogenated hydrocarbons** to α-**elimination**. An important alternate source of **carbenes** is the **decomposition** of **diazo compounds**. **Diazo compounds** are structures bearing a reactive N_2 **leaving group**. The reactivity of this class of compounds is found in the ready release of neutral **nitrogen gas** from a formally charged but net ionic neutral **diazo substrate**. The **stability** of **diazo compounds** is found in the ability of these structures to **delocalize** the **charges** across several atoms. Figure 9.2 illustrates the general structure of **diazo compounds** and associated **resonance forms** with **arrow-pushing**.

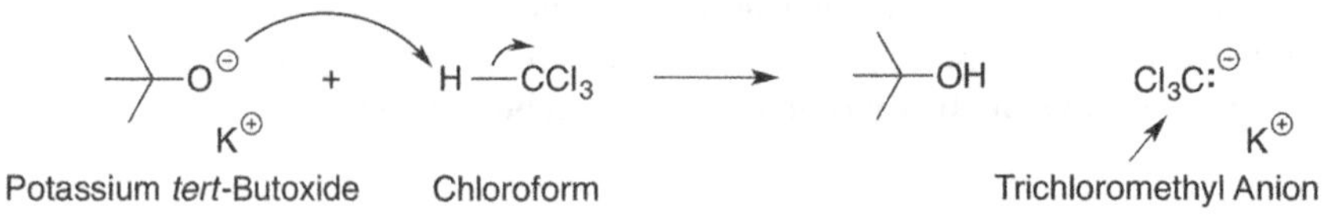

Scheme 9.1 *Chloroform can be deprotonated in the presence of strong bases.*

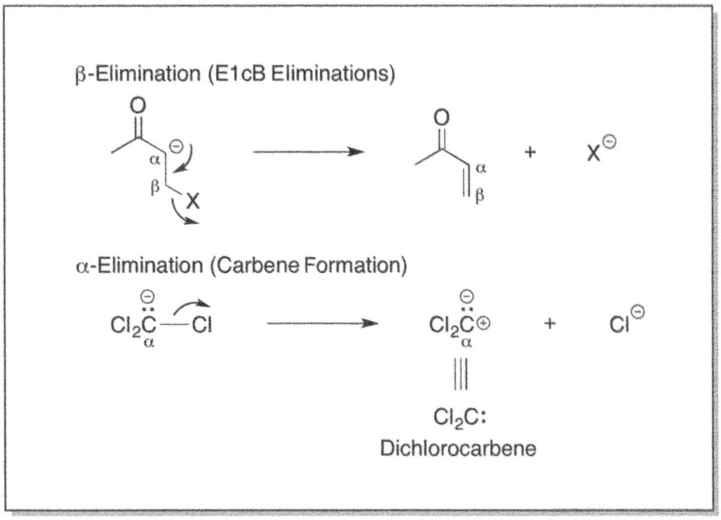

β-Elimination (E1cB Eliminations)

α-Elimination (Carbene Formation)

Scheme 9.2 *α-Elimination versus β-elimination.*

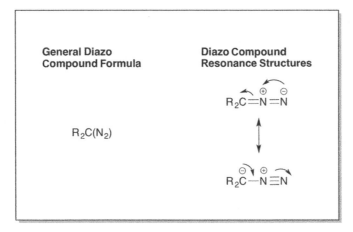

Figure 9.2 *General representation of diazo compounds and resonance forms.*

Decomposition of **diazo compounds** can occur under **photolytic** conditions (treating with light) or under **thermolytic** conditions (treating with heat). Under both cases, the general outcome is the same as treating **halogenated hydrocarbons** with base. Specifically, **α-elimination** occurs liberating a **carbene** and **nitrogen gas**. This process is illustrated in Scheme 9.3 using **arrow-pushing**.

Beyond **α-elimination**, there are additional strategies for the generation of **carbenes** and **carbene equivalents**. As many of these methods generally rely on the involvement of **transition metals** and related mechanistic pathways for **carbene stabilization**, further discussion in this area is beyond the scope of this book. The methods described herein are sufficient for this introductory treatment and for understanding the mechanistic pathways in Sections 9.3.1–9.3.3.

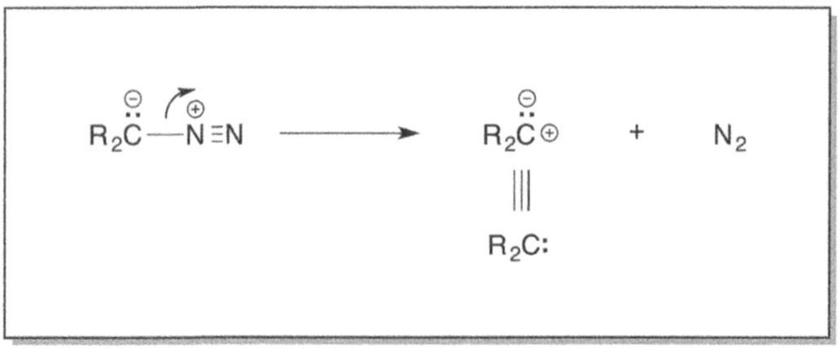

$$R_2\overset{\ominus}{\underset{..}{C}}\!-\!\overset{\oplus}{N}\!\equiv\!N \longrightarrow R_2\overset{\ominus}{\underset{..}{C}}\!{\oplus} \quad + \quad N_2$$

$$\lVert\rVert$$

$$R_2C:$$

Scheme 9.3 Decomposition of diazo compounds leads to carbene formation.

9.3 REACTIONS WITH CARBENES

Considering that the unique nature of **carbenes** effectively allows for the presence of both a **positive charge** and a **negative charge** on the same **carbon atom**, the reactivity of this species is very high. This reactivity can be utilized in a variety of transformations useful for understanding carbene reaction mechanisms and for the preparation of useful organic compounds. In this section, representative classes of **carbene reactions** are presented.

9.3.1 Carbene Dimerization

Considering the Figure 9.1 structure showing a **carbene** with both a **positive charge** and a **negative charge**, one can recognize that **carbene dimerization** as illustrated in Figure 9.3 is possible. However, because the actual **reactivity of carbenes** is so high, the likelihood that two **carbenes** will exist long enough to react with one another is very low. Therefore, observed instances of **carbene dimerization** are more likely to occur through the reaction of a **carbene** with a **carbene precursor**. As illustrated in Scheme 9.4 using **arrow-pushing**, **dichlorocarbene** can react with the **trichloromethyl anion** with subsequent **elimination** of

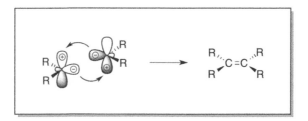

Figure 9.3 Representation of carbene dimerization.

$$Cl_2\overset{\ominus}{\underset{..}{C}}{\oplus} \atop Cl_3C:^{\ominus} \quad \longrightarrow \quad Cl_2\overset{\ominus}{\underset{..}{C}}\!-\!\overset{\overset{\displaystyle Cl}{|}}{\underset{\underset{\displaystyle Cl}{|}}{C}}\!-\!Cl \quad \longrightarrow \quad Cl_2C\!=\!CCl_2$$

Tetrachloroethylene

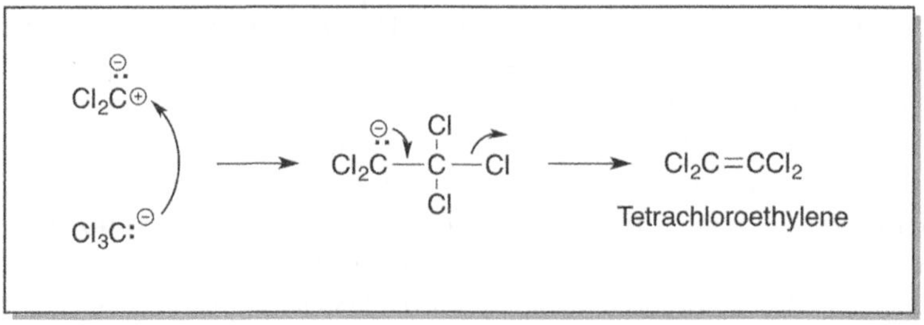

Scheme 9.4 Reaction of dichlorocarbene with the trichloromethyl anion.

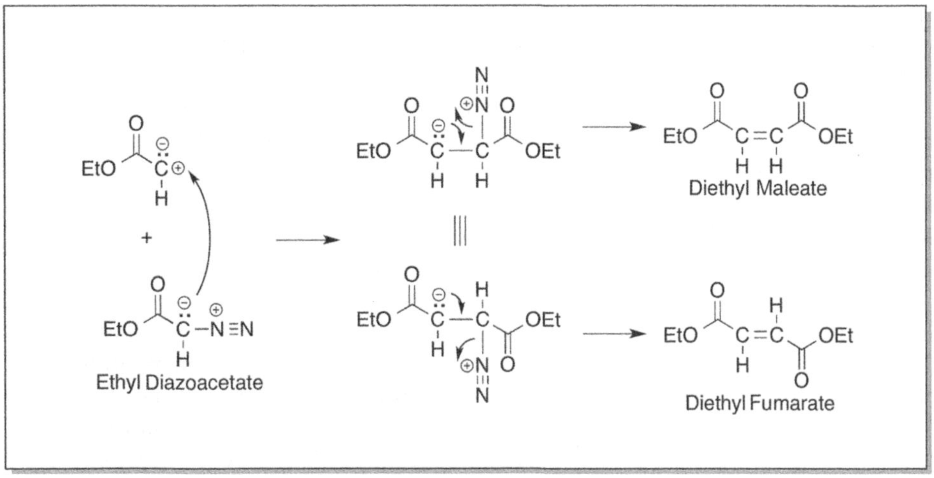

Scheme 9.5 *Reaction of the ethyl acetate carbene with ethyl diazoacetate.*

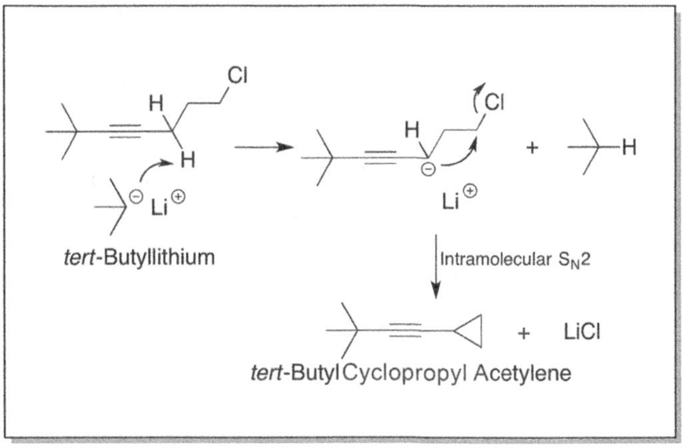

Scheme 9.6 *Example of cyclopropane formation by intramolecular S_N2 reaction.*

a **chloride anion** giving **tetrachloroethylene**. Similarly, as shown in Scheme 9.5, the **ethyl acetate carbene** can react with **ethyl diazoacetate** with subsequent **elimination** of **nitrogen** giving a mixture of **diethyl maleate** and **diethyl fumarate**. In the case of Scheme 9.5, such reactions form mixtures when the final elimination step forms a *cis-* or a *trans-***double bond** as a reflection of **rotation** around the linking **carbon–carbon bond**.

9.3.2 Cyclopropanation Reactions

The generation of **cyclopropane rings** can be both readily achievable and challenging. The fact that **intramolecular S_N2 reactions** (Scheme 9.6) can generate **cyclopropane rings** is reflected by rapid **kinetic effects** presented by the close proximity of a **nucleophilic center** to a **leaving group**. On the other hand, formation of **cyclopropane rings** is challenging due to the high **ring strain** associated with a **three-membered ring** coupled with inherent reactivity. Therefore, the more synthetic options that are available for the preparation of **cyclopropane rings**, the more accessible this valuable structural class becomes. Addressing this need is the direct formation of **cyclopropanes** from **olefins** on reaction with **carbenes**.

Referring to Chapter 6, **hyperconjugation** can be viewed as formation of a "**pseudo double bond**" from a **carbocation**. As illustrated in Figure 9.4, this can be interpreted as a "**protonated**" **double bond**. Recognizing that such **double bond protonation** occurs under acidic conditions by the overlap of a **hydrogen cation's empty s-orbital** with an olefin's electron-rich **double bond**, we can extrapolate that such interactions are possible between **olefins** and other species bearing empty **positively charged orbitals**.

As illustrated in Schemes 9.1 and 9.2, **dichlorocarbene** is available on treatment of **chloroform** with a **strong base**. Furthermore, as shown in Figure 9.1, **carbenes** can be represented as having a positively charged **empty p-orbital**. Mirroring **hyperconjugation**, we recognize that the **dichlorocarbene** empty p-orbital can overlap with an olefin's **double bond**. When this occurs, the **positive charge** is transferred from the **dichlorocarbene** to a carbon atom of the former **olefin**. Joining the former dichlorocarbene's negative charge with the former olefin's positive charge completes the formation of a **dichlorocyclopropane** ring. This process is illustrated in Scheme 9.7 using **arrow-pushing**.

Having established the ability of **carbenes** to form **cyclopropane rings** on reaction with **double bonds**, it is important to evaluate the structural and stereochemical outcomes of these reactions. Referring to Scheme 9.7, one might view the charge-separated intermediate structure as having **rotatable bonds**. Under normal circumstances, this would be a very reasonable assumption. However, due to the close proximity of the **negative charge** to the **positive charge**, ring closure is much faster than **bond rotation**. Therefore, as shown in Scheme 9.8, reactions between **olefins** and **carbenes** proceed in a *syn* manner with both new bonds being established on the same side of the **olefin**. The *anti*-product, which involves **bond rotation**, is not observed. Scheme 9.9 illustrates the products observed on reaction of **dichlorocarbene** with *cis*- and *trans*-**olefins**.

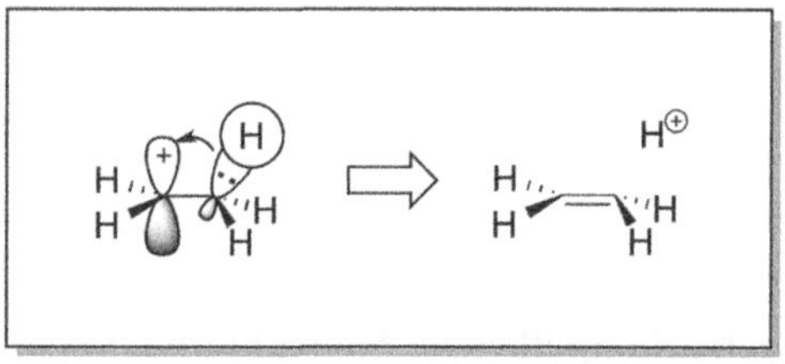

Figure 9.4 Hyperconjugation can be viewed as a "protonation of a double bond."

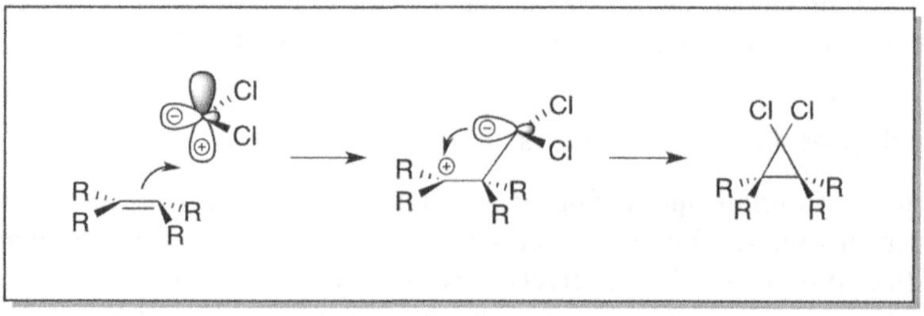

Scheme 9.7 The carbene empty p-orbital can directly interact with an olefin leading to cyclopropane ring formation.

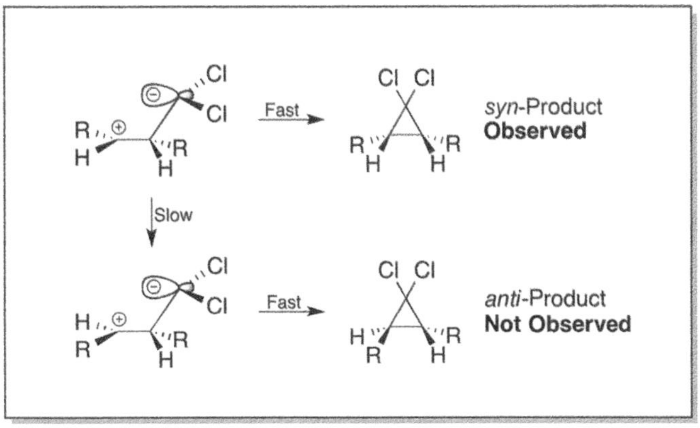

Scheme 9.8 *Carbene additions to olefins generate* syn-*products.*

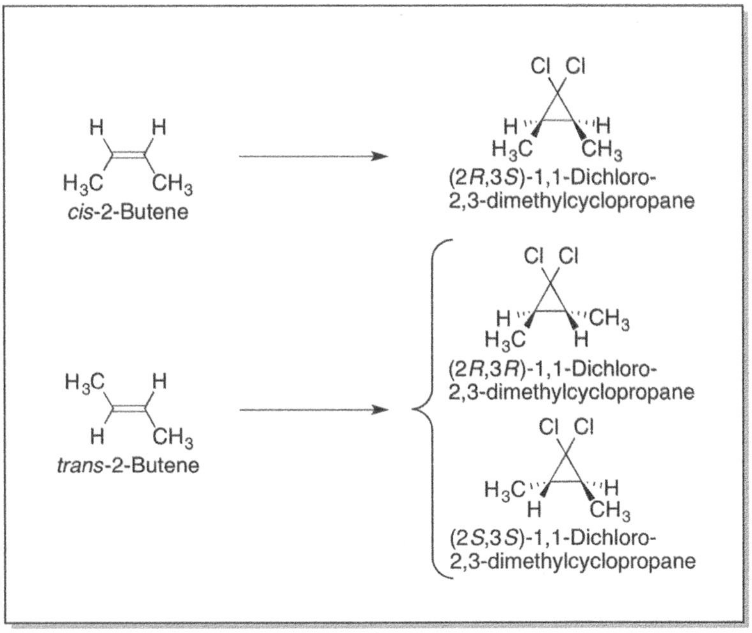

Scheme 9.9 *Dichlorocarbene produces different products from* cis- *and* trans-*olefins.*

While the concept of **stereochemistry** was briefly introduced in previous chapters, discussions of the conventions used in **stereochemical nomenclature** are beyond the scope of this book. However, it is important to understand the various products that form from given reactions. In that context, consider the illustrated products in Scheme 9.9. As shown, only one product is formed on reaction of ***cis*-2-butene** with **dichlorocarbene,** while two products are formed from ***trans*-2-butene**. These products are the result of **stereochemistry** being introduced as a result of the approach of **dichlorocarbene** to the **olefin**.

If we look at an **olefin** from its edge, we recognize that a **carbene** can approach the **olefin** from the top or from the bottom. As illustrated in Figure 9.5, in the case of ***cis*-2-butene**, there is no **stereochemical consequence** because the product that forms from the top

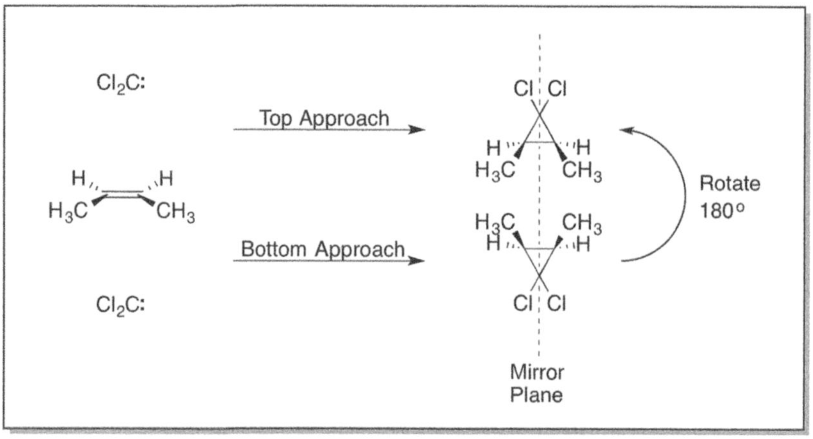

Figure 9.5 *Reaction of cis-2-butene with dichlorocarbene produces the same product from both top and bottom approaches of dichlorocarbene.*

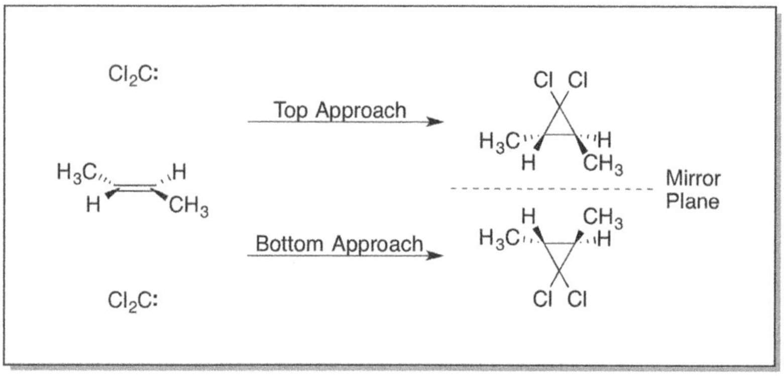

Figure 9.6 *Reaction of trans-2-butene with dichlorocarbene results in formation of enantiomers.*

approach is identical to the product that forms when the **carbene** approaches from the bottom. Furthermore, when examining the structure of (**2R,3S**)-**1,1-dichloro-2,3-dimethyl-cyclopropane**, we recognize that this compound has a **mirror plane** where one side of the molecule is a reflection of the other side.

Comparing the reaction of *cis*-**2-butene** with **dichlorocarbene** to the same reaction with *trans*-**2-butene**, we find a different outcome. As illustrated in Figure 9.6, carbene approach from the top of the **olefin** results in a product with one **stereochemical configuration**, while approach from the bottom results in a different **stereochemical configuration**. Furthermore, when examining the structures of (**2R,3R**)-**1,1-dichloro-2,3-dimethylcyclopropane** and (**2S,3S**)-**1,1-dichloro-2,3-dimethylcyclopropane**, we find that there are no mirror planes present in these molecules. However, these molecules are **mirror images** of each other. When two molecules are nonsuperimposable **mirror images** of each other, such molecules are referred to as **enantiomers**.

When considering top and bottom approaches of **carbenes** to **olefins**, it is important to recognize that reactions with **asymmetrical compounds** will result in outcomes that are different from those illustrated in Figures 9.5 and 9.6. For example, unlike *cis*-**2-butene**, reaction of *cis*-**2-pentene** with **dichlorocarbene** will result in formation of two different

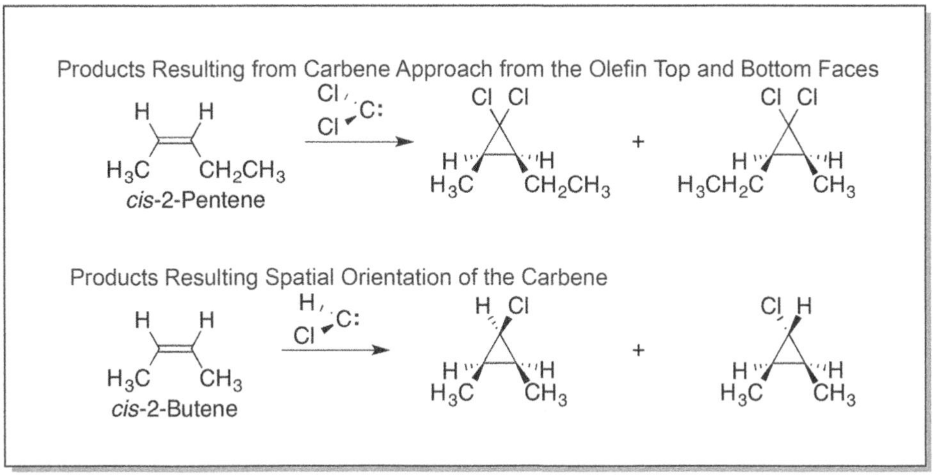

Products Resulting from Carbene Approach from the Olefin Top and Bottom Faces

cis-2-Pentene

Products Resulting Spatial Orientation of the Carbene

cis-2-Butene

Scheme 9.10 *Cyclopropanation products are influenced by the trajectory (top vs. bottom) of the carbene and by the spatial orientation of the carbene.*

products differentiated only by their **stereochemical configuration**. These products form as a result of the **carbene** trajectory approaching from the top and bottom faces of the **olefin**. Likewise, if **chlorocarbene** is reacted with *cis*-**2-butene**, the result is the formation of two different products differentiated only by their **stereochemical configuration**. These products form as a result of the **spatial orientation** of the **carbene** as it reacts with the **olefin**. These examples are illustrated in Scheme 9.10 and are presented to demonstrate that the **spatial positioning** of reaction components has profound impacts on the outcomes of chemical reactions.

9.3.3   O-H Insertion Reactions

Among the most important **carbene reactions** is the **insertion reaction**. In the case of **oxygen–hydrogen bonds**, this reaction plays a particularly important role complementing **nucleophilic** approaches for the preparation of **ethers** discussed in Chapter 5. Scheme 9.11 compares the S_N2 **formation of ethers** (**Williamson ether synthesis**) to the **formation of ethers** resulting from **carbene insertion** into an **oxygen–hydrogen bond**. As illustrated, the **Williamson ether synthesis** proceeds through initial **deprotonation** of an **alcohol** followed by displacement of a **leaving group** by the nucleophilic **oxygen anion**. In the case of **carbene insertion**, the empty *p*-**orbital** of the **carbene** joins with a **lone pair** of electrons present on the **oxygen atom** of an **alcohol**. The intermediate species now bearing a **positively charged oxygen** and a **carbanion** transfers a hydrogen ion from the oxygen to the **negatively charged carbon**. The result is formation of an **ether**.

Perhaps the most significant advantage of **carbene insertion reactions** is that, compared to S_N2 **substitutions**, these reactions do not require **basic reagents** or **deprotonation**. As illustrated in Scheme 9.11, **carbene insertion reactions** generally proceed under **neutral conditions**. This advantage can be utilized in the introduction of ether units to molecules with **base-sensitive functional groups** such as **esters** or **stereocenters** prone to **epimerization**. Examples of **base-mediated side reactions** are illustrated in Figure 9.7 as examples of processes that may be avoided utilizing **carbene insertion** strategies.

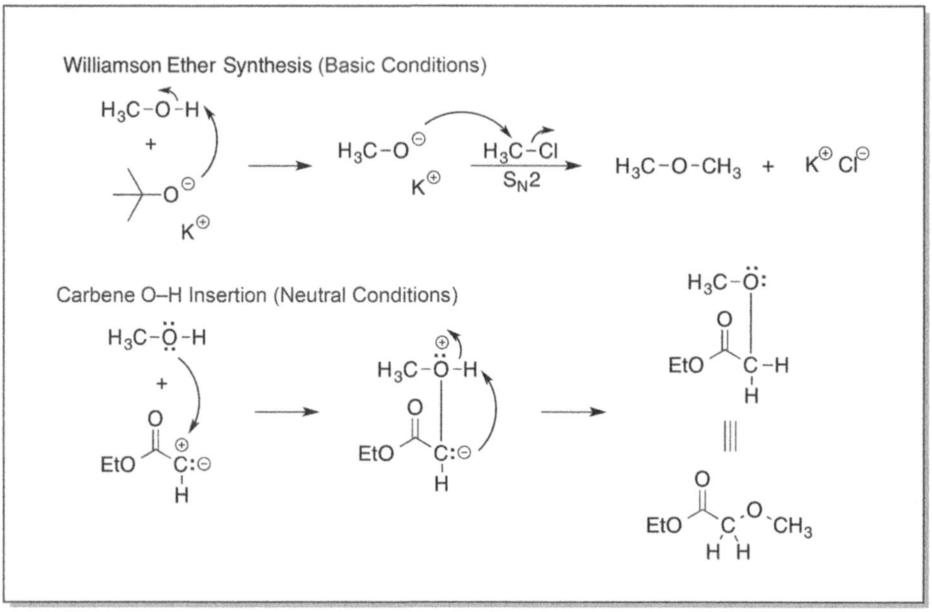

Scheme 9.11 *Carbene O—H insertion reactions are complementary to the Williamson Ether Synthesis.*

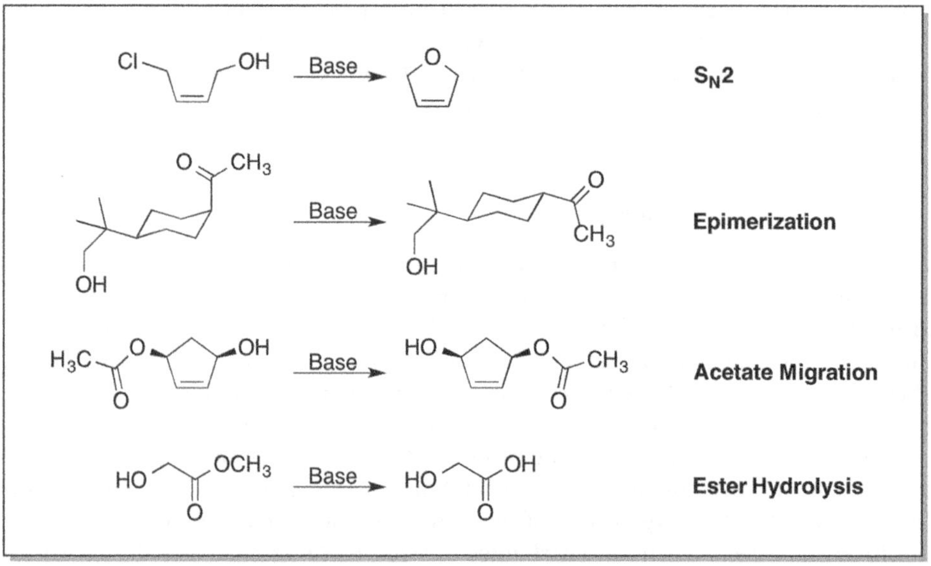

Figure 9.7 *Example base-mediated side reactions avoided using carbene insertion reactions.*

9.4 CARBENES VERSUS CARBENOIDS

Throughout this chapter, the reactive species of focus is the **carbene**. However, throughout the study and practice of **organic chemistry**, the term "**carbenoid**" is routinely found. Simply stated, a **carbenoid** is a reactive species that behaves like a **carbene**. **Carbenoid**

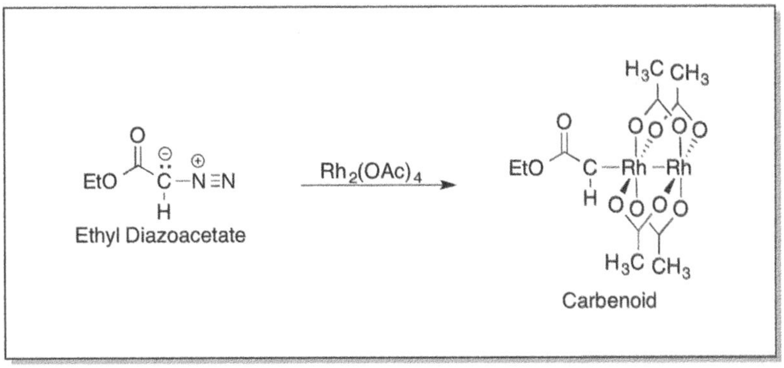

Scheme 9.12   *Formation of a carbenoid on reaction of ethyl diazoacetate with rhodium(II) acetate.*

structures may be easier to work with as their precursors are typically stable and the **carbenoids** are typically formed under **catalytic conditions**.

Expanding upon Scheme 9.11, the illustrated **carbene insertion reaction** is usually executed using a **carbenoid** formed on reaction of **ethyl diazoacetate** with **rhodium(II) acetate**. This transformation is illustrated in Scheme 9.12, and the **carbenoid structure** formed is shown with no charges, conjugation, or lone pairs in order to highlight the general structure. For the purposes of the discussions within this text, **carbenoids** may be treated as **carbenes**.

9.5 SUMMARY

In this chapter, the concept of **carbenes** was discussed as a natural extension of **carbanion** and **carbocation** chemistry. Specifically, **carbenes** were presented as structures that formally possess both a **positive** and a **negative charge** on the same atom. Because **carbenes** have properties of both **carbanions** and **carbocations**, these highly reactive structures are useful for many transformations including the **formation of cyclopropane rings** and the **formation of ethers**. Using **arrow-pushing** principles, mechanistic pathways involving **carbenes** can readily be rationalized.

As discussions in this book have evolved from the **singly charged species** involved in **heterolytic reaction mechanisms** to **net-neutral species** bearing multiple charges as seen in **carbene** chemistry, readers should now be prepared to study reaction mechanisms where no charged species are involved. One class of reactions is known as **pericyclic reactions** and is discussed in Chapter 10.

PROBLEMS

1. Schemes 9.1 and 9.2 illustrate carbene formation through α-elimination. Predict the carbene species formed from the following compounds. Explain your answer using arrow-pushing.

   a. $CHBr_3$

   b. CH_2Cl_2

   c. CH_2I_2

   d. CH_2N_2

   e. $H_2C\begin{smallmatrix}N\\||\\N\end{smallmatrix}$

2. Predict all of the products of the following reactions using arrow-pushing. Show stereochemistry where applicable.

a.    + CHCl₃ $\xrightarrow{\textit{t}\text{BuOK}}$

b.    + CHCl₃ $\xrightarrow{\textit{t}\text{BuOK}}$

c. 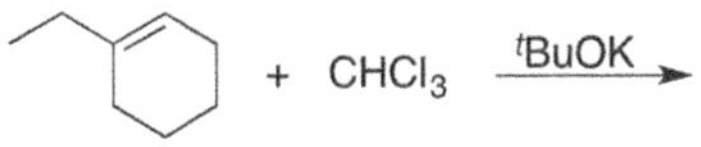   + CHCl₃ $\xrightarrow{\textit{t}\text{BuOK}}$

d. 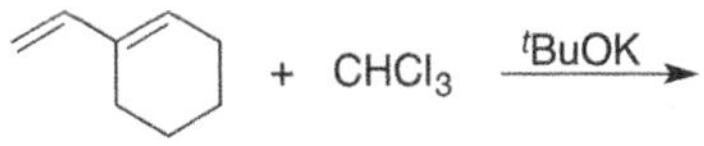   + CHCl₃ $\xrightarrow{\textit{t}\text{BuOK}}$

e. 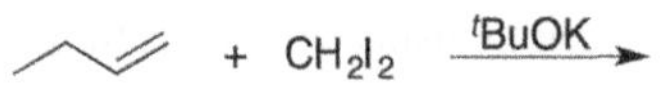   [structure: but-1-ene] + CH$_2$I$_2$ $\xrightarrow{^t\text{BuOK}}$

f. 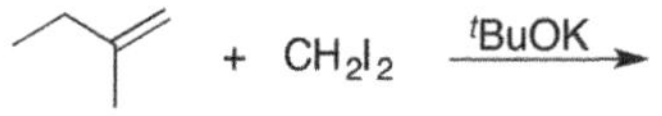   [structure: 2-methylbut-1-ene] + CH$_2$I$_2$ $\xrightarrow{^t\text{BuOK}}$

g. 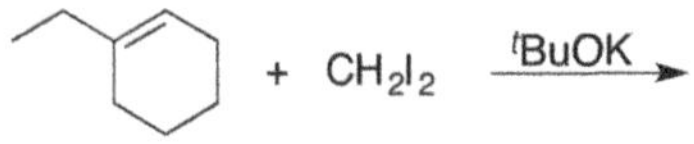   [structure: 1-ethylcyclohexene] + CH$_2$I$_2$ $\xrightarrow{^t\text{BuOK}}$

h. 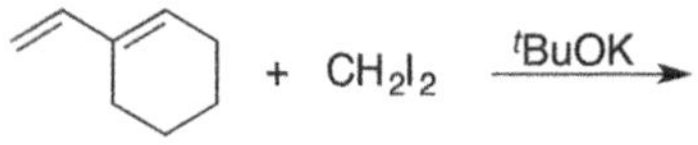   [structure: 1-vinylcyclohexene] + CH$_2$I$_2$ $\xrightarrow{^t\text{BuOK}}$

i. 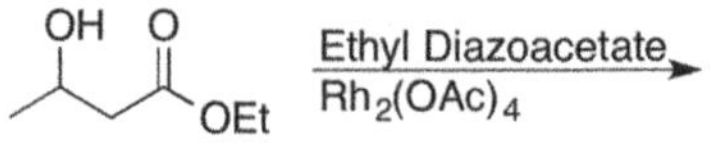   [structure: ethyl 3-hydroxybutanoate] $\xrightarrow[\text{Rh}_2(\text{OAc})_4]{\text{Ethyl Diazoacetate}}$

j.

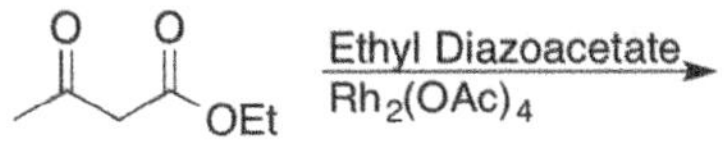

Ethyl Diazoacetate
$Rh_2(OAc)_4$

k.

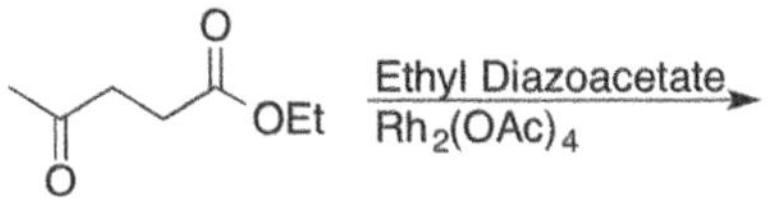

Ethyl Diazoacetate
$Rh_2(OAc)_4$

3. The following transformations involve both carbenes and free radicals. Using arrow-pushing, explain the products of the following reactions.

a.

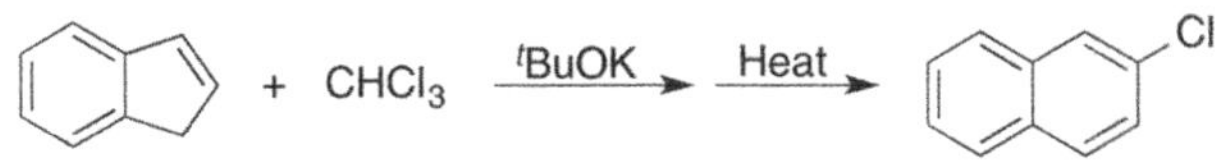

$+$ $CHCl_3$ $\xrightarrow{\ ^tBuOK\ }$ $\xrightarrow{\ Heat\ }$

Cl

b.

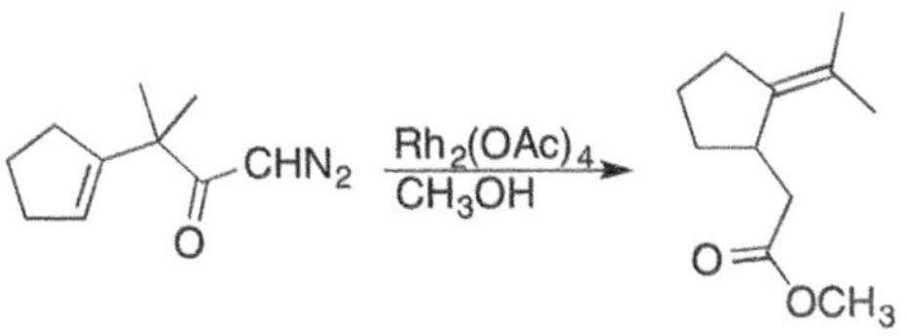

CHN_2 $\xrightarrow[CH_3OH]{Rh_2(OAc)_4}$

OCH_3

4. Referring to Figure 9.7, show the product resulting from reaction of each starting material with ethyl diazoacetate and rhodium(II) acetate. Show all reaction pathways using arrow-pushing and illustrate why the side reaction products shown in Figure 9.7 are not likely to form.

a.

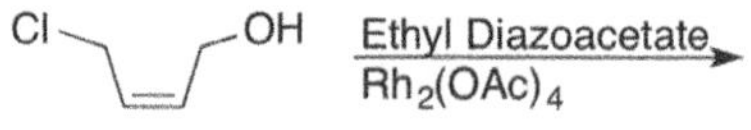

Cl——OH   Ethyl Diazoacetate / Rh₂(OAc)₄ →

b.

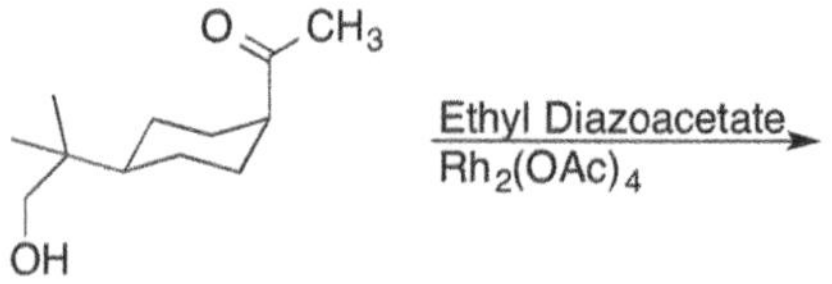

Ethyl Diazoacetate / Rh₂(OAc)₄ →

c.

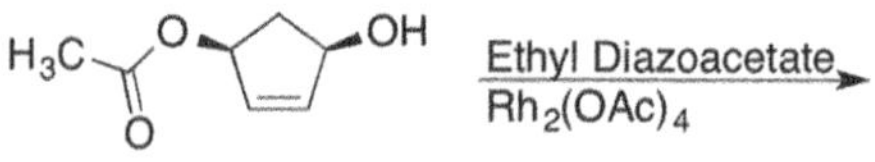

Ethyl Diazoacetate / Rh₂(OAc)₄ →

d.

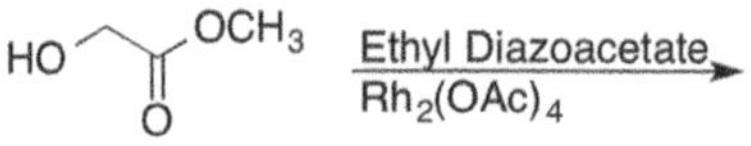

Ethyl Diazoacetate / Rh₂(OAc)₄ →

5. The thermal or photolytic decomposition of azides results in the formation of nitrenes. Nitrenes are chemically similar to carbenes due to their electronic nature and reactivity. One important nitrene reaction is the Curtius rearrangement in which an acyl azide is converted to a carbamate via an isocyanate.

    a.   Using arrow-pushing, show the mechanism for conversion of benzoyl azide to its corresponding nitrene.

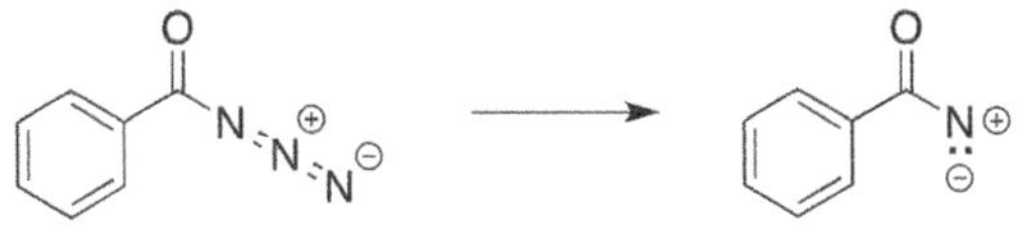

    b.   The Curtius rearrangement of benzoyl nitrene is illustrated in the following scheme. Using arrow-pushing, draw the mechanism for formation of the intermediate isocyanate.

    c.   Using arrow-pushing, draw the mechanism for conversion of the isocyanate in Problem 5b to the illustrated *tert*-butyl carbamate.

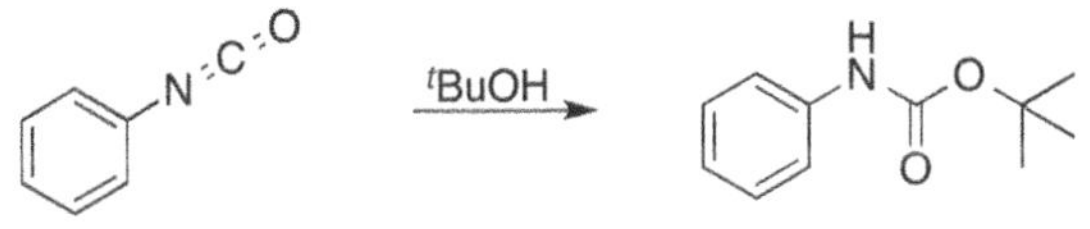

Chapter ***10***

Pericyclic Reactions

The chapters presented thus far focused on understanding the movement of **electrons** in the context of non-charged **free radicals**, **anions**, **cations**, **lone pairs**, and **carbenes**. In each of these cases, either discreet single electrons or electron pairs are used to identify the starting point in given **reaction mechanisms** and their individual mechanistic steps. However, there is a class of reactions that does rely upon the presence of **lone electron pairs** or **charges**. These **charge-neutral reaction** mechanisms usually proceed through **cyclic transition states** and are known as **pericyclic reactions**. Using **arrow-pushing**, this chapter explores various types of **pericyclic reactions** and their associated **mechanistic pathways**.

10.1 WHAT ARE PERICYCLIC REACTIONS?

Simply stated, a **pericyclic reaction** is a type of **organic reaction** in which the reaction **transition state** has a **cyclic geometry** and the reaction proceeds in a **concerted** manner. When we refer to the **cyclic geometry** of a mechanistic **transition state**, we are referring to the spatial arrangement of key atoms into a five-membered ring or a six-membered ring orientation. **Cyclic transition states** versus **acyclic transition states** are illustrated in Figure 10.1 for the **ene reaction**, the **Diels–Alder** reaction, and the **Cope rearrangement**. As shown, unless a **cyclic arrangement of atoms** is established, there can be no **pericyclic reactions**.

The term "**pericyclic reactions**" typically refers to a family of reactions comprising several classifications of transformations. Such classifications include **electrocyclic reactions** in which **conjugated systems** of **double bonds** react to form a new **carbon–carbon single bond**. Another classification of **pericyclic reactions** is the **cycloaddition** in which

Arrow-Pushing in Organic Chemistry: An Easy Approach to Understanding Reaction Mechanisms,
Second Edition. Daniel E. Levy.
© 2017 John Wiley & Sons, Inc. Published 2017 by John Wiley & Sons, Inc.

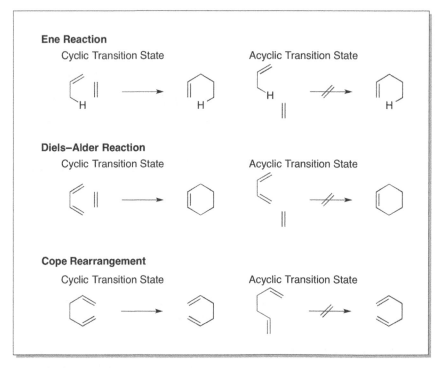

Figure 10.1 *Cyclic transition states enable progression of pericyclic reactions.*

two molecules come together forming a new structure. A third classification of **pericyclic reactions** is the **sigmatropic rearrangement** in which **carbon–carbon single bonds** are simultaneously formed and broken during the reaction. While there are additional classifications of **pericyclic reactions**, this chapter focuses exclusively on the three just described.

In order to facilitate understanding of the mechanistic pathways associated with **pericyclic reactions**, it is important to understand some mechanism-specific bond terminology such as **σ-bonds** and **π-bonds**. Simply put, **σ-bonds** are **single bonds** between two atoms with the "σ" serving as a designation for the **molecular orbital** comprising the **two-electron bond** between the atoms. **σ-Bonds** are formed by overlap of *s*-orbitals, single-lobe overlap of *p*-orbitals, and single-lobe overlap of a *p*-orbital with an *s*-orbital.

Similar to **σ-bonds**, **π-bonds** are formed when both lobes of a first *p*-orbital overlap with both lobes of a second *p*-orbital. The "π" serves as a designation for the **molecular orbital** comprising a second **two-electron bond** between atoms resulting in a **double bond**. While a **π-bond** is weaker than a **σ-bond**, extended **π-bond systems** enhance stability through **conjugation** and allow the progression of **pericyclic reactions**. Figure 10.2 illustrates the atomic orbital combinations leading to formation of **σ-bonds** and **π-bonds**.

10.2 ELECTROCYCLIC REACTIONS

Electrocyclic reactions are transformations in which **conjugated double bond** systems reversibly form new **ring systems**. As stated earlier, **pericyclic reactions** typically proceed through **five-membered transition states** or **six-membered transition states**. However,

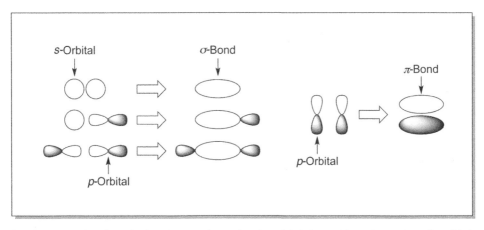

Figure 10.2 *σ-Bonds and π-bonds comprise molecular orbitals formed from the overlap of* s-*orbitals,* p-*orbitals, and combinations thereof.*

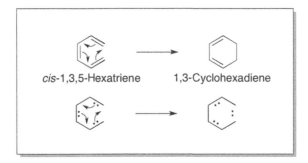

Scheme 10.1 *Electrocyclic conversion of* cis-*1,3,5-hexatriene to 1,3-cyclohexadiene.*

that generalization does not always apply. When considering **electrocyclic reactions**, the important consideration is the ability of extended **conjugated double bond** systems to arrange into a **cyclic configuration** allowing the reaction to proceed. Perhaps the simplest example of an **electrocyclic reaction** is the cyclization of **cis-1,3,5-hexatriene**. As illustrated in Scheme 10.1, under **thermal** or **photolytic conditions**, **cis-1,3,5-hexatriene** cyclizes to form **1,3-cyclohexadiene**. The mechanism, illustrated using **arrow-pushing**, does not involve any charged species. Instead, the mechanism is a cyclic shift of electrons from the extended **π-system** resulting in the formation of a new **carbon–carbon single bond**. This is more easily illustrated by representing the second bond of each **double bond** as **electron pairs**. Using **arrow-pushing**, the **electron pairs** are shown to shift from one **carbon–carbon bond** to an adjacent **carbon–carbon bond**. Remaining consistent with the **octet rule**, the shift of one **electron pair** requires the shift of all **electron pairs** and results in the formation of a new **carbon–carbon single bond**.

Because this transformation depends upon the ability of the three **double bonds** to arrange themselves into a **six-membered ring configuration**, any disruption in this ability hinders the reaction. Factors that can influence the formation of the **six-membered transition state** include additional substitutions that introduce **steric effects**. Figure 10.3 illustrates some substitution patterns that impact the rate and success of this transformation.

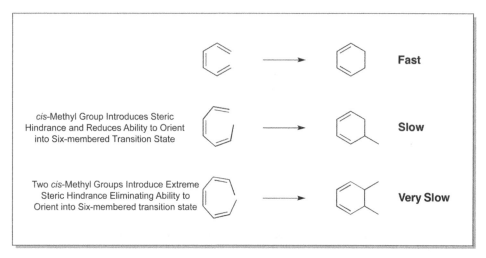

cis-Methyl Group Introduces Steric
Hindrance and Reduces Ability to Orient
into Six-membered Transition State

Two cis-Methyl Groups Introduce Extreme
Steric Hindrance Eliminating Ability to
Orient into Six-membered transition state

Figure 10.3 *Substitution patterns can impact the rate and success of electrocyclic reactions.*

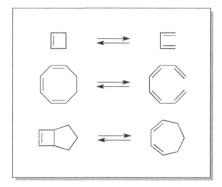

Scheme 10.2 *Electrocyclic reactions involving four-membered rings, eight-membered rings, and bicyclic ring systems.*

Thus far, **electrocyclic reactions** have been illustrated for **six-membered ring systems**. However, these reactions are well known for both smaller and larger rings and their related **conjugated olefins**. As illustrated in Scheme 10.2, **four-membered rings**, **eight-membered rings**, and **bicyclic ring systems** represent additional and non-limiting examples of **electrocyclic reaction** systems. As illustrated for **six-membered rings**, **electrocyclic reactions** involving all ring sizes are dependent upon the ease of formation of the respective **cyclic transition states**. As such, successful transformations are highly dependent upon **substitution patterns** and **steric effects** that may block the ability of the **olefin systems** to adopt the required **cyclic conformations**. In all cases and directly analogous to Scheme 10.1, the **mechanistic pathways** of these reactions can be explained using **arrow-pushing**.

When **substituted olefinic systems** undergo **electrocyclic reactions**, **stereochemical outcomes** result. For example, as illustrated in Scheme 10.3, **(*E,Z,E*)-octa-2,4,6-triene** may conceptually form two **cyclohexadiene** products. One has the two methyl groups in a **cis-orientation** relative to each other, while the other has the two methyl groups in a ***trans-orientation***. Experimentally, the only product resulting from the thermal version of this

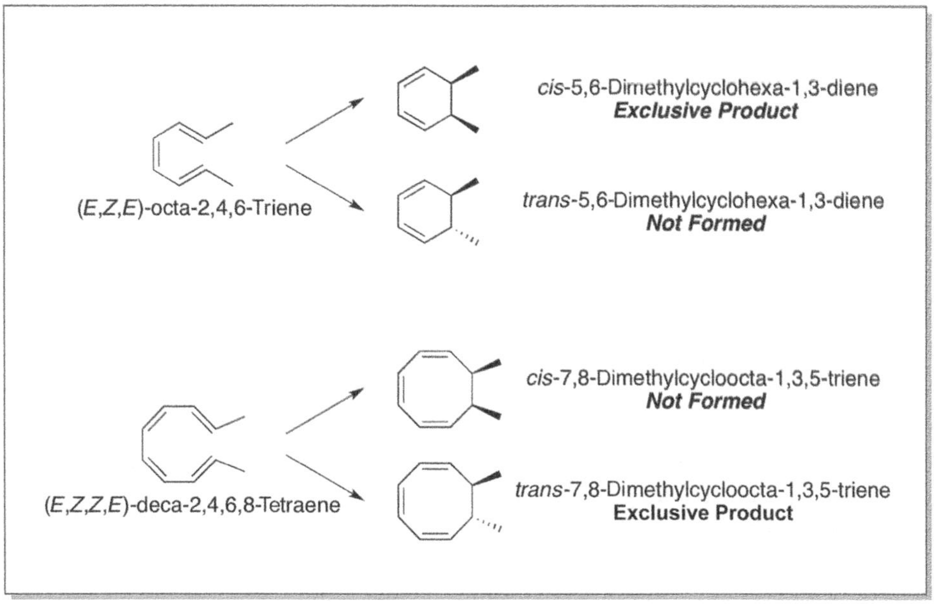

Scheme 10.3 *Stereochemical courses for electrocyclic reactions forming six-membered and eight-membered rings.*

reaction is *cis*-**5,6-dimethylcyclohexa-1,3-diene**. However, the **stereochemical course** for the thermal **electrocyclic reaction** applied to **(*E,Z,Z,E*)-deca-2,4,6,8-tetraene** leads to exclusive formation of ***trans*-7,8-dimethylcycloocta-1,3,5-triene**. While these differences in **stereochemical outcomes** may seem counterintuitive, they are readily explained using the **Woodward–Hoffman rules**—a discussion of which is outside of the scope of this book. For advanced treatments of this reaction type, students are referred to more detailed organic chemistry texts.

10.3 CYCLOADDITION REACTIONS

Cycloaddition reactions are processes where two reactants are brought together through a **cyclic transition state** to form a single product. Such reactions commonly involve **five-membered transition states** and **six-membered transition states** and always proceed through a **concerted reaction mechanism**. There are many types of **cycloaddition reactions** of use in organic chemistry. Among the most common are the **Diels–Alder reaction**, the **ene reaction**, and **1,3-dipolar cycloaddition reactions**. In Sections 10.3.1–10.3.3, each of these **cycloaddition reactions** is discussed using **arrow-pushing**.

10.3.1 The Diels–Alder Reaction

Among the most recognizable **cycloaddition reactions** is the **Diels–Alder reaction**. This reaction, illustrated in Scheme 10.4 using **arrow-pushing**, brings together a **diene** such as **1,3-butadiene** and a **dienophile** such as **ethylene** to form a **cyclohexene** structure. As applied to **electrocyclic reactions**, the **Diels–Alder mechanism** is more easily

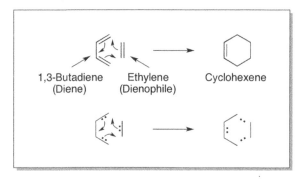

Scheme 10.4   *Diels–Alder reaction with 1,2-butadiene and ethylene.*

illustrated by representing the second bond of each **double bond** as **electron pairs**. Using **arrow-pushing**, the **electron pairs** are shown to shift from one **carbon–carbon bond** to an adjacent **carbon–carbon bond**. Remaining consistent with the **octet rule**, the shift of one **electron pair** requires the shift of all **electron pairs** in the cyclic array and results in the formation of two new **carbon–carbon single bonds**.

The breadth and utility of the **Diels–Alder reaction** allows for the formation of **cyclohexene ring structures** with multiple ring substituents. Furthermore, there are specific and predictable **stereochemical outcomes** associated with the **Diels–Alder reaction**. In the following paragraphs, each of these topics is addressed.

The mechanistic course of the **Diels–Alder reaction** is tied to how the **diene** and the **dienophile** approach one another. Specifically, the *p*-orbitals of the **dienophile** must overlap with the *p*-orbitals of the **diene**. Figure 10.4 illustrates the required **diene–dienophile relationship** for the **Diels–Alder reaction** between **cyclopentadiene** (**diene**) and **maleic anhydride** (**dienophile**) to proceed. As shown, there are two possible orientations these **reactants** may adopt. These orientations are designated "**exo**" where the **transition state** appears extended and "**endo**" where the **transition state** appears folded upon itself. In fact, the **endo** orientation is favored due to additional *p*-orbital overlap available between the **maleic anhydride carbonyls** and the **cyclopentadiene olefins**. These interactions are not possible from the extended **exo orientation**. In general, **Diels–Alder reactions** typically proceed through **endo transition states**.

Due to the preferred **endo** relationship between a **diene** and a **dienophile**, Diels–Alder **reactions** with **substituted dienes** and/or **dienophiles** proceed with specific **stereochemical outcomes**. As illustrated in Scheme 10.5, **cyclopentadiene** reacts with **acrolein** through an **endo transition state** forming the illustrated product. The **exo** product is also formed but only as a minor product component. Furthermore, because there are two possible **endo transition state configurations**, the product is formed as a mixture of **enantiomers**.

Useful **dienes** for **Diels–Alder reactions** are typically **conjugated double bond** systems that may be either **cyclic** or **acyclic**. Strategic **substitutions** on the **diene** system are useful for the generation of **substituted cyclopentene rings** or more complex **ring systems**. Useful **dienophiles** for **Diels–Alder reactions** are typically carbon–carbon double or triple bonds bearing **electron-withdrawing groups**. As with the **diene** component, **dienophile** substitutions are useful for the generation of products that can be further modified as required. Not intended to be an exhaustive list, Figure 10.5 illustrates example **dienes** and **dienophiles** useful for **Diels–Alder reactions**.

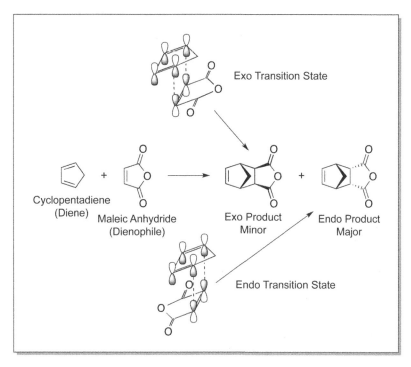

Figure 10.4   *Diene–dienophile orientations for Diels–Alder reaction progression.*

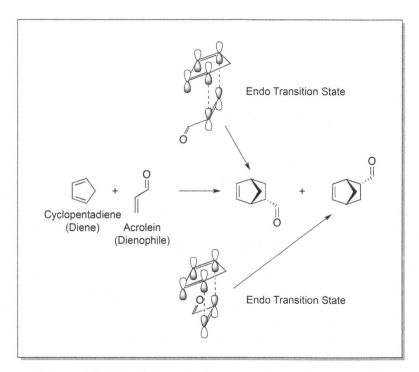

Scheme 10.5   *Diels–Alder reaction between cyclopentadiene and acrolein.*

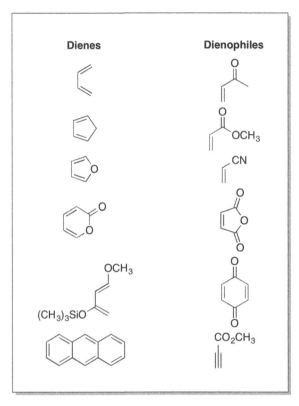

Figure 10.5 Example dienes and dienophiles useful in Diels–Alder reactions.

10.3.2 The Ene Reaction

The **ene reaction** is a **pericyclic reaction** that is mechanistically similar to the **Diels–Alder reaction**. Where the **Diels–Alder reaction** takes place between a **diene** and a **dienophile**, the **ene reaction** requires an **allyl** or "**ene**" component and an "**enophile**." The **ene** component is typically a **carbon–carbon double bond** or a **carbon–carbon triple bond** with at least one hydrogen in an **allylic** or **propargylic** position. **Enophiles** are similar to **Diels–Alder dienophiles** and typically are **carbon–carbon double bonds** or **carbon–carbon triple bonds** bearing an **electron-withdrawing group** (see Fig. 10.5). Unlike the **Diels–Alder reaction**, the **ene reaction** does not generate **ring systems**. Instead, this reaction effectively forms **carbon–carbon single bonds** through a **cyclic transition state**. Scheme 10.6 illustrates the **ene reaction** between **propylene** (the **allyl** component) and **ethylene** (the **enophile**). As applied to **electrocyclic reactions**, the **ene reaction mechanism** is more easily illustrated by representing the second bond of each **double bond** and the **allylic carbon–hydrogen bond** as **electron pairs**. Using **arrow-pushing**, the **electron pairs** are shown to shift from one **carbon–carbon bond** to adjacent sites. Remaining consistent with the **octet rule**, the shift of one **electron pair** requires the shift of all **electron pairs** and results in the breaking of one **carbon–hydrogen bond**, formation of one new **carbon–carbon single bond**, and formation of one new **carbon–hydrogen bond**.

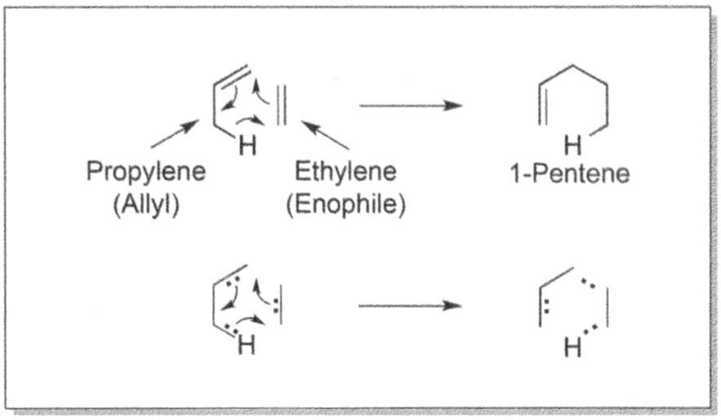

Scheme 10.6 *Ene reaction with propylene and ethylene.*

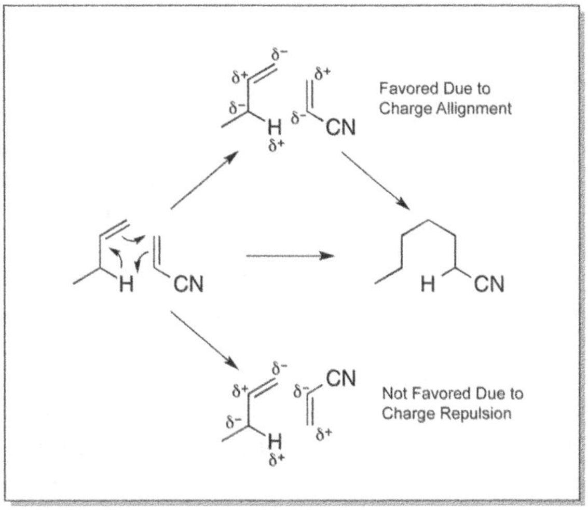

Scheme 10.7 *Ene reaction between 1-butene and acrylonitrile.*

As with the **Diels–Alder reaction**, the **ene reaction** can be applied to **substituted allyl groups** and to substituted **enophiles**. When substituted **enophiles** are involved in the **ene reaction**, **regiochemical** outcomes can be predicted based upon the orientation of reaction components. The regiochemical preference of the **ene reaction** is influenced by the **partial charges** present on the atoms involved in the reaction. Using principles discussed in Chapters 3 and 4, sites of **partial positive charges** and **partial negative charges** can be identified. Keeping in mind that **hydrogen atoms** carry **partial positive charges**, aligning the charged atoms in a fully alternating arrangement allows prediction of the **regiochemical** course of the **ene reaction**. Using **arrow-pushing**, Scheme 10.7 illustrates the progression of an **ene reaction** between **1-butene** and **acrylonitrile**.

Because the **ene reaction** is less **conformationally restricted** compared with other **pericyclic reactions**, it is often used in situations where the reaction components are already constrained relative to one another. Such **intramolecular ene reactions** are useful for the generation of **substituted ring systems**. As illustrated in Scheme 10.8 using **arrow-pushing**, (**E**)-1,7-nonadiene can undergo an **intramolecular ene reaction** to form a **racemic mixture** of **trans-1-methyl-2-vinylcyclohexane**.

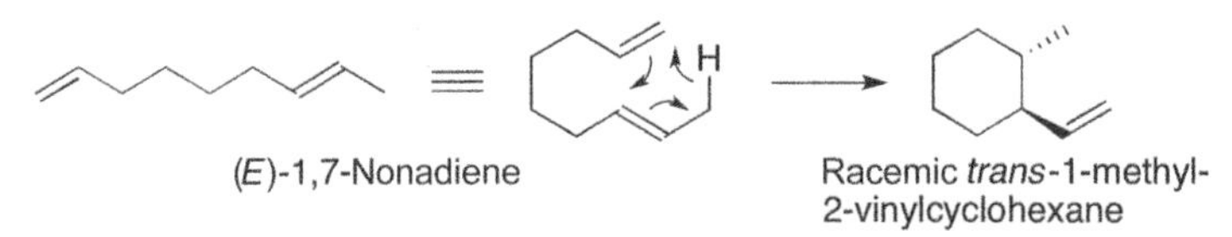

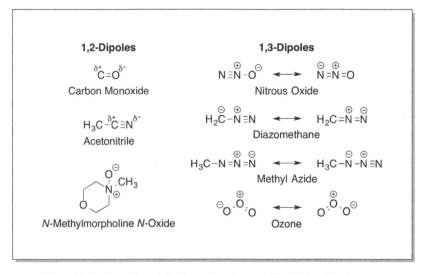

(E)-1,7-Nonadiene

Racemic trans-1-methyl-2-vinylcyclohexane

Scheme 10.8 Intramolecular ene reactions can form substituted ring systems.

10.3.3 Dipolar Cycloaddition Reactions

Where the **Diels–Alder reaction** is useful for the generation of **cyclohexene ring systems** and the **ene reaction** is useful for the formation of **carbon–carbon bonds** through **cyclic transition states**, **dipolar cycloadditions** are useful for the formation of **five-membered rings** bearing **non-carbon atoms**. Among the most common of these transformations is the **1,3-dipolar cycloaddition**.

Before discussing **1,3-dipolar cycloadditions**, one should first be acquainted with a **1,3-dipole**. **Dipoles**, in general, are structures where **charges** are **separated** across **several atoms**. Examples of **dipoles** are shown in Figure 10.6 and include both **functional groups** and **discreet molecules**. **Functional groups** are generally **dipolar** in nature due to **partial charges** being present on the associated **carbon** and **heteroatoms**. Specific molecules that are also **dipoles** include **carbon monoxide**, **nitrous oxide**, and **acetonitrile**. When **charges** are **delocalized** across three atoms in a given **molecule** or **functional group**, the species is referred to as a **1,3-dipole**. In Figure 10.6, **dipoles** are shown with their partial or full **charges** along with relevant **resonance forms**.

As illustrated in Figure 10.6, **1,3-dipoles** can be discreet molecules such as **nitrous oxide** or **ozone**. Similarly, **1,3-dipoles** can be **functional groups** such as **azides**. Recognizing that the **negative charge** of a **1,3-dipole** can act as a **nucleophile**, one can use **arrow-pushing** to describe the formation of **five-membered rings** upon reaction of a **1,3-dipole** with a

Figure 10.6 Examples of dipolar molecules and dipolar functional groups.

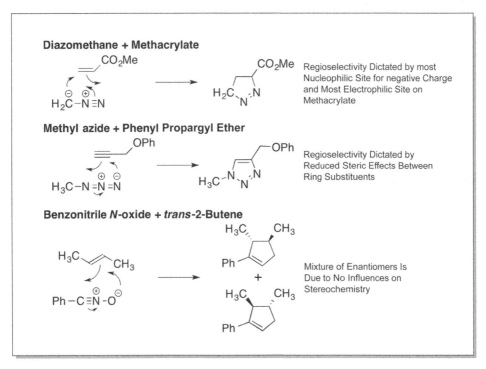

Scheme 10.9 *Examples of 1,3-dipolar cycloadditions.*

carbon–carbon double bond or a **carbon–carbon triple bond**—the component reacting with the **1,3-dipole** being referred to as a **dipolarophile**. Such reactions can generate stable target structures or intermediate structures transiently formed in more complex reaction mechanisms. Using **arrow-pushing**, Scheme 10.9 illustrates several **1,3-dipolar cycloadditions** leading to the formation of **five-membered rings** bearing different arrangements of **heteroatoms**. Mechanistically, the **negative charge** on atom 1 of the **1,3-dipole** adds to a double or triple bond. The newly formed **negative charge** on the **dipolarophile** component then adds to atom 3 of the **dipole** resulting in **ring formation** and negation of all charges.

From Scheme 10.9, we observe that there are factors that influence how the **1,3-dipole** comes together with the **dipolarophile** in order to form observed products. Figure 10.7 illustrates that there are two possible orientations for reactants to approach one another in **1,3-dipolar cycloadditions**. However, the actual course of a given reaction is determined by the **nucleophilicity** of the **anionic atoms** associated with the **1,3-dipole**, the **electrophilicity** of the various atoms associated with the **dipolarophile**, and the formation of **steric hindrance** in the product. Referring to Scheme 10.9, the **nucleophilicity** and **electrophilicity** influences are demonstrated in the reaction of **diazomethane** with **methacrylate**. Furthermore, the **steric influences** are demonstrated in the reaction of **methyl azide** with **phenyl propargyl ether**. When there are no specific influences, it is common to observe two products during **1,3-dipolar cycloadditions**.

With the breadth of combinations available for **1,3-dipolar cycloadditions**, the **ozonolysis** reaction stands out as a special case. As shown in Scheme 10.10, **ozone** reacts with *cis* or *trans* **double bonds** through a **1,3-dipolar cycloaddition** forming a **1,2,3-trioxolane** known as a **molozonide**. Once formed, this species undergoes a **retro 1,3-dipolar cycloaddition**

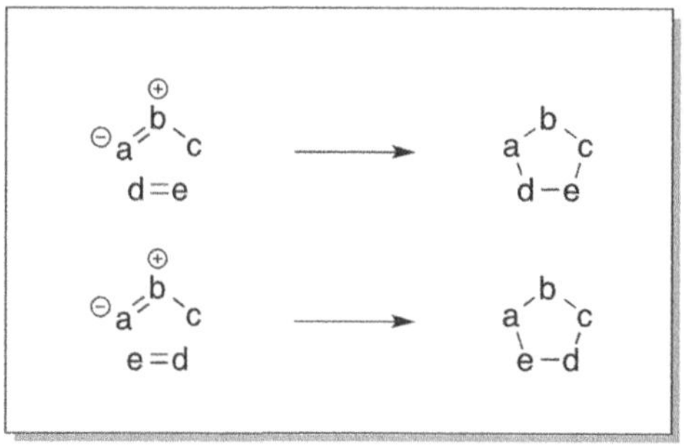

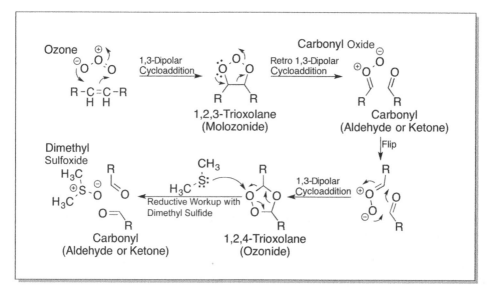

Figure 10.7 *1,3-Dipoles can approach dipolarophile in two possible orientations.*

Scheme 10.10 *Mechanistic pathway for ozonolysis reactions.*

forming a **carbonyl** and a **carbonyl oxide**. The **carbonyl oxide** then flips and reacts with the **carbonyl** in a second **1,3-dipolar cycloaddition** forming a **1,2,4-trioxolane** known as an **ozonide**. This species can then undergo reaction with **dimethyl sulfide** forming two **carbonyl** compounds and **dimethyl sulfoxide**.

Summarizing this discussion of **1,3-dipolar cycloadditions**, the **ozonolysis** example illustrates that such **cycloadditions** can occur with **enophiles** comprising **heteroatoms** such as oxygen. The utility of **heteroatoms** in **1,3-dipolar cycloadditions** broadly extends into other **pericyclic transformations** including, but not limited to, **Diels–Alder reactions, ene reactions**, and **sigmatropic reactions** discussed in Section 10.4.

10.4 SIGMATROPIC REACTIONS

Sigmatropic reactions are characterized by **cyclic transition states** leading to the breaking of one **σ-bond** (**carbon–carbon single bond**) and the formation of another **σ-bond**. Because these reactions proceed through **cyclic transition states**, they fall under the classification of **pericyclic reactions**. In this section, **sigmatropic reactions** are intro-duced in the context of the **Cope rearrangement** and the **Claisen rearrangement**.

10.4.1 The Cope Rearrangement

The **Cope rearrangement**, illustrated in Scheme 10.11 using **arrow-pushing**, is charac-terized by the **thermal isomerization** of **1,5-dienes**. Through a **cyclic transition state**, one **σ-bond** is broken while another **σ-bond** is formed. By counting the atoms between the **broken σ-bond** and the **formed σ-bond**, we observe that the new **σ-bond** is formed at the third atom counted from each end of the **broken σ-bond**. Therefore, the **Cope rear-rangement** is referred to as a **[3,3] sigmatropic rearrangement**.

While the example presented in Scheme 10.10 reflects the general mechanism of the **Cope rearrangement**, the reaction applied to **1,5-hexadiene** results in the formation of **1,5-hexadiene**. In order to utilize the **Cope rearrangement** as a synthetically useful reaction, **asymmetric dienes** can be utilized. As illustrated in Scheme 10.12 using **arrow-pushing**, the **Cope rearrangement** of **3-methyl-1,5-hexadiene** produces **5-*trans*-1,5-heptadiene** via the **[3,3] sigmatropic mechanism**.

Like **1,5-diene Cope rearrangement substrates**, **Cope rearrangement products** are also **1,5-dienes**. Therefore, these **[3,3] sigmatropic rearrangements** are generally **reversible** resulting in mixtures of starting materials and products. In practice, the major component of such reaction mixtures is the more **thermodynamically stable product** unless there is another factor influencing the **Cope rearrangement pathway**. Such factors include strategic substitutions on the starting **1,5-diene**.

As discussed earlier, Scheme 10.12 illustrates the **Cope rearrangement** of **3-methyl-1,5-hexadiene** resulting in formation of **5-*trans*-1,5-heptadiene**—a product that is structurally different compared to the starting material. If a **hydroxyl group** replaces the **methyl group**, the **Cope rearrangement** proceeds identically to that illustrated in Scheme 10.12 with the exception of the product being an **enol**. Through **tautomerization**, the **enol** spontaneously

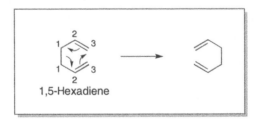

Scheme 10.11 *Cope rearrangement of 1,5-hexadiene.*

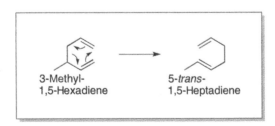

Scheme 10.12 *Cope rearrangement of 3-methyl-1,5-hexadiene.*

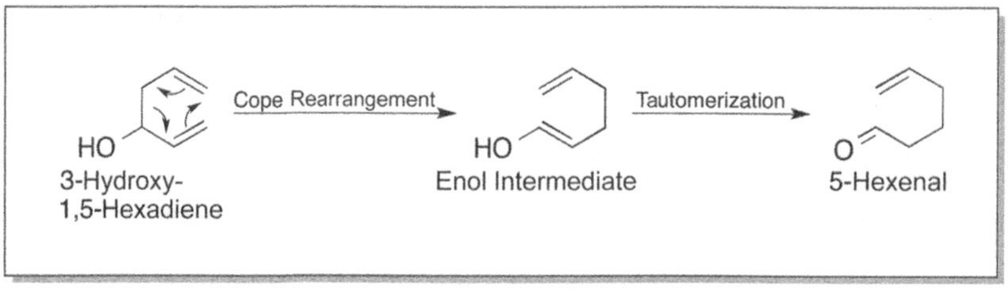

Scheme 10.13   *Oxy-Cope rearrangement of 3-hydroxy-1,5-hexadiene.*

converts to an **aldehyde**. The resulting **5-hexenal** is not a **1,5-diene** and is therefore incapable of undergoing a **Cope rearrangement**. Consequently, **5-hexenal** is the only identifiable product formed from **3-hydroxy-1,5-hexadiene**. This version of the **Cope rearrangement**, illustrated in Scheme 10.13 using **arrow-pushing**, is known as the **oxy-Cope rearrangement**.

10.4.2 The Claisen Rearrangement

The **Claisen rearrangement** allows **migration** of an **allyl group** from an **oxygen atom** of an **allyl phenyl ether** to an **ortho position** on the **phenyl ring**. This reaction, illustrated in Scheme 10.14 using arrow-pushing, is a **sigmatropic rearrangement** mechanistically similar to the **Cope rearrangement**. Specifically, if the position of the newly **formed σ-bond** is counted from the atoms associated with the **broken σ-bond**, we recognize that the **Claisen rearrangement** is a **[3,3] sigmatropic rearrangement**. As illustrated in Scheme 10.14, this **pericyclic reaction** proceeds by **movement of electrons** beginning with a **lone pair** from the **oxygen atom** and progressing through the **aromatic phenyl ring**. Once the **allyl group** has migrated, the intermediate **ketone tautomerizes** to restore the **aromatic ring** and reveal the **phenol hydroxyl group**.

One advantage associated with the **Claisen rearrangement** is the ability to introduce **multiple substituents** on a **phenol** structure through the use of **sequential Claisen rearrangements**. As illustrated in Scheme 10.15, this strategy can be used to generate both symmetrically and asymmetrically **substituted phenols** based upon the nature of the **allyl groups** utilized. In general a **phenol** structure is reacted with **allyl chloride** or a **substituted allylating agent** under basic conditions. The resulting **ether** is then subjected to the **Claisen rearrangement**. Repeating this two-step process with a second **allylating agent** results in formation of a **doubly substituted phenol**.

Although the **Claisen rearrangement** relies on the presence of an **allyl ether**, an actual **allyl group** is not required for this reaction to proceed. For example, if a **functional group** that can

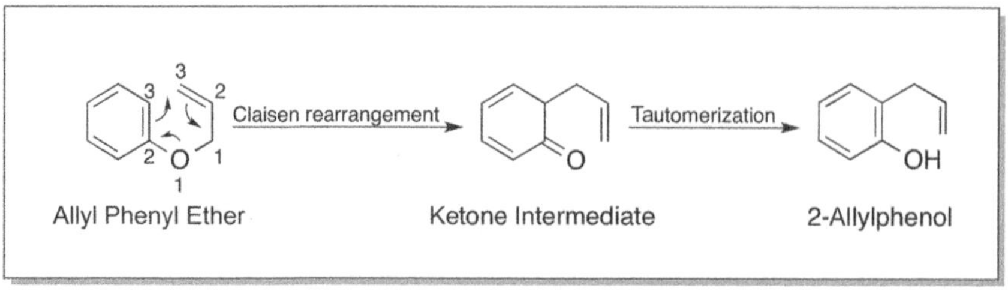

Scheme 10.14   *Claisen rearrangement of allyl phenyl ether.*

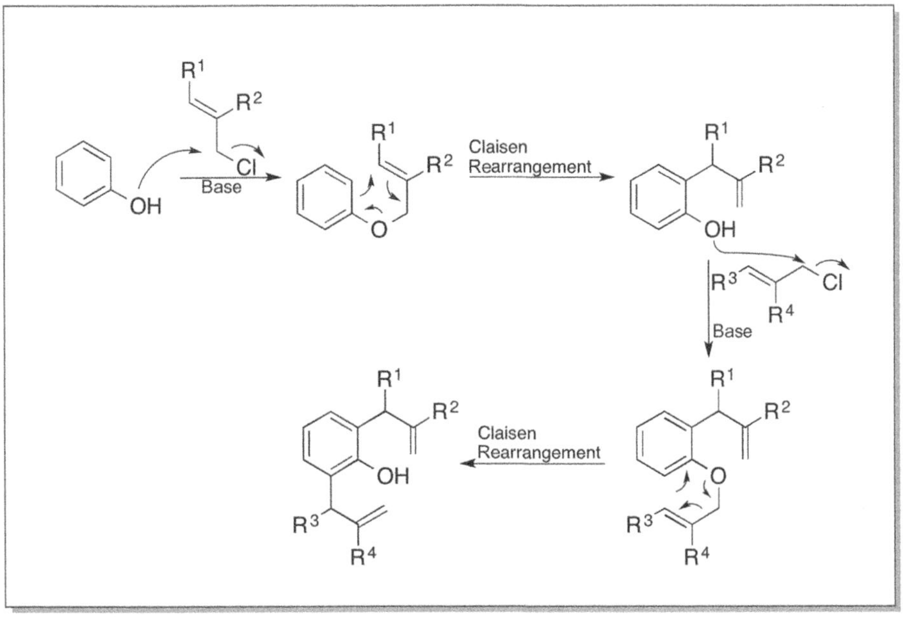

Scheme 10.15 *Multiple Claisen rearrangements can be used to generate phenol structures with multiple substitutions.*

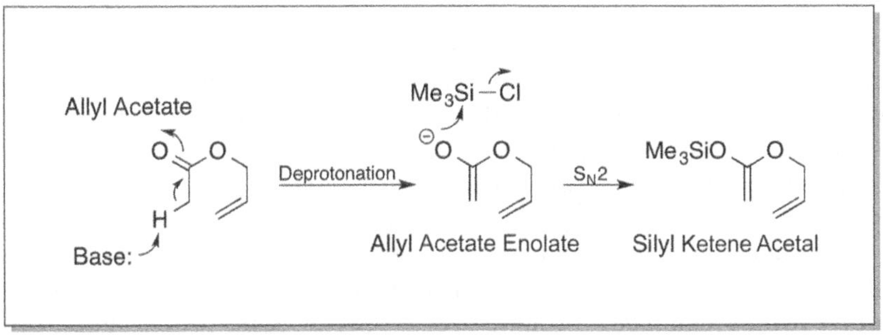

Scheme 10.16 *Allyl acetate can be converted into a silyl ketene acetal precursor for the Ireland–Claisen rearrangement.*

be converted into an **allyl-like functionality** is present, **[3,3] sigmatropic rearrangements** can proceed. Such a strategy is utilized in the **Ireland–Claisen rearrangement**.

As illustrated in Scheme 10.16, the **Ireland–Claisen rearrangement** begins with an **allyl ester**. As discussed in Chapter 3, under **basic conditions**, **esters** can act as **acids** and become **deprotonated**. Furthermore, as discussed in Chapter 5, **deprotonated acids** can act as **nucleophiles** and participate in S_N2 **reactions**. Such an S_N2 **reaction** with **trimethylchlorosilane** generates a **silyl ketene acetal** bearing the **allyl** double bond configuration required for a standard **Claisen rearrangement**.

As illustrated in Scheme 10.17 using **arrow-pushing**, **silyl ketene acetals** of **allyl esters** undergo **[3,3] sigmatropic rearrangements** according to the **Ireland–Claisen rearrangement**. While the **silyl** component is required for the **rearrangement**, the resulting **silyl ester** is not stable enough to be isolated and undergoes spontaneous **hydrolysis**.

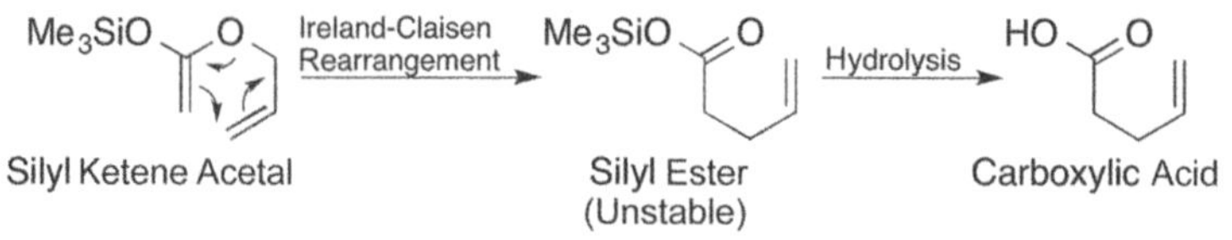

Scheme 10.17 *The Ireland–Claisen rearrangement generates carboxylic acids with terminal double bonds.*

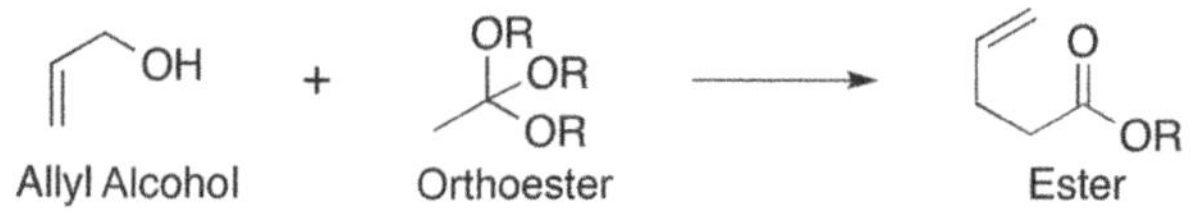

Scheme 10.18 *Example of the Johnson–Claisen rearrangement.*

Thus, the isolated products from **Ireland–Claisen rearrangements** are **carboxylic acids** bearing functionally useful **terminal double bonds** that can be modified according to methods described throughout this book and in **general organic chemistry** textbooks.

Having established that **esters** can be manipulated to undergo **[3,3] sigmatropic rearrangements**, another variation of the **Claisen rearrangement** deserves mention. The **orthoester Claisen rearrangement**, also known as the **Johnson–Claisen rearrangement**, is illustrated in Scheme 10.18. Unlike the **Claisen rearrangement**, which retains **allyl** components, and the **Ireland–Claisen rearrangement**, which generates **carboxylic acids**, the **Johnson–Claisen rearrangement** produces **esters**.

An **orthoester** is a functional group with the same **oxidation state** of an **ester** but bearing no **carbonyl** component. Examples of **orthoesters** and their parent **carboxylic acids**, shown in Figure 10.8, include **trimethyl orthoformate, trimethyl orthoacetate**, and **triethyl orthopropionate**. As illustrated, a **carbon atom** bearing three **carbon–oxygen bonds** characterizes the **orthoester functional group**. **Orthoesters** are readily **hydrolyzed** to form **esters** and **carboxylic acids** and are sometimes used to keep reaction mixtures anhydrous.

Due to the unique chemistry of **orthoesters**, these groups react with **allylic alcohols** under mildly **acidic conditions** to generate **esters** with **olefin tails** similar to those generated from the **Ireland–Claisen rearrangement**. Using **arrow-pushing**, Scheme 10.19 illustrates the mechanism for this transformation beginning with initial **protonation** of one of the **orthoester alkoxy groups**. The **protonated alkoxy group** is eliminated with assistance from a **lone electron pair** on a neighboring **oxygen atom**. The resulting **cation** residing on the **carbonyl oxygen** is neutralized when an **allylic alcohol** adds to this system. Still under **acidic conditions**, a second **alkoxy group** is **protonated** and **eliminated**. Following the second **alkoxy elimination**, the resulting **cation** is **neutralized** on the loss of a **hydrogen atom** from the adjacent carbon with simultaneous formation of a **carbon–carbon double bond**. The resulting species is a **ketene acetal** and is analogous to the **silyl ketene acetals** required for the **Ireland–Claisen rearrangement**. With the required **double**

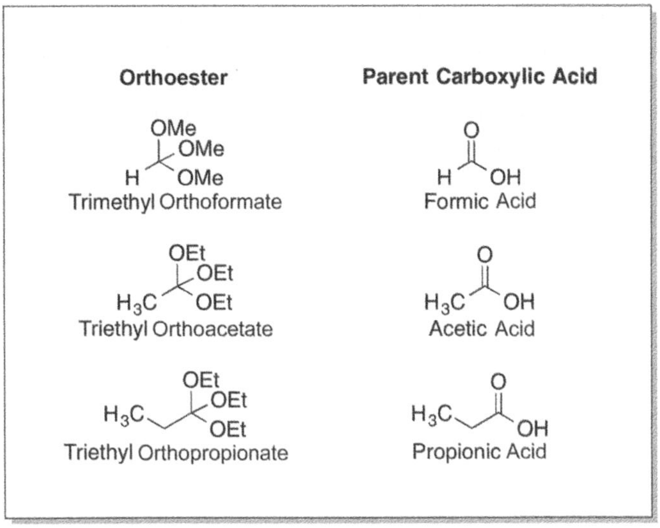

Figure 10.8 *Examples of orthoesters.*

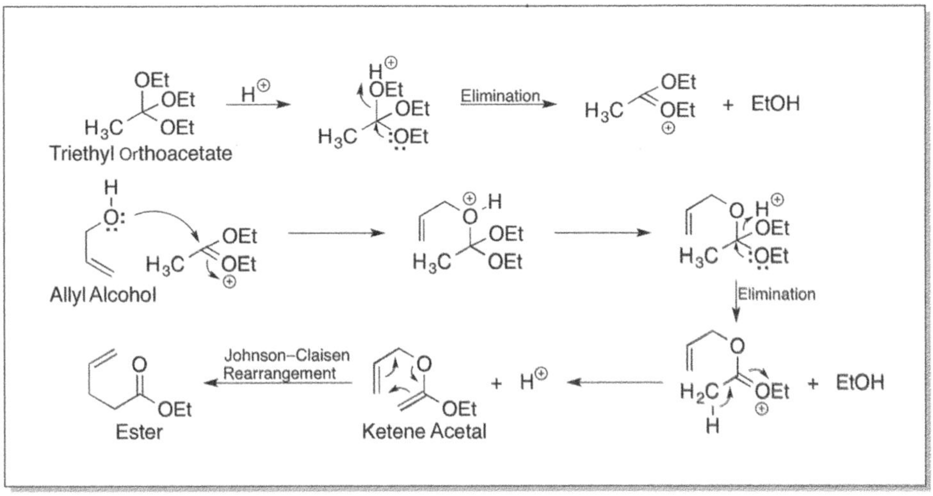

Scheme 10.19 *Mechanism for the Johnson–Claisen rearrangement.*

bond configuration established, the **ketene acetal** undergoes a **[3,3] sigmatropic rearrangement** with the **allyl group** forming the observed **ester**.

10.5 SUMMARY

In this chapter, various **pericyclic reactions** were presented and mechanistically explained using **arrow-pushing**. Furthermore, the examples selected for this chapter illustrate the versatility of products that can be obtained. Such versatility can be structural—based on the different products obtained from each of these reactions. Alternately, versatility can be functional as noted from reaction variations including the **oxy-Cope rearrangement**, the

Ireland–Claisen rearrangement, and the **Johnson–Claisen rearrangement**. While structural versatility in products allows for the generation of diverse target molecules, versatility in the functional groups generated allows chemists to efficiently plan for follow-up reactions with greater control over how next-stage products are formed. Through the study of this chapter, in conjunction with the material discussed throughout this book, students can begin planning synthetic strategies toward more complex target molecules.

PROBLEMS

1. The Carroll rearrangement is mechanistically similar to a variation of the Claisen rearrangement. Using arrow-pushing, propose a mechanism for the conversion of allyl acetoacetate to 5-hexene-2-one.

Allyl Acetoacetate   5-Hexene-2-one

2. The Fischer indole synthesis brings together phenylhydrazine and acetone to form 2-methylindole. The mechanism involves a sigmatropic rearrangement. Using arrow-pushing, propose a mechanism for this reaction.

Phenylhydrazine   Acetone   2-Methylindole

3. As demonstrated by the Fischer indole synthesis, sigmatropic rearrangements can occur with heteroatoms such as nitrogen and oxygen. Furthermore, the use of heteroatoms is broadly applicable to pericyclic reactions. Predict the products of the following pericyclic reactions. For each reaction, show the mechanism and predict the major product formed. Justify your answers using arrow-pushing and electronic/steric influences. For each example, state the type of pericyclic reaction (electrocyclic, cycloaddition, sigmatropic rearrangement). Predict the intermediate structures when not shown.

a.

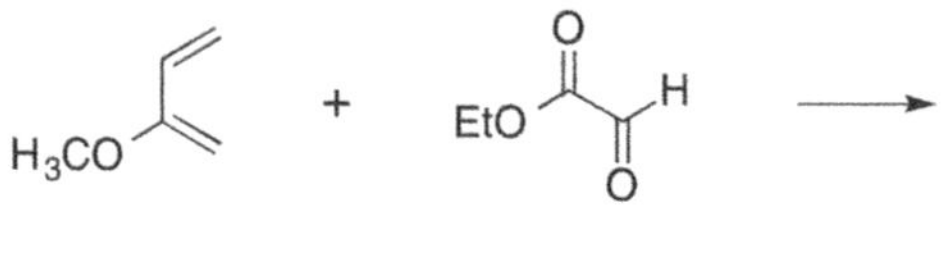

b.

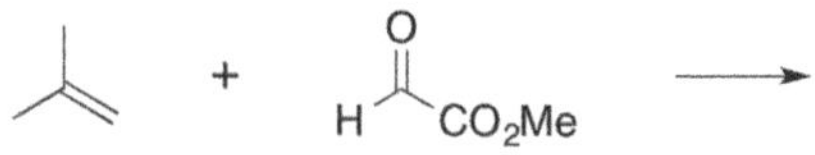

c.

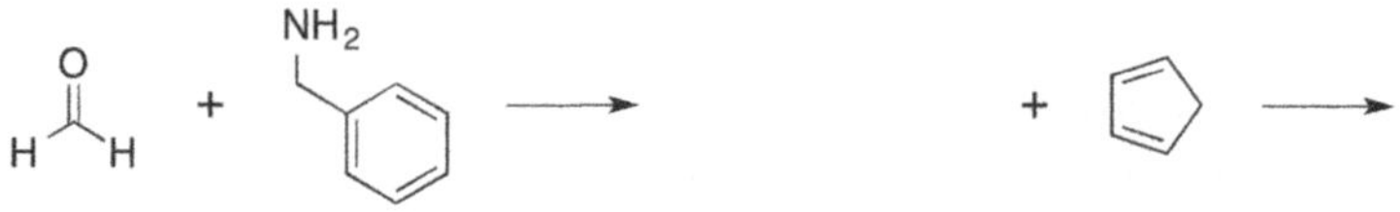

d.

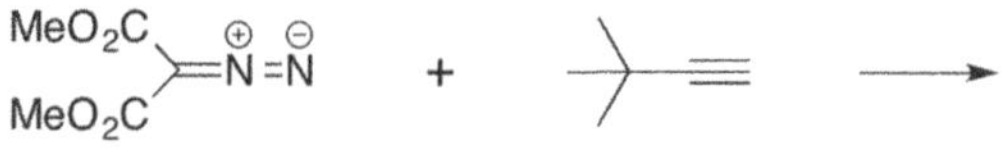

e.

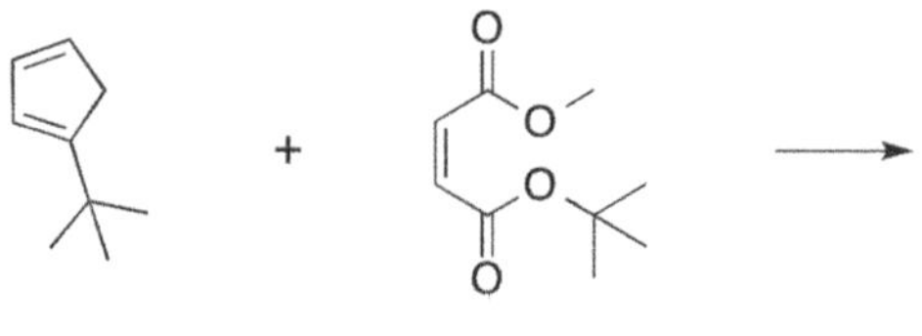

f.

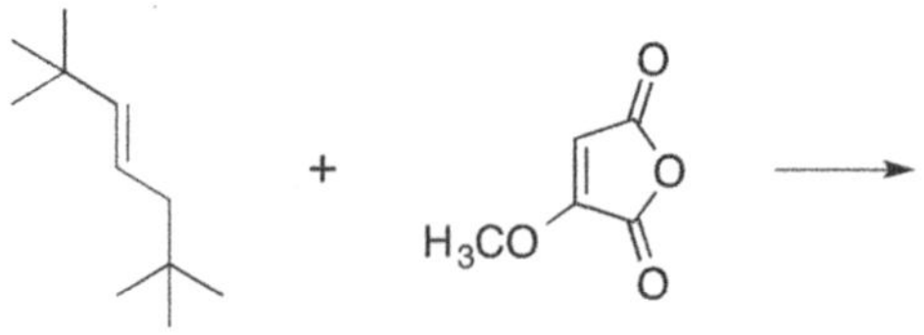

4. The aza-Cope rearrangement follows the same general mechanistic pathway as the standard Cope rearrangement. Using principles discussed throughout this book, predict the structures formed at each stage of the illustrated reaction and leading to the formation of 3-butenylamine and benzaldehyde.

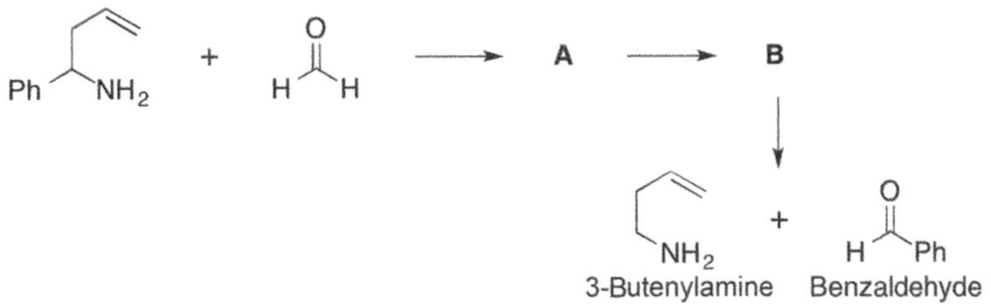

3-Butenylamine   Benzaldehyde

5. As stated in this chapter, orthoesters are readily hydrolyzed to esters and carboxylic acids. Propose a mechanism for the hydrolysis of trimethyl orthoacetate to methyl acetate. Justify your answer using arrow-pushing. Explain why trimethyl orthoacetate may be useful in keeping reaction mixtures anhydrous.

Trimethyl Orthoacetate   Methyl Acetate

6. Scheme 10.10 illustrates how sigmatropic rearrangements are classified by counting the atoms between the broken σ-bond and the newly formed σ-bond. Specify the type of sigmatropic rearrangement for the following reactions. Show your work by numbering the relevant atoms.

a.

b.

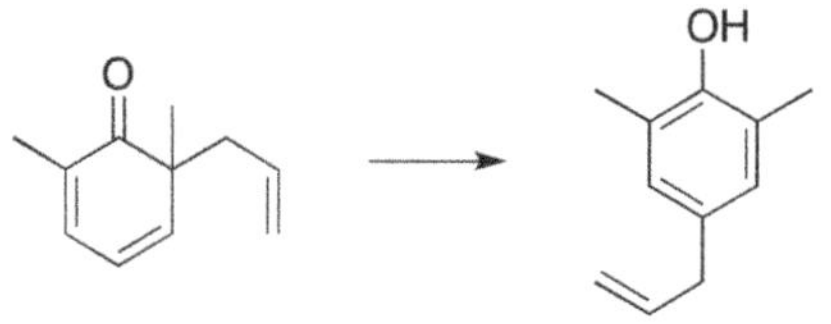

c.

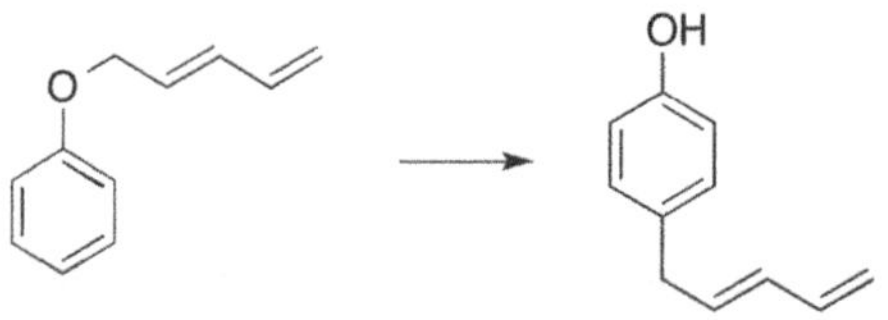

d.

e.

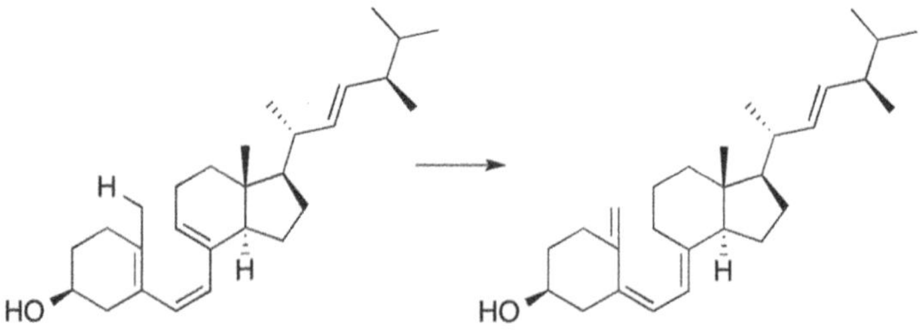

7. Pericyclic reactions are useful for the generation of more complex ring systems. Propose a pericyclic reaction strategy for the formation of the following structures from completely acyclic starting materials. Show product formation using arrow-pushing and specify the type of pericyclic reaction applied. In some cases, multiple pericyclic reactions are required.

a.

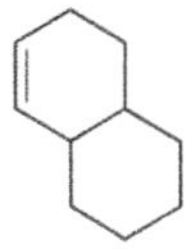

b.

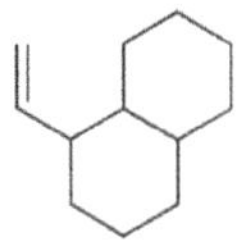

c.

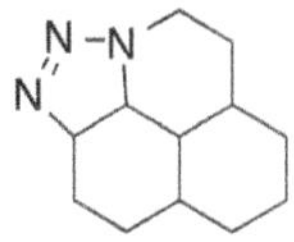

d.

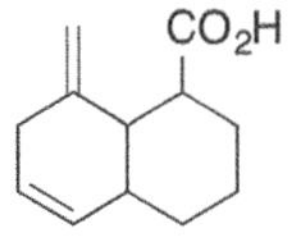

Chapter *11*

Moving Forward

Organic chemistry is a very mature science upon which numerous disciplines depend, ranging from pharmaceuticals and food science to agrochemicals and material science. In approaching organic chemistry, discussions within this book focus on utilizing the **acid/ base** properties of organic molecules, in conjunction with the **electronic** properties of associated **functional groups**, to rationalize chemical reactions through the movement of **electrons**. Similar strategies are applied to **free radicals**, **carbenes**, and **pericyclic reactions**. This technique of **arrow-pushing** is presented as an adjunct to the memorization of the numerous **name reactions** available to organic chemists today. However, along with the treatments of various **mechanistic components** of organic reactions, this book includes introductions to many of the fundamental reactions studied in introductory organic chemistry courses. In this chapter, these reactions are revisited in order to emphasize that, through the application of **arrow-pushing**, a broader and deeper understanding of organic chemistry can be achieved.

11.1 FUNCTIONAL GROUP MANIPULATIONS

Functional group manipulations involve the transformation of one functional group into another usually with no additional changes to the core molecular structure. Throughout this book, many different **functional groups** were presented beginning with those illustrated in Figure 1.3 and continuing through each chapter and their associated problem sets. Considering **olefins**, among the simplest of **functional groups**, transformations into **alkyl halides** were presented in Chapter 8. Specific examples, reiterated in Schemes 11.1 and 11.2, included both the **addition** of **halogens** across **double bonds** and the application of **Markovnikov's rule** when adding **acids** across **double bonds**.

Arrow-Pushing in Organic Chemistry: An Easy Approach to Understanding Reaction Mechanisms,
Second Edition. Daniel E. Levy.
© 2017 John Wiley & Sons, Inc. Published 2017 by John Wiley & Sons, Inc.

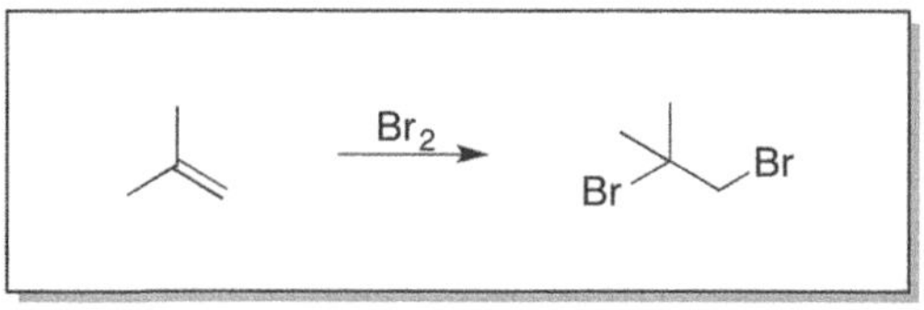

Scheme 11.1 *Addition of bromine across a double bond.*

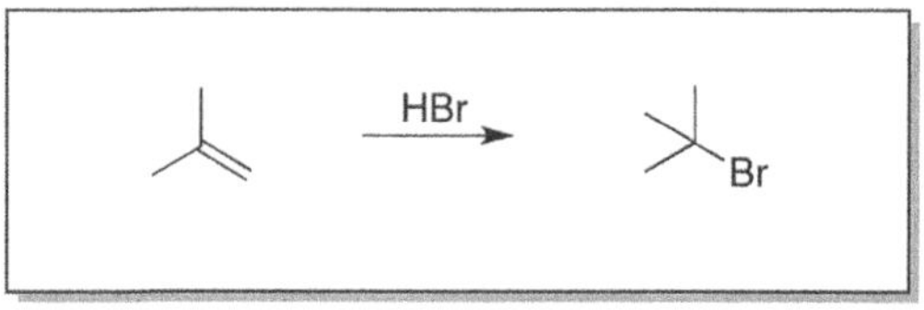

Scheme 11.2 *Markovnikov addition of hydrobromic acid across a double bond.*

While the examples presented in Schemes 11.1 and 11.2 illustrate only the formation of **alkyl bromides**, it is important to recognize that **halogens** can be replaced through **nucleophilic displacements**. These displacements can occur via either S_N1 or S_N2 mechanisms. Regarding S_N1 reactions, ionization generally occurs under **solvolytic** conditions limiting the **nucleophile** to the **solvent** used. In the case of S_N2 reactions, the only limiting factors relate to the relative **nucleophilicities** of the incoming **nucleophiles** compared with those of the **leaving groups**. Thus, as illustrated in Figure 11.1, **alkyl halides** can be converted into a wide variety of useful **functional groups**.

Upon further examination of the **functional group transformations** summarized in Figure 11.1, there are a number of possible conversions applicable to the product **functional groups**. Among these are the conversions of **alcohols** to **ethers** illustrated in Scheme 11.3. Additionally, transformation of **carboxylic acids** to **esters** and **amides** is illustrated in Figure 11.2. The related conversions of **esters** to **acids** and **amides** are shown in Figure 11.3. Finally, transformations of **aldehydes** and **ketones** to **imines**, **oximes**, and **enamines** are summarized in Figure 11.4.

In addition to the **functional group transformations** discussed in this book, there are many more that depend on **oxidative** and **reductive** mechanisms. These mechanisms are covered in depth in introductory organic chemistry courses and will not be presented here in detail. As an introduction, Figure 11.5 summarizes such transformations that include the **oxidation** of **alcohols** to **aldehydes**, **ketones**, and **carboxylic acids**. Likewise, Figure 11.5 introduces the **reductive** transformations of **aldehydes**, **ketones**, and **carboxylic acids** to **alcohols** as well as **amides** to **amines**. As will be revealed through further coursework, additional **functional group manipulations** are available and rationalized utilizing the principles of **arrow-pushing** discussed throughout this book.

11.2 NAME REACTIONS

While the focus of this book is to introduce the technique of **arrow-pushing** as a strategy for understanding the general principles of **organic chemistry** without broad memorization of reactions, some **name reactions** are mentioned. These **name reactions** are presented for two reasons. First, their underlying mechanisms highlight the principles relevant to the

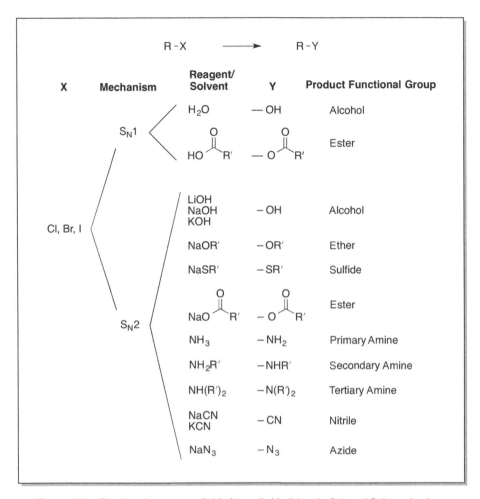

Figure 11.1 *Functional groups available from alkyl halides via S_N1 and S_N2 mechanisms.*

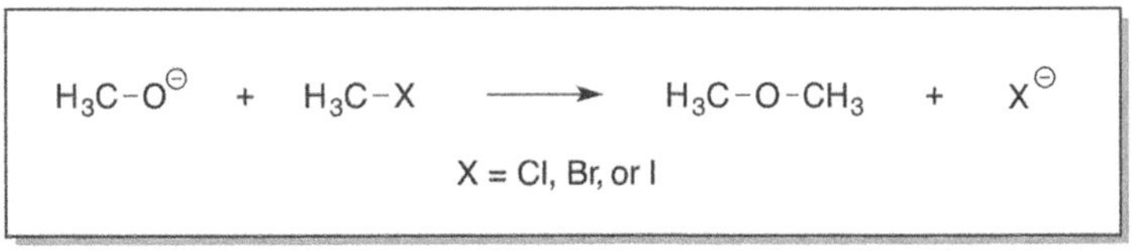

$$H_3C-O^{\ominus} \ + \ H_3C-X \ \longrightarrow \ H_3C-O-CH_3 \ + \ X^{\ominus}$$

X = Cl, Br, or I

Scheme 11.3 *Conversion of alcohols to ethers—the Williamson ether synthesis.*

chapters in which they are presented. Second, they represent important and fundamental tools for general **organic chemistry** transformations. While the focus of this book advocates development of a full understanding of **organic reaction mechanisms** as a means of learning the subject, once this understanding is achieved, recognition of these reactions by name presents a significant shortcut to the description of and communication of **synthetic processes**. The name reactions presented in this book are reviewed in the following paragraphs.

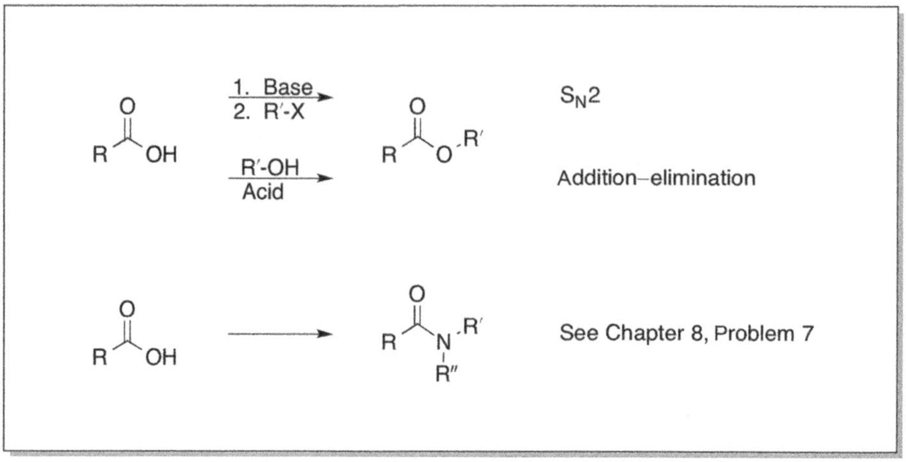

Figure 11.2 *Transformations of carboxylic acids to esters and amides.*

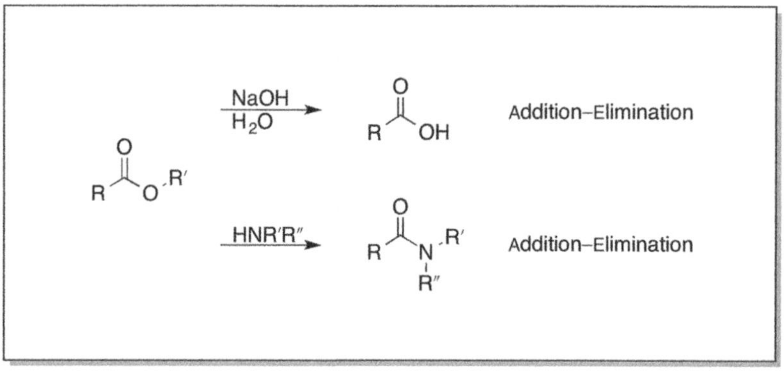

Figure 11.3 *Transformations of esters to carboxylic acids and amides.*

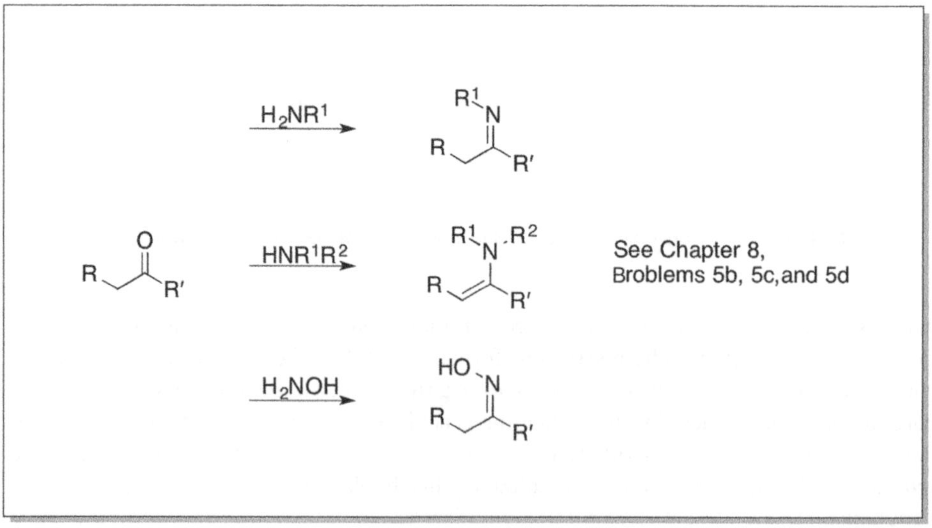

Figure 11.4 *Transformations of aldehydes and ketones to imines, oximes, and enamines.*

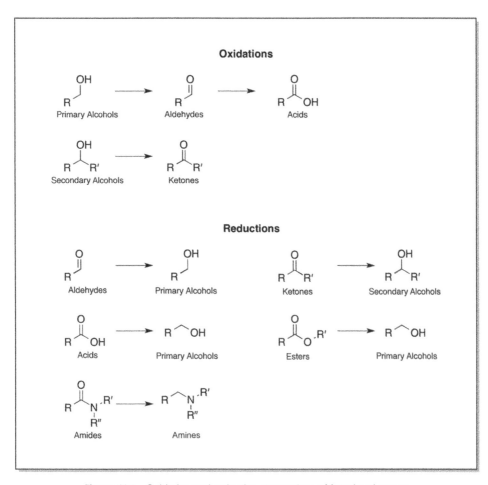

Figure 11.5 *Oxidative and reductive conversions of functional groups.*

In Chapter 1, **electrocyclic** reactions were presented as early examples utilizing **arrow-pushing** techniques. These were selected because of their simplicity relating to the **nonionic** character of the reactions. Specifically, the **acid–base properties** of the starting molecules are of lesser importance as the reactions illustrated proceed through the movement of **electrons** through the existing olefinic systems. The reactions illustrated include the **Diels–Alder reaction** (Scheme 11.4), the **Cope rearrangement** (Scheme 11.5), and the **Claisen rearrangement** (Scheme 11.6). These and related **pericyclic reactions**, depending upon the same mechanistic principles, were presented in greater detail in Chapter 10 and are covered in depth in introductory organic chemistry coursework.

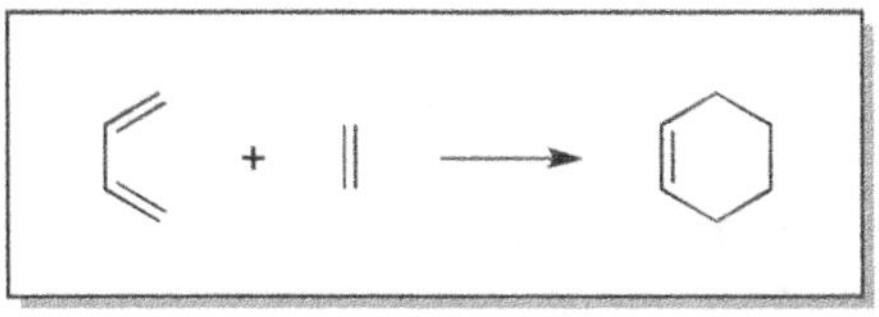

Scheme 11.4 *The Diels–Alder reaction.*

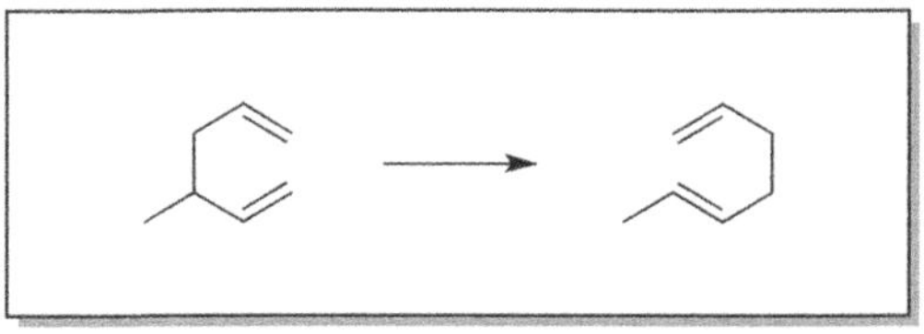

Scheme 11.5   *The Cope rearrangement.*

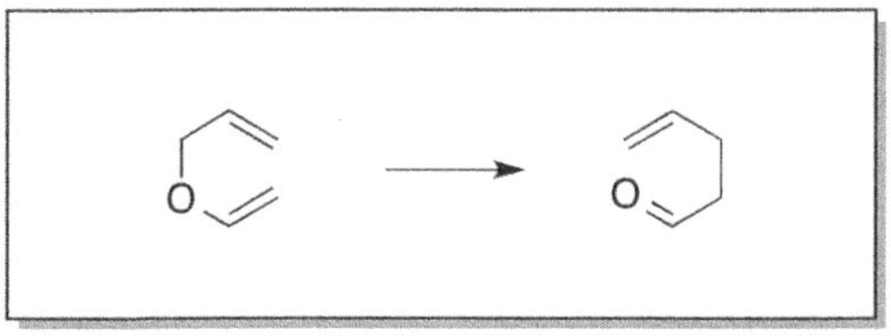

Scheme 11.6   *The Claisen rearrangement.*

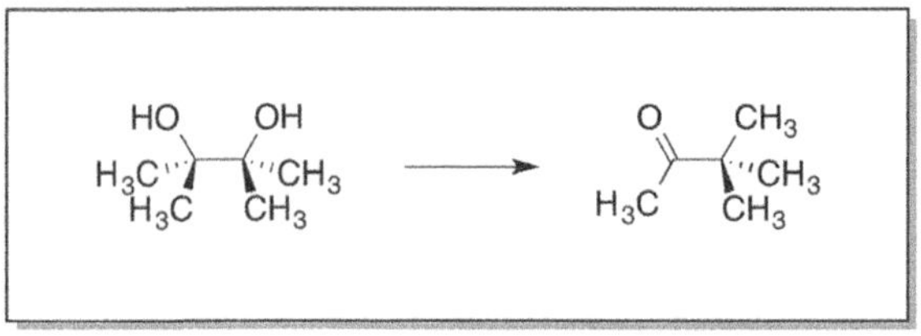

Scheme 11.7   *The pinacol rearrangement.*

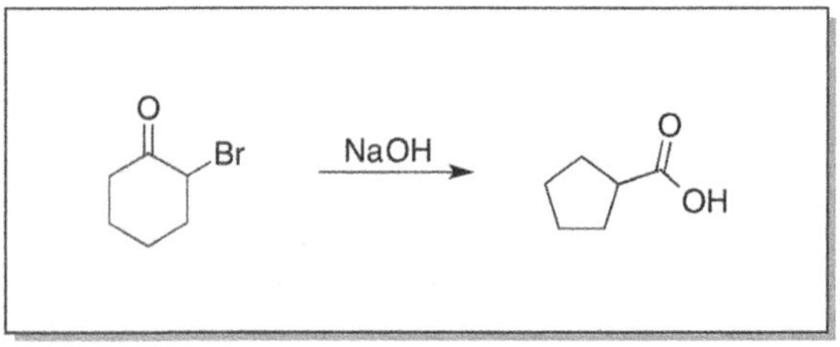

Scheme 11.8   *The Favorskii rearrangement.*

The earlier described **rearrangement** reactions are not the only ones presented within this book. In addition to **pericyclic reactions**, some reactions dependent upon **ionic** mechanisms were presented. These include the **pinacol rearrangement** (Scheme 11.7) and the **Favorskii rearrangement** (Scheme 11.8). These examples were presented within the context of **alkyl shifts** and the related **hydride shifts**. Through these examples, the concepts of **ionic stability** and **spontaneous ionic transformations** to more stable **ionic** species were explored. These concepts are especially prevalent when examining **solvolysis**-mediated processes where S_N1 and **E1** mechanisms are involved.

Moving from **rearrangements**, **condensation reactions** were also presented. **Condensation reactions** occur when two reactive species join to form a new compound. The first of these was the **aldol condensation** (Scheme 11.9). Later, a more complex application of the **aldol condensation** was presented in the form of the **Robinson annulation**

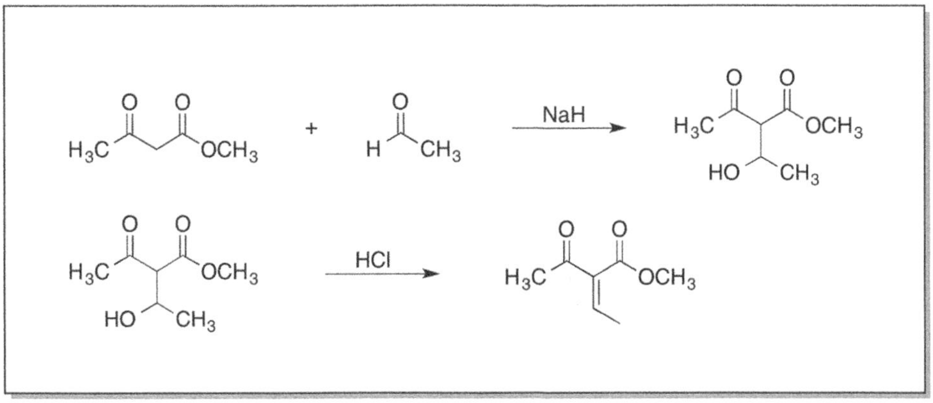

Scheme 11.9 The aldol condensation.

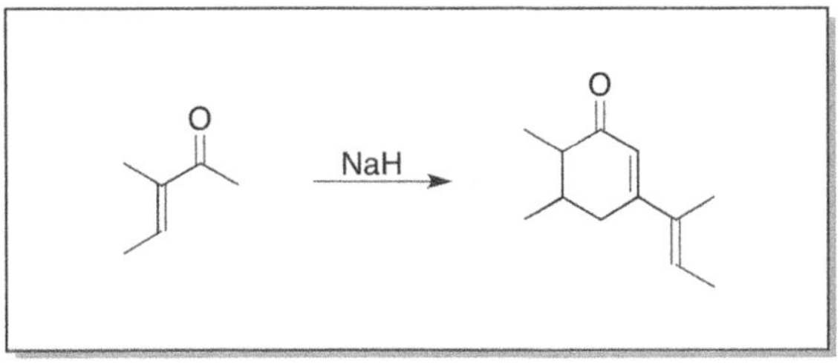

Scheme 11.10 The Robinson annulation.

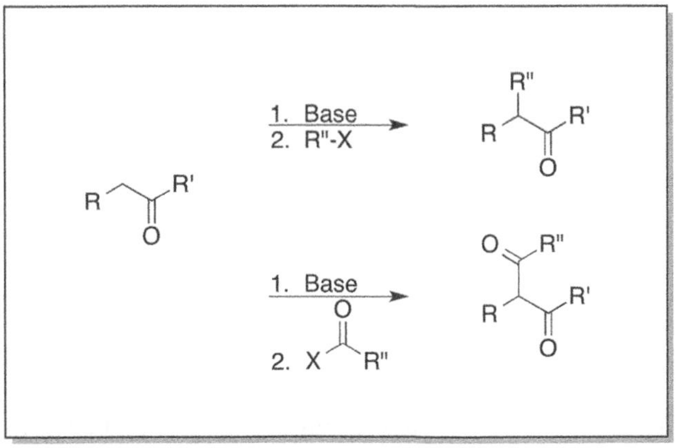

Scheme 11.11 Alkylation and acylation reactions adjacent to carbonyls.

(Scheme 11.10). For both of these reactions, the underlying lessons relate to the ability to induce reactions and incorporate substitutions at carbon atoms adjacent to **carbonyl groups**. Similar reactivities of such carbon atoms can be utilized for **alkylation** (S_N2) and **acylation** (**addition–elimination**) reactions as illustrated in Scheme 11.11.

Regarding **acylation** reactions, **acylation** of **alcohols** produces **esters** and **acylation** of **amines** produces **amides**. Both of these transformations are illustrated in Figure 11.2.

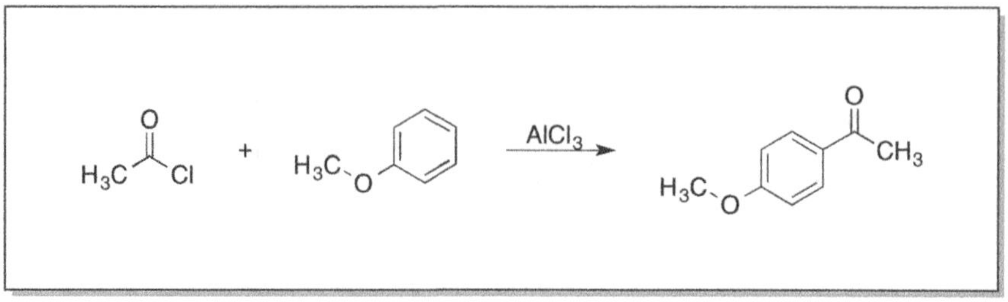

Scheme 11.12 *The Friedel–Crafts acylation.*

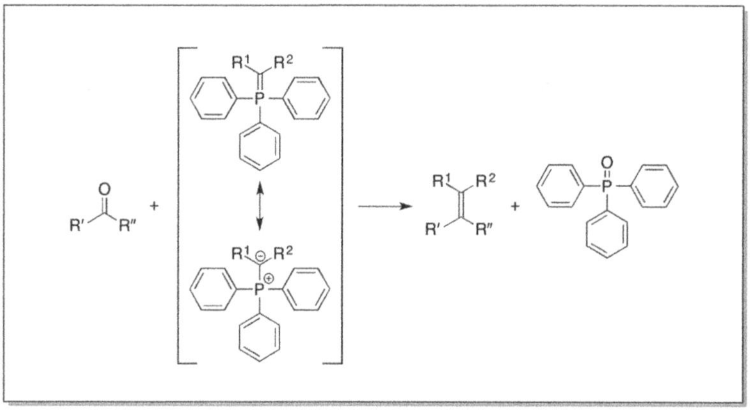

Scheme 11.13 *The Wittig reaction.*

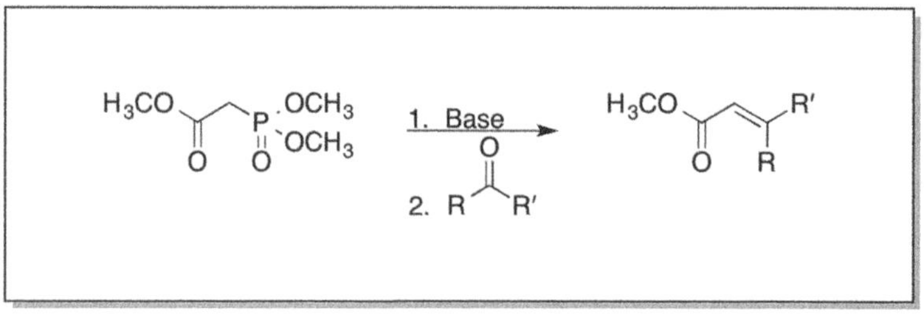

Scheme 11.14 *The Horner–Emmons reaction.*

These, in addition to the introduction of **acyl groups** adjacent to **carbonyls** (Scheme 11.11), only hint at the breadth of related **acylation** reactions available and useful in **organic synthesis**. One additional reaction is the **Friedel–Crafts acylation** illustrated in Scheme 11.12. Through this transformation, extended functionalization of **aryl** groups becomes accessible.

Thus far, the **aldol condensation** was presented as a method for adding **carbon atoms** adjacent to **carbonyl groups**, and the **Friedel–Crafts acylation** was presented as useful for the addition of **carbon atoms** to **aromatic rings**. In addition to these reactions, the **Wittig reaction** (Scheme 11.13) and the **Horner–Emmons reaction** (Scheme 11.14) were presented as capable of replacing the **carbon–oxygen double bond** of **aldehydes** and **ketones** with **carbon–carbon double bonds**. The new extensions can be simple or

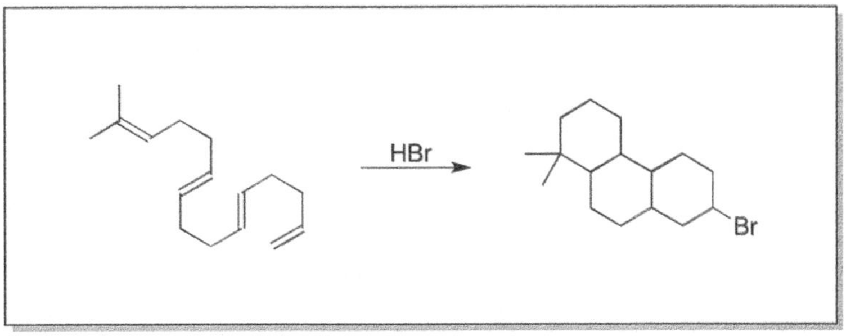

Scheme 11.15 *A cation–π cyclization.*

functionalized. Additionally, the newly formed **double bonds** can be modified through addition of **halogens** or **acids**.

One final example of a **name reaction** presented within the text of this book is the **cation–π cyclization**. This reaction, illustrated in Scheme 11.15, returns to the previously described reaction classes that include **pericyclic reactions** and **rearrangements**. Inclusion of this reaction complements the various **nucleophiles** used throughout the examples of this book by highlighting the **nucleophilic** nature of **double bonds**.

While mechanistically distinct, the **aldol condensation** and the **Friedel–Crafts acylation** result in the incorporation of additional carbon atoms to the starting structure. Such is the case with all of the **name reactions** specified within this book. As will become apparent through study and practice, the ability to expand and extend organic molecular structures is extremely important when planning the synthesis of complex molecules. To this end, the greater the number of available reactions useful for expanding molecular structures, the greater the versatility in synthetic route planning and diversity in available target compounds. Table 11.1 summarizes the transformations described within this book in context of their utilities in building organic structures.

While most of the reactions mentioned in Table 11.1 have already been described, there are a number of reactions mentioned that are the subject of the problems presented at the end of this chapter. As will become apparent, the mechanistic pathways of those reactions follow the principles of **arrow-pushing** presented within this book. However, one class of reactions, **organometallic reactions**, has not yet been discussed. The primary reason for this is that the mechanisms associated with these reactions are beyond the scope of this book and are typically reserved for more advanced treatments of mechanistic organic chemistry. Because of the synthetic utility of this class of reactions, they are introduced here using simplified rationalizations of their mechanistic pathways.

The **Grignard reaction**, illustrated in Scheme 11.16, is actually a **1,2-addition reaction** between a **Grignard reagent** and an **aldehyde** or a **ketone** leading to formation of an **alcohol**. The **organometallic** aspect of this reaction is associated with generation of the **Grignard reagent**. As illustrated in Scheme 11.17, **Grignard reagents** are prepared on reaction of **alkyl halides** with **magnesium metal**. Without going into the specifics of this mechanism, a **magnesium atom** essentially inserts itself into the **carbon–halide bond**. Recognizing that **magnesium metal** has no net charge and that the **magnesium atom** of a **Grignard reagent** has a +2 net charge, the **magnesium metal** becomes **oxidized** during this reaction. Therefore, the insertion of **magnesium** into a **carbon–halide bond** is referred to as an **oxidative addition**. A simplified representation of the **oxidative addition** mechanism is shown in Scheme 11.17 using **arrow-pushing**. Once formed, **Grignard reagents** can be used in **Grignard reactions**.

TABLE 11.1 Name reactions and reaction types useful for modification and expansion of organic structures.

Reaction	Described	Utility
Rearrangement reactions	*Structural rearrangements with no net gain or loss in the number of atoms*	
Cope rearrangement	Schemes 1.6, 10.10	Rearrangement of dienes to alternate dienes
Oxy-Cope rearrangement	Scheme 10.12	Rearrangement of dienes with formation of ketones
Aza-Cope rearrangement	Chapter 10, Problem 4	Through intermediate imines, pericyclic rearrangement of olefinic amines to alternate olefinic amines
Claisen rearrangement	Scheme 10.13	Pericyclic migration of an allyl group from an oxygen extending the carbon chain length with tethered olefin
Ireland–Claisen rearrangement	Schemes 10.15, 10.16	Pericyclic migration of silyl ketene acetal from an oxygen extending the carbon chain length with tethered carboxylic acid
Johnson–Claisen rearrangement	Scheme 10.17	Pericyclic migration resulting from reaction of an allylic alcohol with an orthoester extending the carbon chain length with tethered ester
Carroll rearrangement	Chapter 10, Problem 1	Pericyclic reaction related to the Ireland–Claisen rearrangement with a tandem decarboxylation and forming a ketone tethered to an olefin
Fischer indole synthesis	Chapter 10, Problem 2	Rearrangement of phenyl hydrazones leading to formation of indole ring systems
Favorskii rearrangement	Scheme 11.8; Chapter 7, Problem 4c	Ring contraction of a cyclohexenone forming a cyclopentene ring with an exocyclic acyl group
Pinacol rearrangement	Scheme 6.10	Rearrangement of a 1,2-vicinal diol forming a ketone
Cycloadditions	*Condensation of two molecules through a cyclic transition state leading to a cyclic product containing the total number of atoms present in the two starting materials*	
Diels–Alder reaction	Schemes 1.2, 10.4	Pericyclic condensation of a diene with a dienophile forming a cyclohexene structure
Robinson annulation	Scheme 11.10; Chapter 11, Problem 2e	Condensation of a ketone with an α,β-unsaturated ketone forming a cyclohexenone
Functional group transformations	*Introduction of a new functional group or conversion of one functional group to another functional group with no net change in the core molecular structure*	
Swern oxidation	Chapter 8, Problem 3d	Conversion of primary alcohols to aldehydes and secondary alcohols to ketones
Williamson ether synthesis	Scheme 11.3	S_N2 reaction between deprotonated alcohols and alkyl halides forming ethers
Friedel–Crafts acylation	Scheme 11.12; Chapter 8, Problem 6h; Chapter 11, Problem 4; Chapter 11, Problem 5	Addition of an acyl (ketone) functional group or substituent to a phenyl ring
Grignard reaction	Schemes 8.12, 11.16, 11.17	Nucleophilic addition to carbonyls converting aldehydes and ketones to alcohols

(Continued)

TABLE 11.1   (Continued)

Reaction	Described	Utility
Condensations	*Condensation of two molecules leading to a product containing the total number of atoms present in the two starting materials*	
Aldol condensation	Scheme 11.9; Chapter 1, Problem 1e; Chapter 5, Problem 8a	Condensation between an aldehyde and a ketone forming a β-hydroxyketone
Robinson annulation	Scheme 11.10; Chapter 11, Problem 2e	Condensation of a ketone with an α,β-unsaturated ketone forming a cyclohexenone (final product is the result of dehydration resulting in loss of one oxygen atom and two hydrogen atoms)
Ene reaction	Figure 10.1, Scheme 10.6	Pericyclic condensation of two olefins extending the carbon chain length
Olefination reactions	*Reaction between two molecules joining carbon atoms and forming a new carbon–carbon double bond*	
Wittig reaction	Schemes 1.1, 11.13; Chapter 11, Problem 2a	Conversion of aldehydes and ketones to substituted or unsubstituted double bonds
Horner–Emmons reaction	Scheme 11.14; Chapter 11, Problem 2c	Conversion of aldehydes and ketones to substituted or unsubstituted double bonds
Aldol condensation	Scheme 11.9; Chapter 1, Problem 1e; Chapter 5, Problem 8a	Condensation between an aldehyde and a ketone initially forming a β-hydroxyketone (the olefin is formed via subsequent dehydration of the β-hydroxyketone leading to an α,β-unsaturated ketone)
Organometallic reactions		
Grignard reaction	Schemes 8.12, 11.16, 11.17	Nucleophilic addition to carbonyls converting aldehydes and ketones to alcohols
Suzuki reaction	Schemes 11.18, 11.19	Coupling of aryl/vinyl halides with aryl/vinyl boronic acids generating new carbon–carbon bonds
Sonogashira reaction	Chapter 11, Problem 3e	Coupling of aryl/vinyl halides with acetylenes generating new carbon–carbon bonds
Buchwald–Hartwig amination	Chapter 11, Problem 3f	Coupling of aryl halides with amines generating new carbon–nitrogen bonds

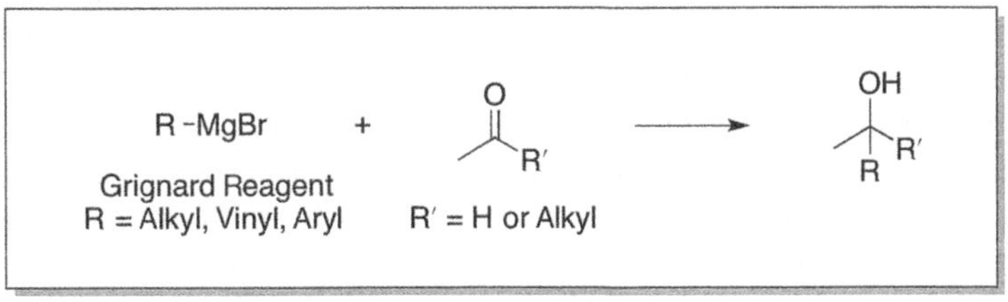

R –MgBr   +   (O=)R′   ⟶   (OH)R,R′

Grignard Reagent
R = Alkyl, Vinyl, Aryl   R′ = H or Alkyl

Scheme 11.16   *The Grignard reaction.*

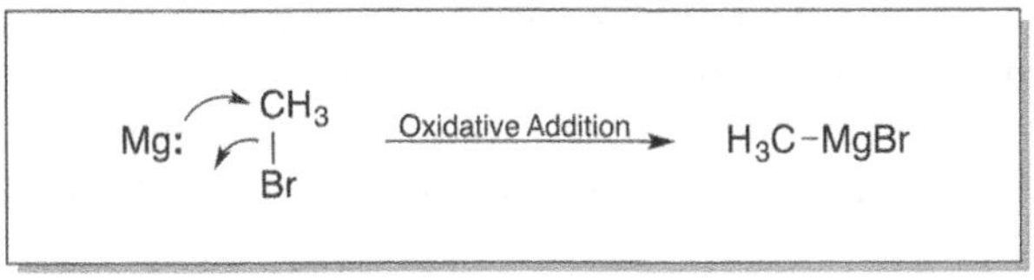

Mg: CH$_3$–Br $\xrightarrow{\text{Oxidative Addition}}$ H$_3$C –MgBr

Scheme 11.17   *Formation of Grignard reagents involves oxidative addition.*

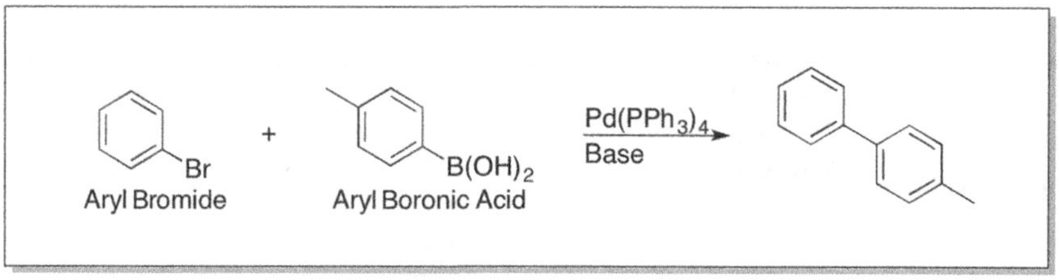

Scheme 11.18 *The Suzuki reaction.*

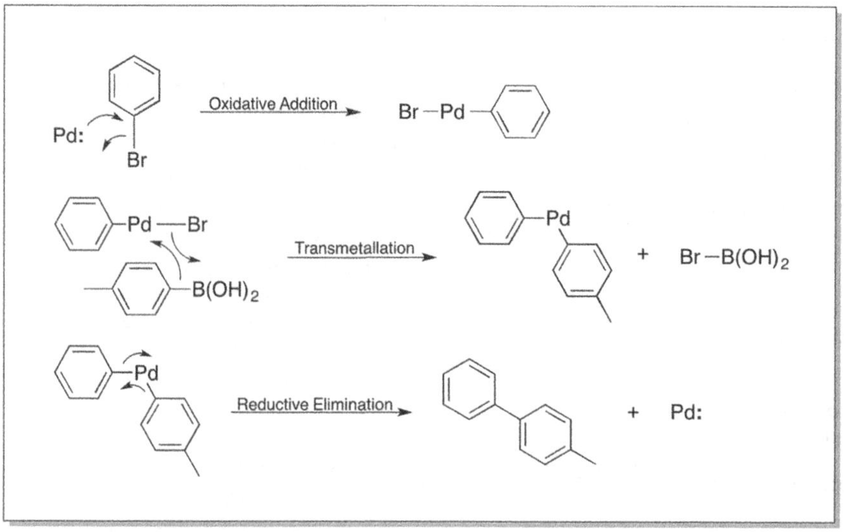

Scheme 11.19 *Simplified Suzuki reaction mechanism.*

While the **Grignard reaction** and generation of **Grignard reagents** involve relatively simple mechanisms, their introduction at this time is relevant when considering the more complex **Suzuki reaction**. As illustrated in Scheme 11.18, the **Suzuki reaction** allows for the direct coupling of a **vinyl or aryl halide** with a **vinyl or aryl boronic acid**. **Suzuki reactions** generally require the presence of a **palladium catalyst** such as **tetrakis(triphenylphosphine)palladium**. Such catalysts present the **palladium metal** as a neutral species. In addition to the **palladium catalyst**, a base such as **sodium ethoxide** or **potassium *tert*-butoxide** is required for this reaction to proceed.

Extending from the mechanisms presented for the formation of **Grignard reagents**, Suzuki reactions begin with an initial **oxidative addition** where the **neutral palladium atom** inserts itself into the **carbon–bromide bond** of the **aryl bromide** while taking on a +2 charge. Next, a **transmetallation** reaction occurs in which the palladium atom replaces the **boronic acid** with no net change in the **palladium atom** charge. Finally, a new **carbon–carbon** bond is formed when the two aryl groups join and expel the **palladium atom**. Because the process of releasing the palladium atom reverts the palladium charge from +2 to neutral, this step is referred to as a **reductive elimination**. Like the mechanisms presented for the formation of **Grignard reagents**, this process is simplified for the purposes of introduction. Scheme 11.19 illustrates these individual steps using **arrow-pushing**. The **palladium atom** is illustrated without the **triphenylphosphine ligands** for clarity of the illustrations.

TABLE 11.2   Reagent classes and associated properties.

Reagent class (class name)	Examples	Properties	Uses
R–Li (alkyllithium)	Methyllithium Butyllithium *sec*-Butyllithium *tert*-Butyllithium	Strong base	Deprotonation of weak organic acids, E2 eliminations
		Strong nucleophile when R is not bulky	S_N2/S_N2' displacements, addition reactions, addition–elimination reactions
R–MgBr (alkylmagnesium bromide, alkyl Grignard)	Methylmagnesium bromide (methyl Grignard)	Strong nucleophile when R is not bulky	S_N2/S_N2' displacements, addition reactions, addition–elimination reactions
R_2CuLi (dialkyllithiocuprate)	Dimethyllithiocuprate	General reactive nucleophile	1,4-addition reactions
R_2N–Li (lithium dialkylamide)	Lithium diisopropylamide	Strong base, nonnucleophilic when R is bulky	Deprotonation of weak organic acids with acidities as high as $pK_a = 35$
M–H (metal hydride)	Sodium hydride Potassium hydride	Strong nonnucleophilic base	Deprotonation of weak organic acids with acidities as high as $pK_a = 25$
RO–K (potassium alkoxide)	Potassium *tert*-butoxide	Nonnucleophilic base	Deprotonation of organic acids with acidities as high as $pK_a = 18$
M–OH (metal hydroxide)	Sodium hydroxide Potassium hydroxide	Nucleophilic bases	Deprotonation of organic acids with acidities as high as $pK_a = 16$, hydrolysis of esters, amides, and nitriles
R_3N (trialkylamine)	Triethylamine Diisopropylethylamine	Nonnucleophilic base	Deprotonation of organic acids, acid scavenger

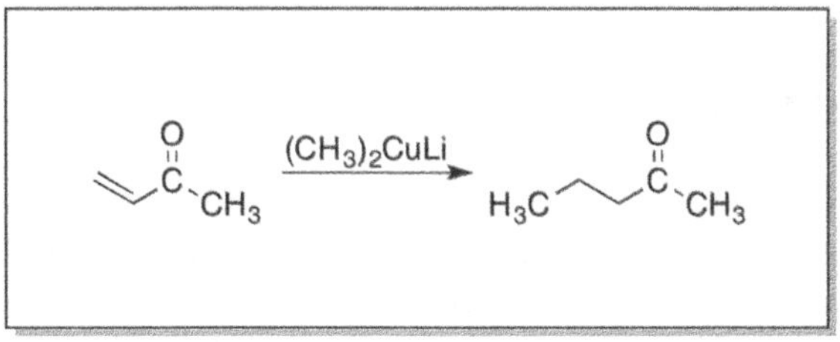

Scheme 11.20 *The Michael addition.*

The role of the **base**, referred to in Scheme 11.18, is omitted from the simplified **Suzuki reaction mechanism** illustrated in Scheme 11.19 in order to focus on the key mechanistic components of this reaction. In fact, the **base** serves three purposes. First, the **base** is involved in formation of the initial **palladium complex**. Second, the **base** reacts with the expelled **boron** unit to form a **borate**. Finally, the **base** facilitates the **reductive elimination** step. All of these aspects of the **Suzuki reaction** will be addressed in more advanced organic chemistry coursework.

As will be revealed through further coursework, many more **name reactions** are available. Furthermore, new reactions, yet to be named, are continually being discovered. In approaching all of these reactions, it is imperative to develop **mechanistic understandings** in order to correctly apply the reactions within the scope of their utilities and limitations. In this respect, **arrow-pushing** presents a valuable approach to the derivation of mechanistic understanding prior to committing the reaction names to memory.

11.3   REAGENTS

Throughout this book, and in association with the reactions presented, various **reagents** were presented that, due to their specific properties, react in very specific ways. These **reagents** differ in their **basicity**, **nucleophilicity**, and preferred sites of reaction. Table 11.2 summarizes the various properties of the reagent classes presented.

Of the **reagents** listed in Table 11.2, **dialkyllithiocuprates** stand out because of their unique ability to participate in **1,4-addition reactions**. Such reactions, also known as **conjugate additions**, are generally referred to as **Michael additions**. This **name reaction** is illustrated in Scheme 11.20 with the reaction of **dimethyllithiocuprate** with **methyl vinyl ketone**.

When considering the **reagents** listed in Table 11.2, it is important to remember that this table is not inclusive. There are many permutations of the **reagents** listed in the table as well as innumerable additional **reagents** that have been made useful to various aspects of **organic chemistry**. In fact, many research groups focus exclusively on the design and preparation of novel **reagents** capable of solving difficult synthetic problems. It is through this aspect of **organic chemistry** that some of the most significant advances have been realized.

11.4   FINAL COMMENTS

By now, having worked through the material in this book, readers should be well acquainted with the fundamental principles of **arrow-pushing**. Furthermore, through the examples presented herein, the reader should have acquired an understanding of how to apply **arrow-pushing** to explain **reaction processes** and to predict **reaction products**. While this book was intended to serve solely as a supplement to introductory organic chemistry texts, the content was designed to move from the basic foundation of organic chemistry to the direct application of **arrow-pushing** techniques, thus enabling the reader to begin to advance through the study of **organic chemistry**. Finally, in closing, readers should endeavor to understand the underlying principles of **organic chemistry** in order to embrace the full substance of this mature and continually relevant discipline.

PROBLEMS

1. Describe the following functional group transformation in mechanistic terms. Show arrow-pushing.

a.

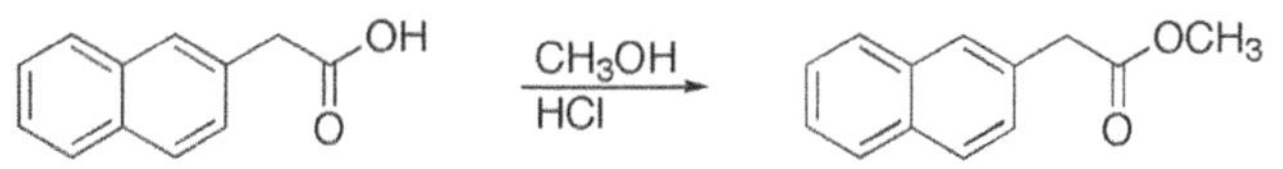

b.

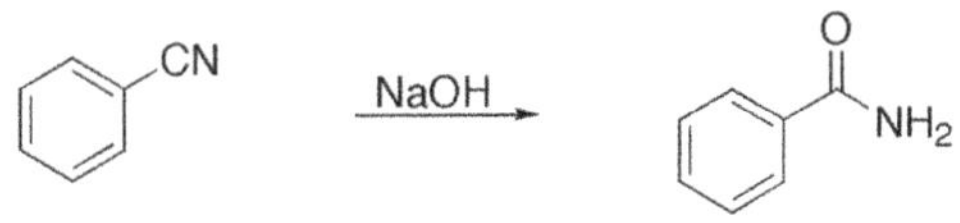

c.

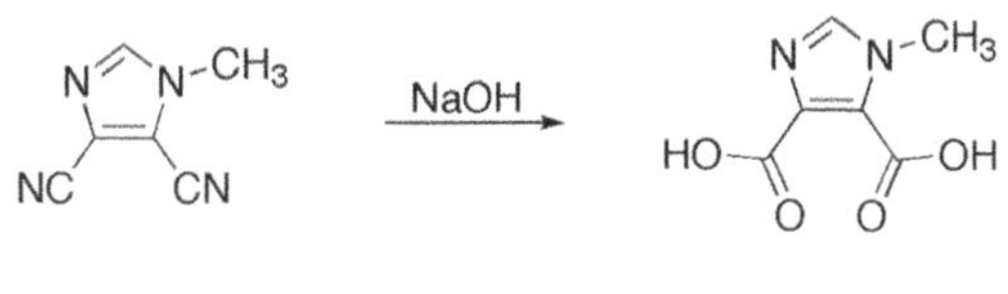

2. Explain the following reactions in mechanistic terms. Show arrow-pushing and describe the reaction as a name reaction.

a.

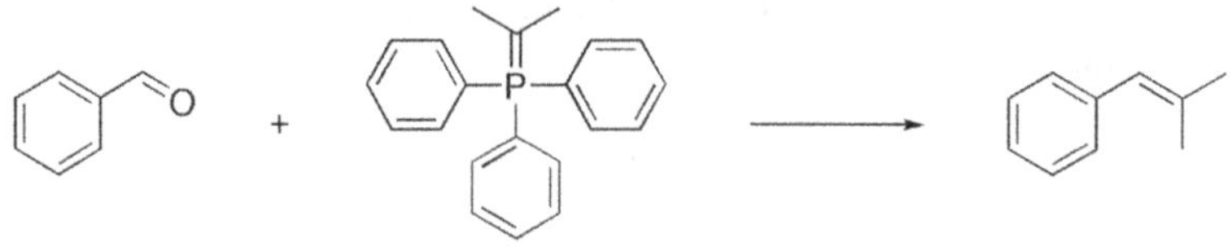

b.

c.

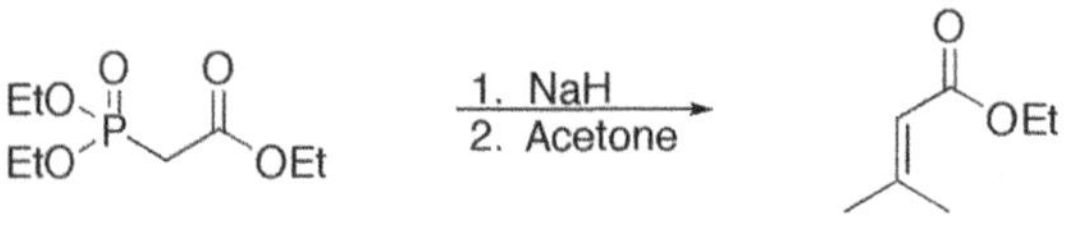

d.

e.

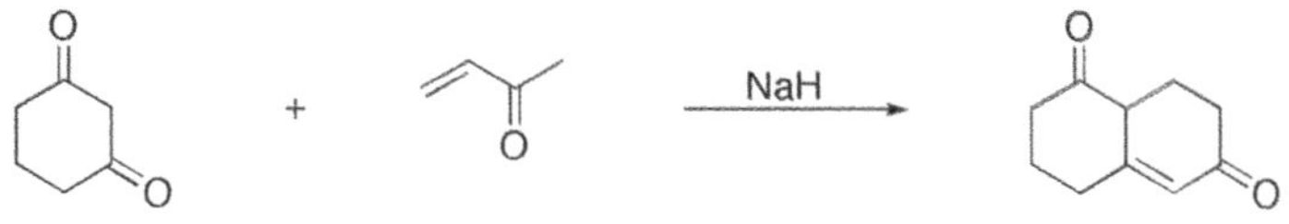

f.

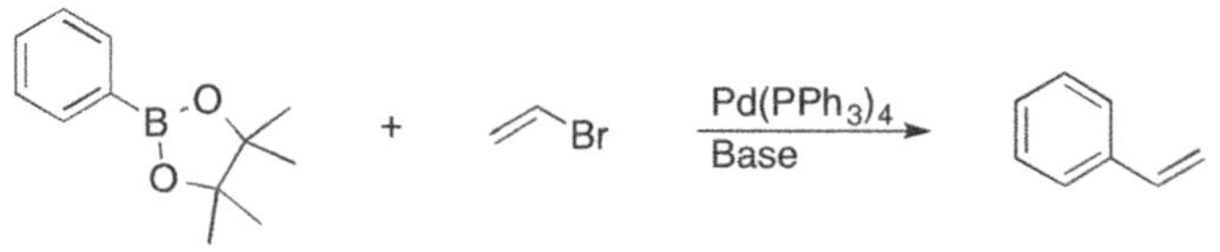

3. Explain the following name reactions in mechanistic terms. Show arrow-pushing.

   a.   The ene reaction

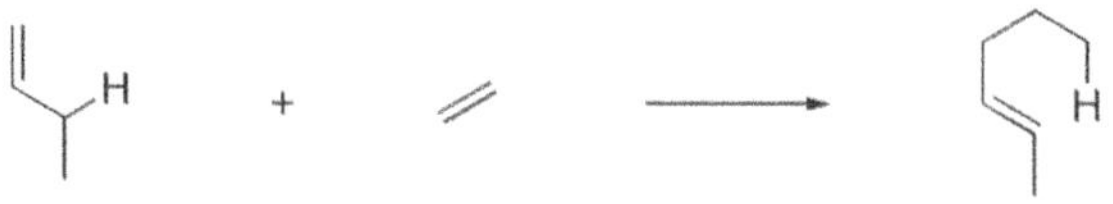

   Note: Only the hydrogen involved in the reaction is shown.

   b.   The McLafferty rearrangement

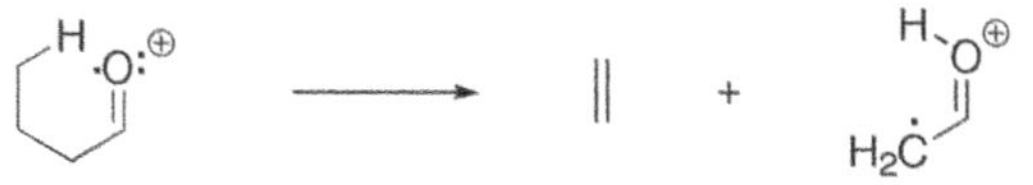

   Note: The radical cation present in the starting material is the result of the carbonyl oxygen losing a single electron. This reaction is generally observed during electron impact mass spectrometry.

c.   1,3-Dipolar cycloaddition

$$H_3C-C\overset{\oplus}{\equiv}N-O^{\ominus} \quad + \quad \equiv \quad \longrightarrow$$

d.   The Swern oxidation

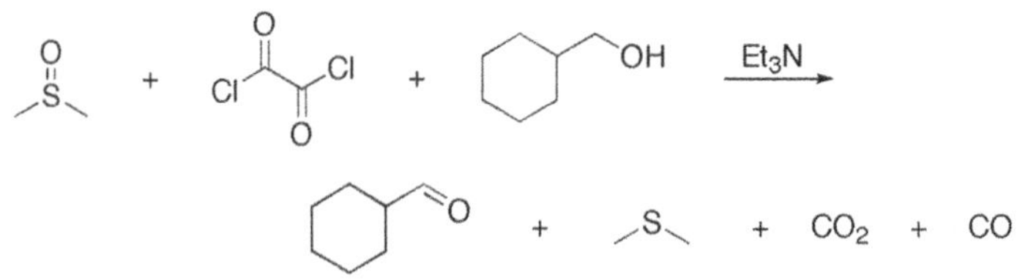

Hint: The oxygen atom in dimethyl sulfoxide is nucleophilic.

e.   The Sonogashira reaction

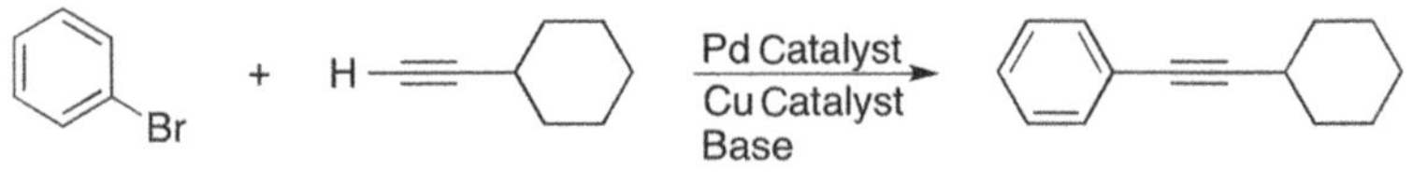

$$\xrightarrow[\text{Base}]{\text{Pd Catalyst} \atop \text{Cu Catalyst}}$$

Hint: The copper catalyst metallates the acetylene.

f.   The Buchwald–Hartwig amination

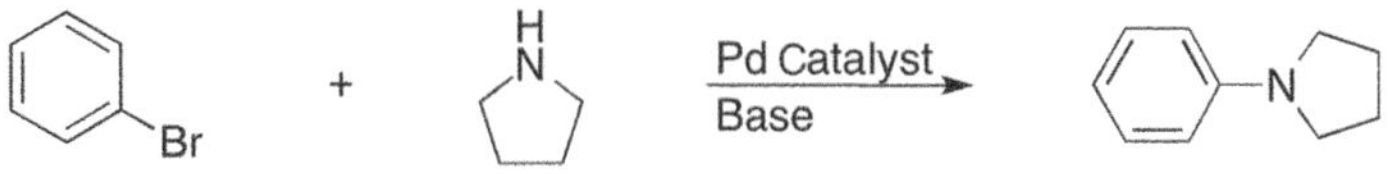

4. The Friedel–Crafts acylation, illustrated in Scheme 11.12, shows the formation of one product. However, the reaction, as illustrated, actually forms a mixture of two products. Using the arguments presented in the solution set for Chapter 8, identify the second product. Show partial charges and arrow-pushing.

5. Predict all products formed from a Friedel–Crafts acylation on the following compounds with acetyl chloride. Rationalize your answers using partial charges.

a.

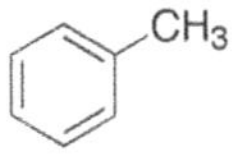

b.

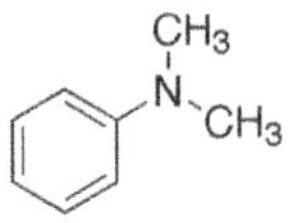

c.

d.

e.

f.

6. From the following list of compounds, propose a synthetic strategy for the specified compounds. Up to three synthetic steps may be required. Any chemical reagents described in this book or any general organic chemistry text may be used. Show all arrow-pushing.

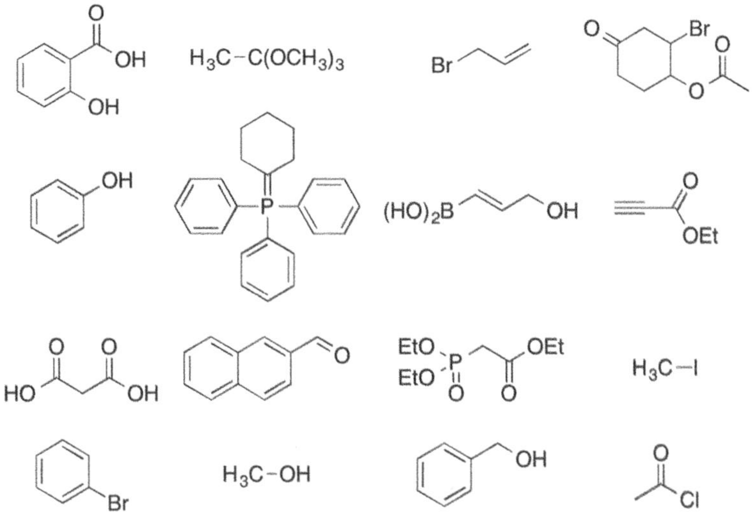

a.

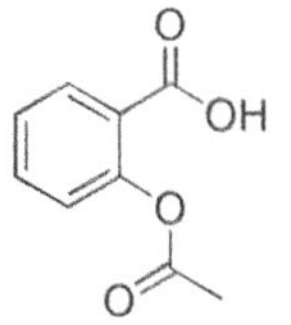

(acetylsalicylic acid, aspirin)

b.

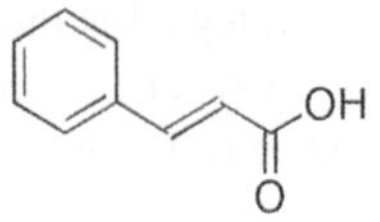

(cinnamic acid)

c.

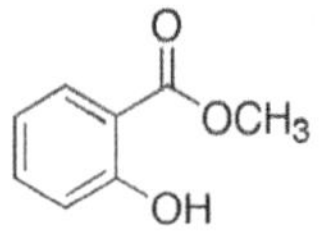

(methyl salicylate, oil of wintergreen)

d.

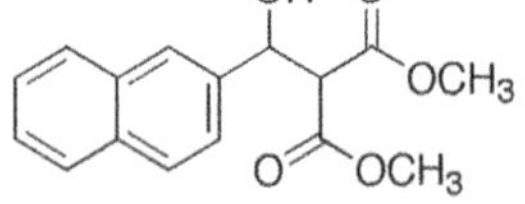

e.

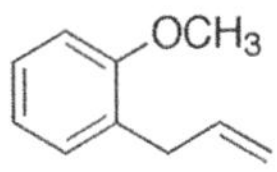

f.

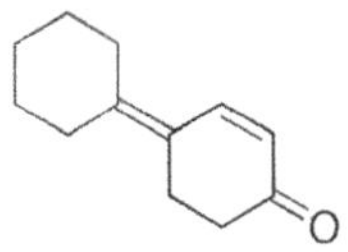

g.

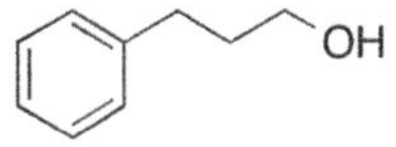

via a route not related to that used in Problem 6b

h.

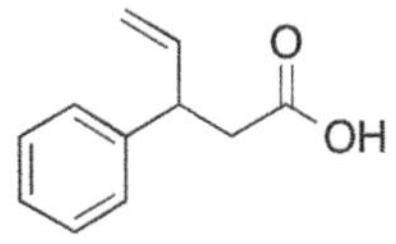

via a route not related to those used in Problems 6b and 6(g)

Appendix *1*

pK_a *Values of Protons Associated with Common Functional Groups*

While this book teaches that organic chemistry can be learned without relying upon memorization of a multitude of chemical reactions, familiarity with pK_a values associated with various functional groups is essential. The pK_a values listed in the following table provide a general calibration of the acidities of protons associated with common functional groups. In advancing through organic chemistry, accurate recollection of these values is indispensable.

Arrow-Pushing in Organic Chemistry: An Easy Approach to Understanding Reaction Mechanisms,
Second Edition. Daniel E. Levy.
© 2017 John Wiley & Sons, Inc. Published 2017 by John Wiley & Sons, Inc.

Common protic acids

H–F Hydrofluoric acid	**3.18**	H–I Hydroiodic acid	**–10**
H–Cl Hydrochloric acid	**–2.2**	H–CN Hydrocyanic acid	**9.3**
H–Br Hydrobromic acid	**–4.7**	H–N$_3$ Hydrazoic acid	**4.6**

Neutral functional groups
Carboxylic acids and amides

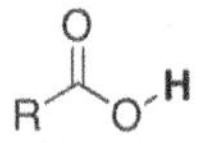

Carboxylic acids   **4–6**

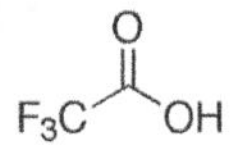

Trifluoroacetic acid   **0.30**

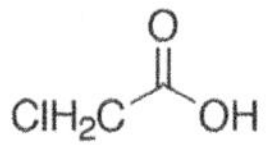

Chloroacetic acid   **2.85**

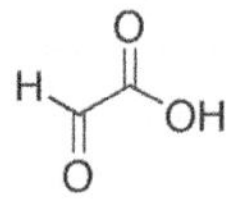

Glyoxylic acid   **3.18**

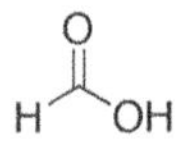

Formic acid   **3.75**

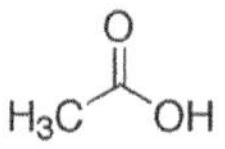

Acetic acid   **4.75**

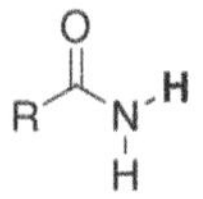

Amides   **15–17**

Neutral Functional Groups—Continued
Alcohols, amines, and thiols

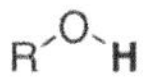

R–O–H
Alcohols
15–19

F_3C–CH(H)(H)–O–H
2,2,2-Trifluoroethanol
11–12

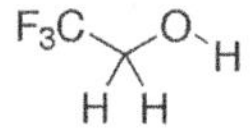

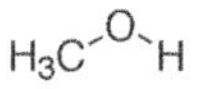

H_3C–O–H
Methanol
15

H_3C–CH(H)(H)–O–H
Ethanol
15–16

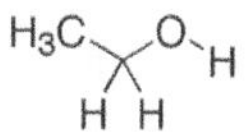

H_3C–CH(H_3C)(H)–O–H
2-Propanol
(isopropanol)
16–17

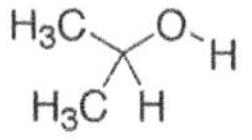

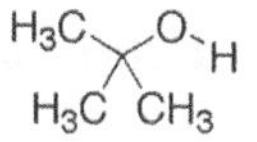

H_3C–C(H_3C)(CH_3)–O–H
2-Methyl-2-propanol
(*tert*-butanol)
18–19

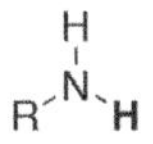

H–N(H)–R–H
Amines
33–38

H_2N–H
Ammonia
35

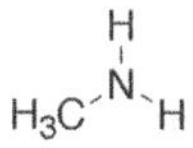

H_3C–N(H)–H
Methylamine
35

H_3C–S–H
Methanethiol
10.4

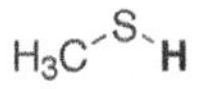

Aldehydes, ketones, esters, and amides

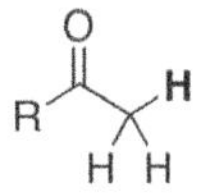

R–C(=O)–CH(H)(H)–H
Aldehydes and
ketones
20–25

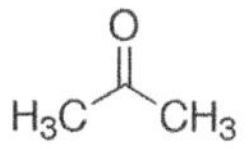

H_3C–C(=O)–CH_3
Acetone
20

H–CH(H)(H)–C(=O)–O–R
Esters
25–30

H_3C–O–C(=O)–CH(H)(H)–C(=O)–O–CH_3
Dimethyl malonate
13

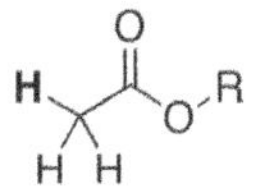

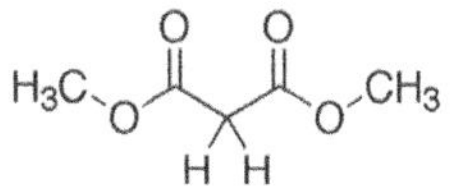

H–CH(H)(H)–C(=O)–N(R)–R
Amides
30–35

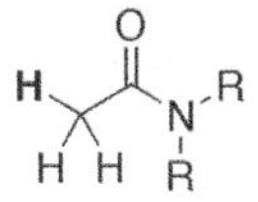

Neutral Functional Groups—Continued
Nitriles and nitro compounds

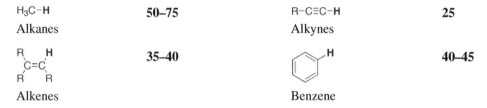

Nitriles	**20–25**	H₃C–CN Acetonitrile	**25**
		Nitromethane	**10–15**

Alkanes, alkenes, and alkynes

H_3C-H Alkanes	**50–75**	$R-C\equiv C-H$ Alkynes	**25**
Alkenes	**35–40**	Benzene	**40–45**

Protonated functional groups
Alcohols and ethers

Alcohols	**–2.2**	Ethers	**–2.2**

Amines

$(CH_3CH_2)_3\overset{\oplus}{N}-H$ Amines (triethylamine)	**10**

Carboxylic acids and esters

Carboxylic acids	**–6**	Esters	**–6**

Amides and nitriles

Amides	**0**	$H_3C-C\equiv\overset{\oplus}{N}-H$ Nitriles (acetonitrile)	**–10**

Aldehydes and ketones

Aldehydes	**–7 to –9**	Ketones	**–7 to –9**

Appendix *2*

Answers and Explanations to Problems

In each chapter of this book, problem sets are presented in order to support and enhance the reader's understanding of the principles discussed. These problem sets frequently extend beyond the topics presented and are included for the following reasons:

- Problems directly in line with presented topics serve to support the understanding of the basic principles of arrow-pushing applied to organic chemistry reaction mechanisms.
- Problems extending beyond the scope of topics presented in each chapter serve to encourage readers to extrapolate from principles discussed and to apply those principles to more advanced but relevant reaction mechanisms.
- Problems based on hypothetical examples or examples that would never be practical in real-world settings encourage readers to think outside of the box and to use their imaginations in order to come up with creative solutions.

Through working the problems in each chapter and through the detailed explanations presented in this appendix, readers will:

- Develop a broader understanding of the applications of arrow-pushing to complex mechanistic scenarios.
- Begin to develop foresight for the design of appropriate synthetic pathways.
- Learn to approach difficult problems through recognition of possible outcomes, multiple outcomes, and solutions.

Arrow-Pushing in Organic Chemistry: An Easy Approach to Understanding Reaction Mechanisms,
Second Edition. Daniel E. Levy.
© 2017 John Wiley & Sons, Inc. Published 2017 by John Wiley & Sons, Inc.

CHAPTER 1 SOLUTIONS

1. *Add arrow-pushing to explain the following reactions.*

   When drawing arrows to illustrate movement of electrons, it is important to remember that electrons form the bonds that join atoms. The following represent heterolytic-type reaction mechanisms:

   a. $N \equiv C^{\ominus}$ + $H_3C - I$ ⟶ $N \equiv C - CH_3$ + $I^{\ominus}$

   This is an example of an S_N2 reaction mechanism converting an alkyl iodide (iodomethane) to an alkyl nitrile (acetonitrile). Arrow-pushing is illustrated in the following:

   $N \equiv C^{\ominus}$ + $H_3C - I$ ⟶ $N \equiv C - CH_3$ + $I^{\ominus}$

   b. 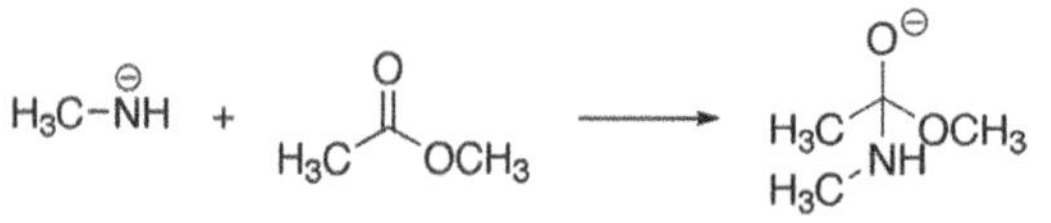 $H_3C - \overset{\ominus}{N}H$ + (methyl acetate structure: H_3C—C(=O)—OCH_3) ⟶ (tetrahedral intermediate: H_3C—C($O^{\ominus}$)(OCH_3)—NH—CH_3)

   This is an example of the first step of an addition–elimination reaction mechanism converting an ester (methyl acetate) to an amide (*N*-methylacetamide). For clarity, the anion was repositioned in the scheme. Arrow-pushing is illustrated in the following:

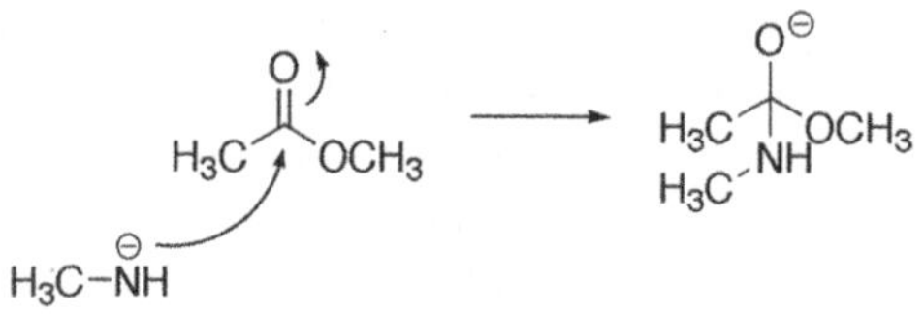

   (reaction scheme with H_3C—C(=O)—OCH_3 and $H_3C - \overset{\ominus}{N}H$ ⟶ tetrahedral intermediate H_3C—C($O^{\ominus}$)(OCH_3)(NH—CH_3))

   c.   (tetrahedral intermediate H_3C—C($O^{\ominus}$)(OCH_3)(NH—CH_3)) ⟶ H_3C—C(=O)—$\overset{CH_3}{\underset{H}{N}}$ + $^{\ominus}O - CH_3$

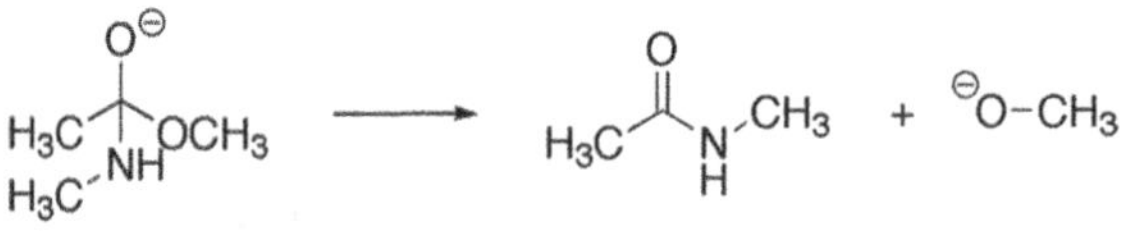

   This is an example of the second step of an addition–elimination reaction mechanism converting an ester (methyl acetate) to an amide (*N*-methylacetamide). Arrow-pushing is illustrated in the following:

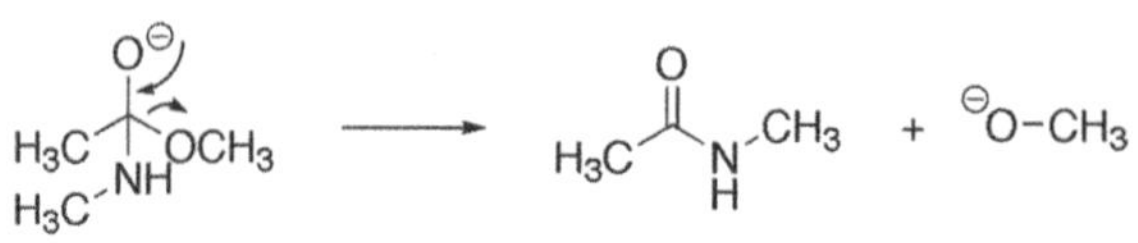

   (reaction scheme: H_3C—C($O^{\ominus}$)(OCH_3)(NH—CH_3) ⟶ H_3C—C(=O)—$\overset{CH_3}{\underset{H}{N}}$ + $^{\ominus}O - CH_3$)

   d.  $H_3C - \overset{\cdot\cdot}{N}H_2$ + H_3C⌒⌒Cl ⟶ H_3C⌒⌒$\overset{H_2}{\underset{\oplus}{N}}^{CH_3}$ + $Cl^{\ominus}$

This is an example of an S$_N$2 reaction mechanism converting an alkyl chloride (chloro-propane) to an ammonium salt (*N*-methyl-*N*-propylammonium chloride). For clarity, the amine was repositioned in the scheme. Arrow-pushing is illustrated in the following:

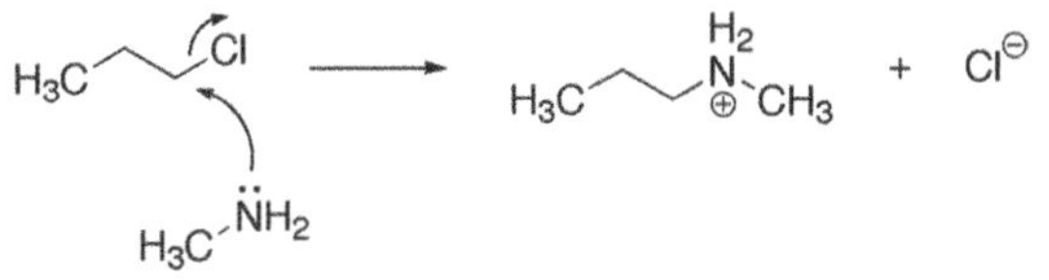

e.

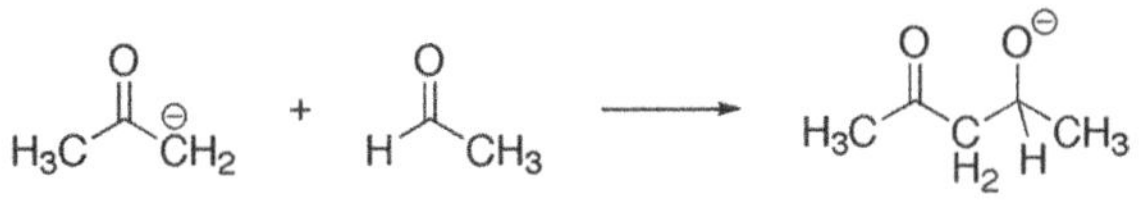

This is an example of an aldol condensation between an acetone anion and acetalde-hyde. Note that the mechanism proceeds through addition of an anion to an aldehyde carbonyl. Arrow-pushing is illustrated in following:

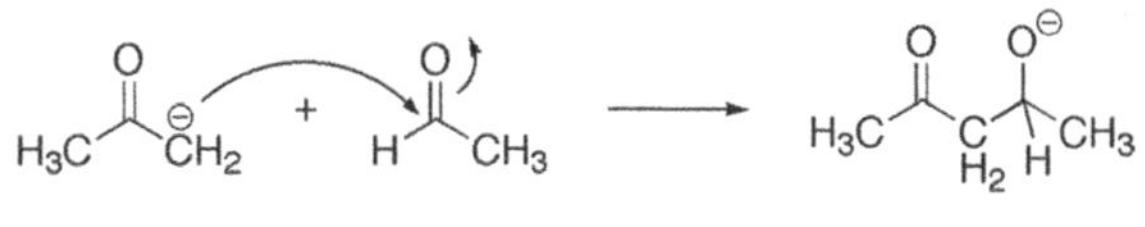

f.

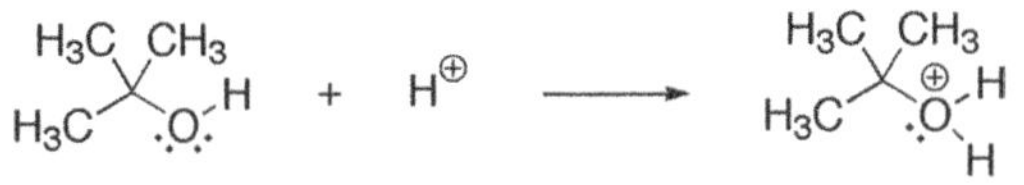

This is an example of the first step in the acid-mediated solvolysis of a tertiary alco-hol. Note that protonation of the alcohol occurs under strongly acidic conditions with the lone pair of electrons moving toward the positive charge residing on the proton. Arrow-pushing is illustrated in the following:

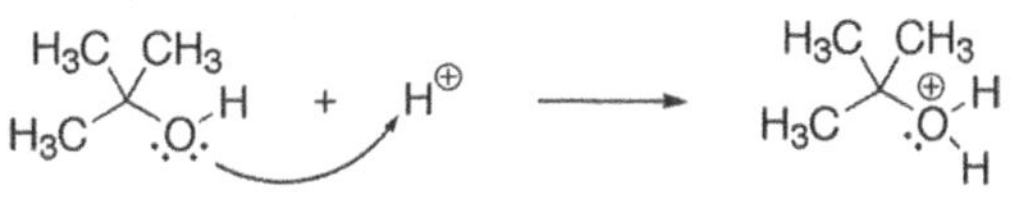

g.

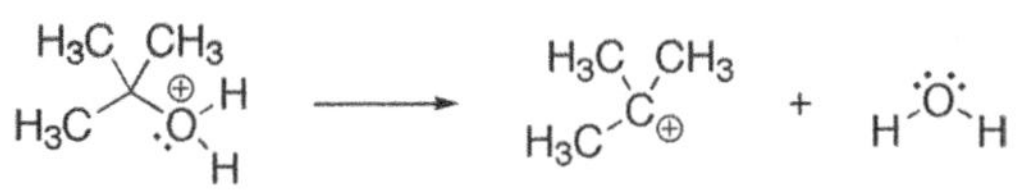

This is an example of the second step in the acid-mediated solvolysis of an alcohol. Note that the protonated alcohol separates as water and leaves the positive charge on the carbon atom. For clarity, the bond was lengthened to allow space for the arrow. Note that the electrons in the bond move toward the positive charge residing on the oxygen. Arrow-pushing is illustrated in the following:

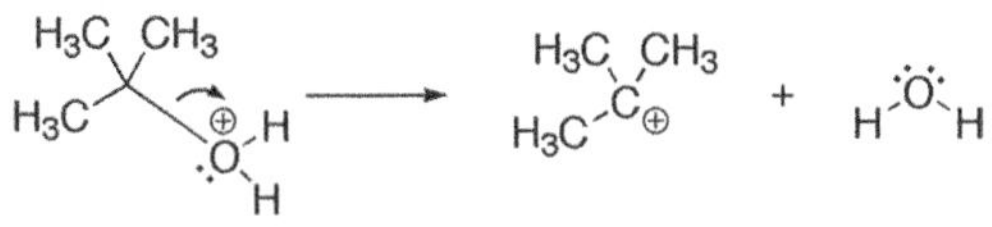

h.

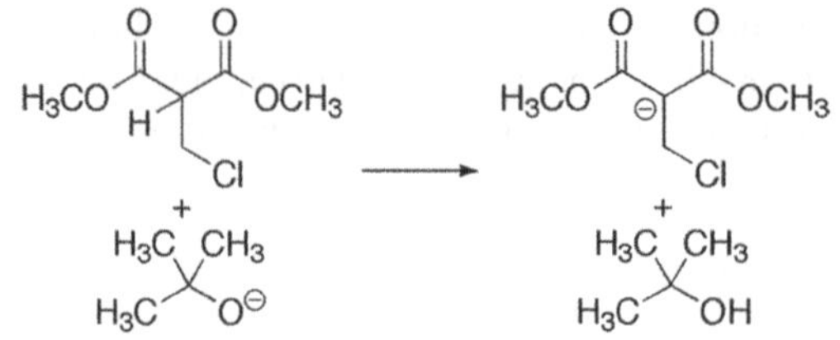

This is an example of the first step of an E1cB reaction mechanism. Note the base-mediated deprotonation of the diester converting the *tert*-butoxide anion to *tert*-butanol. For clarity, the anion was repositioned and the bond was lengthened. Arrow-pushing is in the following:

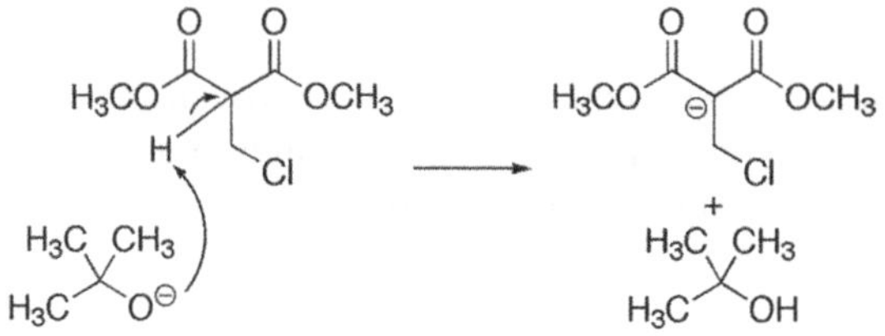

i.

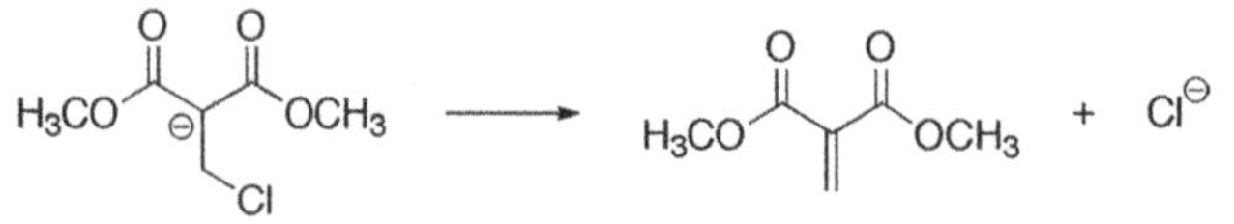

This is an example of the second step of an E1cB reaction mechanism. Note that the displacement of the chloride anion is the result of an anion present on an adjacent carbon atom. Arrow-pushing is illustrated in the following:

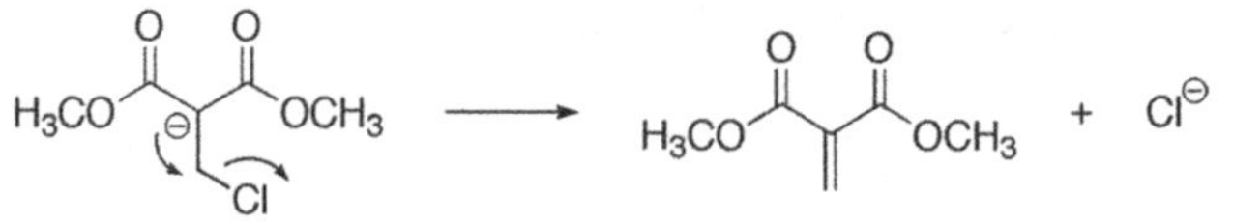

It is important to recognize that, in actuality, the steps illustrated in Problems 1h and 1i occur almost simultaneously. The "individual steps" are illustrated in order to simplify the presentation of this mechanistic type. E1cB eliminations are discussed in greater detail in Chapter 7.

The following represent reaction mechanisms involving free radicals:

j. $Br-Br \longrightarrow Br\cdot + Br\cdot$

This is an example of the homolytic cleavage of a bromine molecule to form two bromide radicals. Note the use of single-barbed arrows to describe radical-based mechanisms resulting in movement of single electrons. For clarity, the bond is elongated. Arrow-pushing is illustrated in the following:

k. 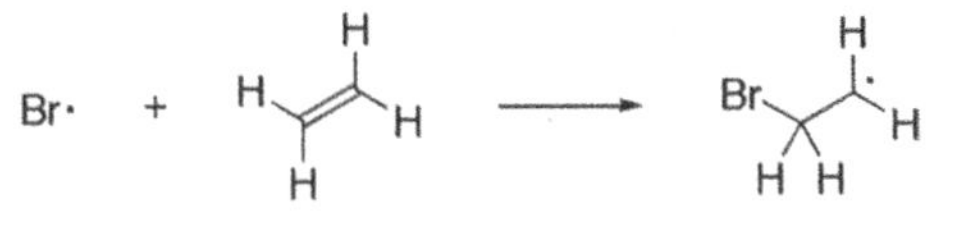   Br· +

This is an example of the addition of a bromide radical to an olefin. Note that a single-barbed arrow is used for each electron that is moving. Arrow-pushing is illustrated in the following:

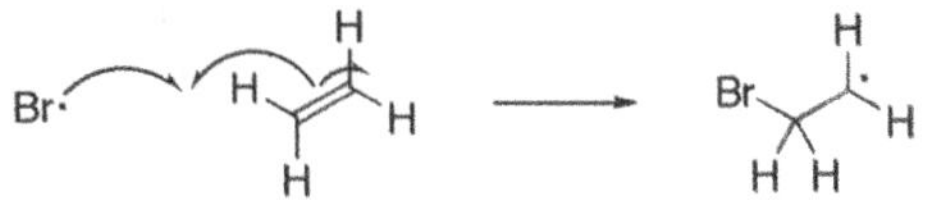

l.

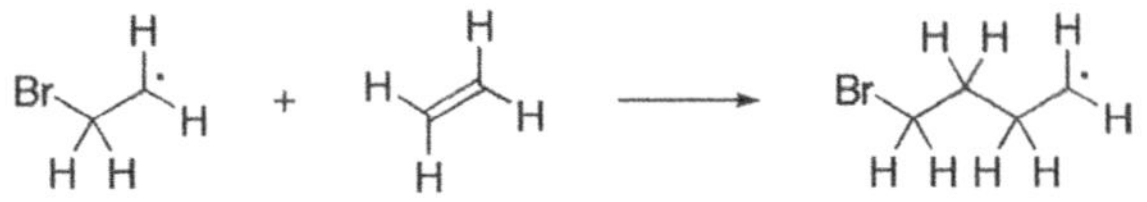

This is an example of a step in the free radical polymerization of ethylene forming polyethylene. As in the previous example, note that a single-barbed arrow is used for each electron that is moving. Arrow-pushing is illustrated in the following:

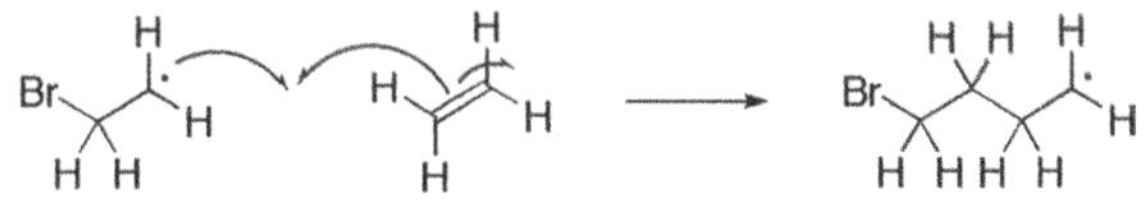

The following represents a concerted reaction mechanism:

m.

This is an example of a Claisen rearrangement. As illustrated, concerted mechanisms can be described either by movement of electron pairs or by movement of single electrons. However, these mechanisms are generally represented by the movement of electron pairs using double-barbed arrows as is done for heterolytic reaction mechanisms. While, mechanistically, the movement of electron pairs is preferred over the movement of single electrons, both processes are illustrated in the following using arrow-pushing:

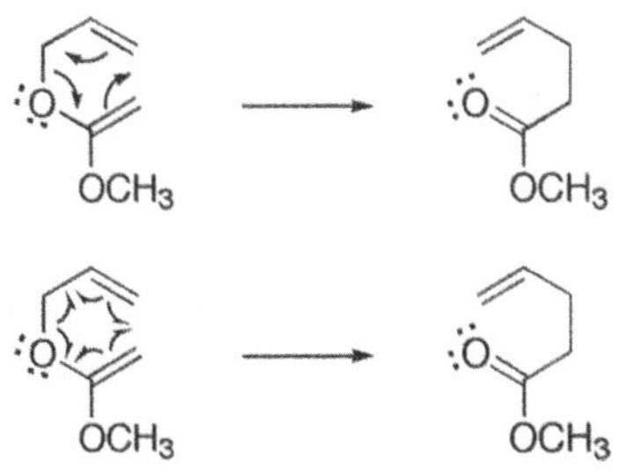

The following represents a heterolytic-type reaction mechanism:

n.

This is an example of a cation–π cyclization. Note that unlike the previously described heterolytic reaction mechanisms, this reaction is influenced by a positive charge. Also, please note that this reaction shares some characteristics with concerted mechanisms in that formation of the new bonds occurs almost simultaneously. Arrow-pushing is illustrated in the following:

2. *Place the partial charges on the following molecules:*

a.

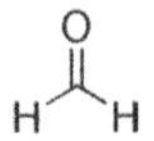

Carbonyls are polarized such that a partial negative charge resides on the oxygen and a partial positive charge resides on the carbon.

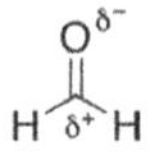

b.

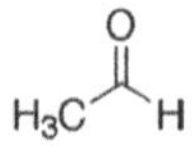

Because of the polarity of the carbonyl, adjacent groups are also polarized. In general, where a partial positive charge rests, an adjacent atom will bear a partial negative charge.

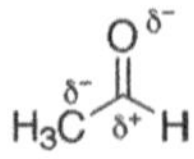

c.

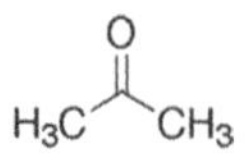

Because of the polarity of the carbonyl, adjacent groups are also polarized. In general, where a partial positive charge rests, an adjacent atom will bear a partial negative charge. This can occur on more than one adjacent atom.

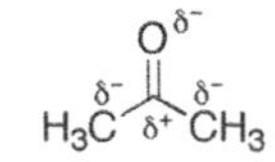

d.

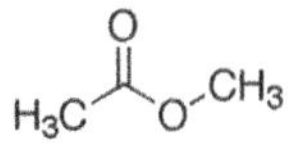

Because of the polarity of the carbonyl, adjacent groups are also polarized. In general, where a partial positive charge rests, an adjacent atom will bear a partial negative charge. This can occur on more than one adjacent atom or heteroatom.

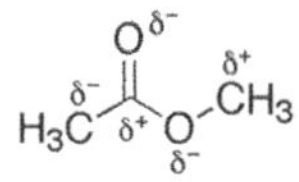

e.

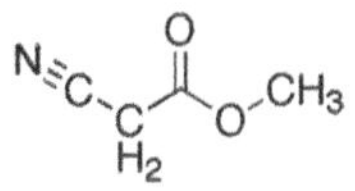

Nitriles, like carbonyls, are polarized with the nitrogen bearing a partial negative charge and the carbon possessing a partial positive charge.

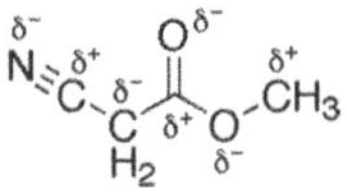

f.

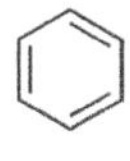

Benzene has no localized positive or negative charges because of its symmetry. The two illustrated resonance forms are equivalent rendering benzene a non-polar molecule.

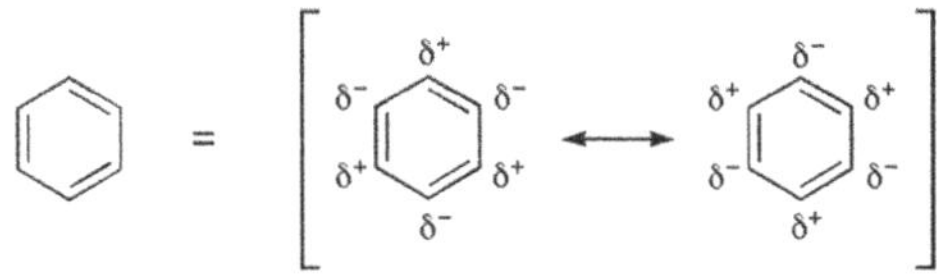

g.

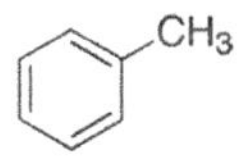

As will be discussed in Chapter 3, methyl groups are electron donating. This is not due to any defined positive charges on the carbon atom and is more the result of **hyperconjugation**. **Hyperconjugation**, in this case, relates to the ability of the carbon–hydrogen σ-bonds of the methyl group to donate electrons into the conjugated system of benzene. While this effect will be discussed in more detail later, for now let us define methyl groups as possessing a partial negative charge. This resulting negative charge thus polarized each double bond in the ring.

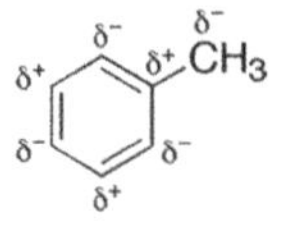

h.

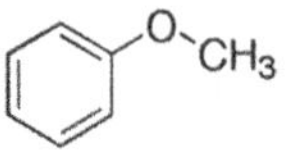

As with the previous example, groups possessing partial negative charge characteristics donate electrons into conjugated systems and polarize the double bonds. This effect is generally noted with heteroatoms such as oxygen. Also, while in the previous example a methyl group was argued to possess a partial negative charge, the partial positive charge illustrated here is due to the overriding partial negative characteristics of the oxygen atom.

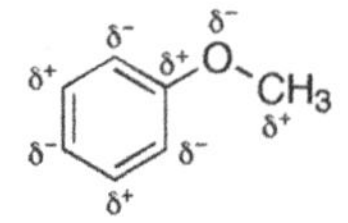

i.

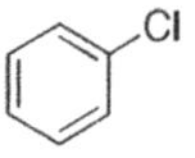

As with the previous example, heteroatoms such as chlorine possess partial negative charge characteristics and donate electrons into conjugated systems polarizing the double bonds.

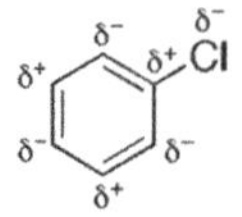

j.

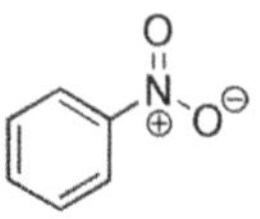

As with groups possessing negative charge characteristics, when a positive charge is present on an atom connected to a conjugated system, the double bonds are polarized. This polarization is opposite of that observed for negatively charged groups.

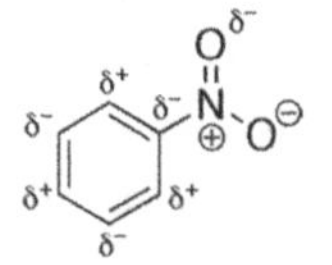

k.

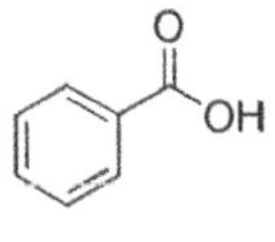

As with groups possessing negative charge characteristics, when a partial positive charge is present on an atom connected to a conjugated system, the double bonds are polarized. This polarization is opposite to that observed for negatively charged groups.

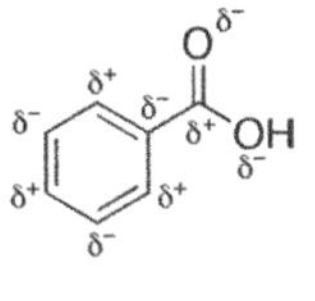

Please note for Problems 2l through 2r: When multiple groups are present on conjugated systems, their charged characteristics can work together or oppose each other depending on where they are placed relative to each other. The following problems address this point:

l.

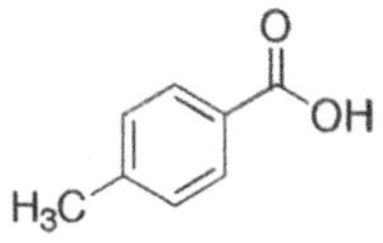

In this case, the carboxylic acid being electron withdrawing induces a partial positive charge at the *para* position. This is the same position where an electron-donating methyl group is placed. *Consider what impact the methyl group has on the acidity of the carboxylic acid.*

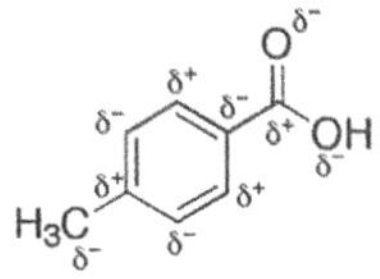

m.

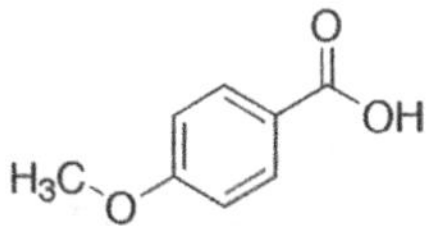

In this case, the carboxylic acid being electron withdrawing induces a partial positive charge at the *para* position. This is the same position where an electron-donating methoxy group is placed. Also, while in a previous example a methyl group was argued to possess a partial negative charge, the partial positive charge illustrated here is due to the overriding partial negative characteristics of the oxygen atom. *Consider what impact the methoxy group has on the acidity of the carboxylic acid.*

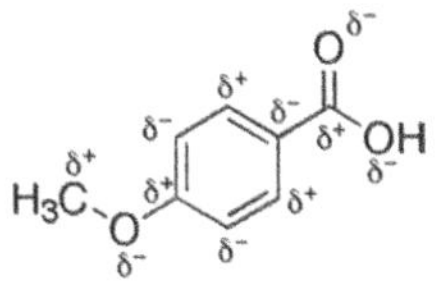

n.

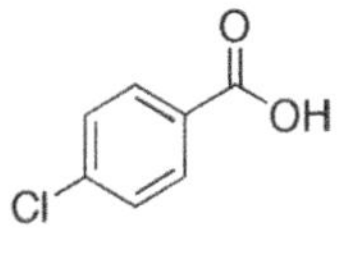

In this case, the carboxylic acid being electron withdrawing induces a partial positive charge at the *para* position. This is the same position where an electron-donating chloride is placed. *Consider what impact the chloro group has on the acidity of the carboxylic acid.*

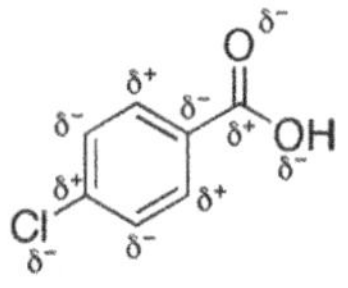

o.

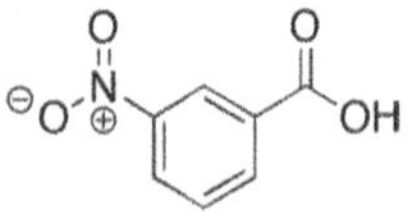

In this case, the carboxylic acid being electron withdrawing induces a partial negative charge at the *meta* position. This is the same position where an electron-withdrawing nitro group is placed. *Consider what impact the nitro group has on the acidity of the carboxylic acid.*

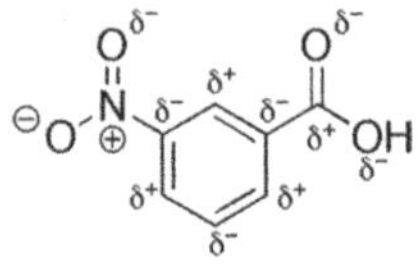

p.

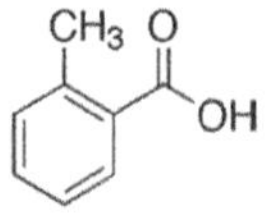

In this case, the carboxylic acid being electron withdrawing induces a partial positive charge at the *ortho* position. This is the same position where an electron-donating methyl group is placed. *Consider what impact the methyl group has on the acidity of the carboxylic acid.*

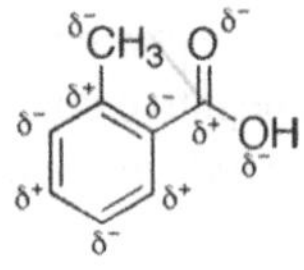

q.

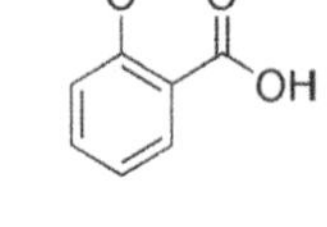

In this case, the carboxylic acid being electron withdrawing induces a partial positive charge at the *ortho* position. This is the same position where an electron-donating methoxy group is placed. Also, while in a previous example a methyl group was argued to possess a partial negative charge, the partial positive charge illustrated here is due to the overriding partial negative characteristics of the oxygen atom. *Consider what impact the methoxy group has on the acidity of the carboxylic acid.*

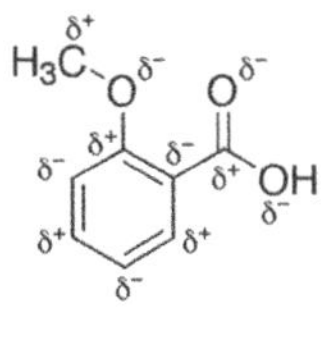

r.

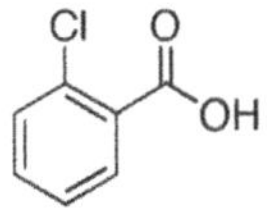

In this case, the carboxylic acid being electron withdrawing induces a partial positive charge at the *ortho* position. This is the same position where an electron-donating chloride is placed. *Consider what impact the chloro group has on the acidity of the carboxylic acid.*

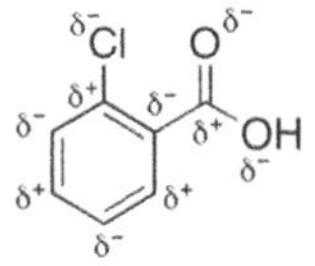

CHAPTER 2 SOLUTIONS

1. *Arrange the free radicals in each group from highest to lowest stability.*

As discussed in this chapter, free radical stability is influenced by conjugation (the ability to delocalize across several atoms) and hyperconjugation (the ability to share electrons with neighboring bonds and lone electron pairs).

a.

$$
\begin{array}{cccc}
\text{H}-\overset{\text{H}}{\underset{\text{CH}_2}{\overset{|}{\underset{|}{\text{C}}}}}\cdot & \text{H}-\overset{\text{H}}{\underset{\text{H}_3\text{C}-\text{CH}}{\overset{|}{\underset{|}{\text{C}}}}}\cdot & \text{H}_3\text{C}-\overset{\text{CH}_3}{\underset{\text{CH}_3}{\overset{|}{\underset{|}{\text{C}}}}}\cdot & \text{H}_3\text{C}-\overset{\text{H}}{\underset{\text{CH}_2}{\overset{|}{\underset{|}{\text{C}}}}}\cdot
\end{array}
$$

This set of structures illustrates the importance of hyperconjugation based on the number of carbon–hydrogen bonds adjacent to the free radical center. More neighboring bonds lead to additional hyperconjugation opportunities and greater stability.

$$
\text{H}_3\text{C}-\overset{\text{CH}_3}{\underset{\text{CH}_3}{\text{C}}}\cdot \;>\; \text{H}_3\text{C}-\overset{\text{H}}{\underset{\text{CH}_2\text{CH}_3}{\text{C}}}\cdot \;>\; \text{H}-\overset{\text{H}}{\underset{\text{H}_3\text{C}-\text{CH}, \text{CH}_3}{\text{C}}}\cdot \;>\; \text{H}-\overset{\text{H}}{\underset{\text{CH}_2\text{CH}_2\text{CH}_3}{\text{C}}}\cdot
$$

b.

$$
\begin{array}{ccc}
\text{H}_3\text{C}-\overset{\text{CH}_3}{\underset{\text{N}-\text{CH}_3}{\text{C}}}\cdot & \text{H}_3\text{C}-\overset{\text{CH}_3}{\underset{\text{O}-\text{CH}_3}{\text{C}}}\cdot & \text{H}_3\text{C}-\overset{\text{CH}_3}{\underset{\text{CH}_2\text{CH}_3}{\text{C}}}\cdot
\end{array}
$$

This set of structures illustrates the importance of lone electron pairs residing on heteroatoms adjacent to free radical centers. While the order of electronegativity is $O > N > C$, oxygen has two nonbonded electron pairs and nitrogen has one nonbonded electron pair. These nonbonded electron pairs are more easily shared with free radicals compared with the electron pairs associated with carbon–hydrogen bonds. Thus, the order of stability is illustrated in the following with the two heteroatomic structures being roughly equivalent in stability.

$$
\left[\; \text{H}_3\text{C}-\overset{\text{CH}_3}{\underset{\text{N}-\text{CH}_3}{\text{C}}}\cdot \;\approx\; \text{H}_3\text{C}-\overset{\text{CH}_3}{\underset{\text{O}-\text{CH}_3}{\text{C}}}\cdot \;\right] \;>\; \text{H}_3\text{C}-\overset{\text{CH}_3}{\underset{\text{CH}_2\text{CH}_3}{\text{C}}}\cdot
$$

c.

$$
\begin{array}{ccc}
\text{H}_3\text{C}-\overset{\text{CH}_3}{\underset{\text{CH}_2\text{CH}_3}{\text{C}}}\cdot & \text{H}_3\text{C}-\overset{\text{CH}_3}{\underset{\text{C}_6\text{H}_5}{\text{C}}}\cdot & \text{H}_3\text{C}-\overset{\text{CH}_3}{\underset{\text{O}-\text{CH}_3}{\text{C}}}\cdot
\end{array}
$$

This set of structures illustrates the importance of delocalization of free radical centers across multiple centers. In the case of the isopropylphenyl radical, the radical center is delocalized into the phenyl ring, making this structure the most stable of these three.

2. *Figure 2.6 listed several common free radical initiators including azobisisobutyronitrile (AIBN). Using arrow-pushing, explain the mechanism of AIBN decomposition.*

In the case of AIBN, initial homolytic cleavage of a carbon–nitrogen bond occurs either thermally or photolytically, generating an isobutyronitrile radical and a nitrogen radical. The resulting radical residing on the nitrogen then facilitates homolytic cleavage of a second carbon–nitrogen bond resulting in the formation of nitrogen gas and a second isobutyronitrile radical.

An alternate mechanism involves the simultaneous thermal or photolytic cleavage of two carbon–nitrogen bonds, leading to the simultaneous formation of two isobutyronitrile radicals and nitrogen gas.

Since these are free radical mechanisms involving the movement of single electrons, arrow-pushing is illustrated using single-barbed arrows.

3. *Using arrow-pushing, explain how AIBN as a free radical initiator assists reactions using tributyltin hydride—$(C_4H_9)_3SnH$.*

As discussed in Problem 2, thermal or photolytic decomposition of AIBN leads to formation of the isobutyronitrile radical. As illustrated in the following, the Sn–H bond of tributyltin hydride is a fairly weak bond susceptible to homolytic cleavage in the presence of species capable of extracting a hydrogen atom. In this case, such a species is the isobutyronitrile radical which converts to isobutyronitrile with the formation of a tin radical. The tin radical then becomes involved in the propagation of the free radical reaction.

Since this reaction mechanism involves the movement of single electrons, arrow-pushing is illustrated using single-barbed arrows.

4. *Predict the major bromination product for each compound when reacted with N-bromo-succinimide. Explain your answers.*

a.
$$H_3C \quad H \quad CH_3$$
$$H-C-C-C-CH_3$$
$$H_3C \quad H \quad CH_3$$

For this example, the hydrogen atom shown in boldface in the following is adjacent to the most carbon–hydrogen bonds compared with all other hydrogen atoms. Therefore, the free radical formed on homolytic cleavage of this carbon–hydrogen bond has the most opportunities for stabilization by hyperconjugation and will therefore be the most stable free radical site. Free radical bromination will therefore produce the illustrated bromide as the major product.

$$H_3C \quad H \quad CH_3 \qquad\qquad H_3C \quad H \quad CH_3$$
$$\textbf{H}-C-C-C-CH_3 \qquad\qquad Br-C-C-C-CH_3$$
$$H_3C \quad H \quad CH_3 \qquad\qquad H_3C \quad H \quad CH_3$$

b.
$$H_3C \quad H \quad CH_3$$
$$H_3C-C-C-C-CH_3$$
$$H_3C \quad H \quad CH_3$$

For this example, no hydrogen atoms are adjacent to carbon–hydrogen bonds. However, there are hydrogen atoms that are adjacent to carbon–carbon bonds. Like carbon–hydrogen bonds, carbon–carbon bonds are also capable of sharing electrons through hyperconjugation. Therefore, homolytic cleavage of one of the two carbon–hydrogen bonds at the center of the molecule (one hydrogen atom shown in boldface) will lead to the formation of the most stable free radical. Free radical bromination will therefore produce the illustrated bromide as the major product.

$$H_3C \quad H \quad CH_3 \qquad\qquad H_3C \quad H \quad CH_3$$
$$H_3C-C-C-C-CH_3 \qquad\qquad H_3C-C-C-C-CH_3$$
$$H_3C \quad \textbf{H} \quad CH_3 \qquad\qquad H_3C \quad Br \quad CH_3$$

c.

For this example, homolytic cleavage of a carbon–hydrogen bond in the allylic position results in the formation of an allyl radical. Allyl free radicals are stabilized by conjugation as illustrated using arrow-pushing. Because delocalization of the free radical results in the formation of two potential bromination sites, one may anticipate that two products will form. However, one of the allylic radicals is further stabilized by hyperconjugation due to the presence of adjacent carbon–hydrogen bonds present on the terminal methyl group. Therefore, the illustrated bromide will be the major product.

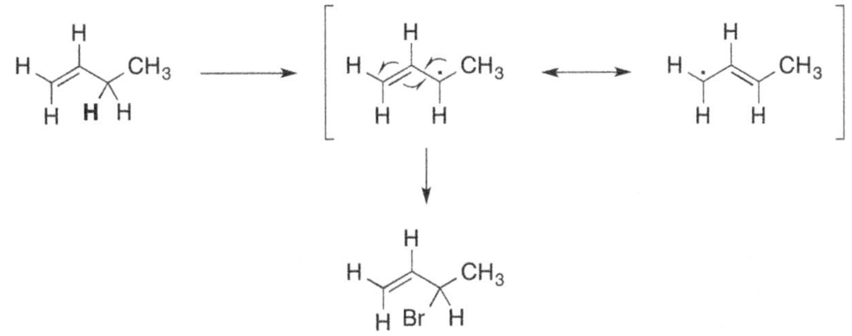

d.

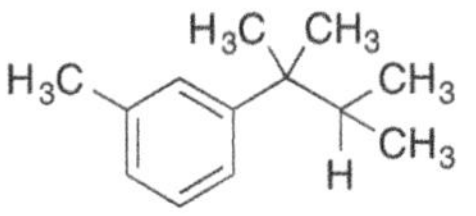

For this example, homolytic cleavage of a carbon–hydrogen bond can lead to the formation of either a free radical stabilized by hyperconjugation or a free radical stabilized by delocalization into the phenyl ring. Because resonance stabilization leads to greater stability, the hydrogen shown in boldface is the most likely site of reaction, and the illustrated bromide will be the major product.

5. *Predict the products resulting from free radical "degradation" of each structure. Show your work using arrow-pushing.*

a.

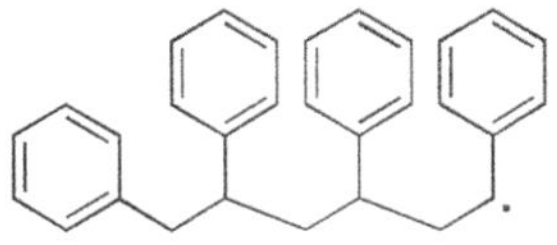

The illustrated free radical starting material is a polystyrene fragment. Due to the relative stability of benzylic free radicals, sequential cleavage of styrene fragments results in the formation of two styrene molecules and a stilbene radical. Loss of a hydrogen atom results in the formation of stilbene.

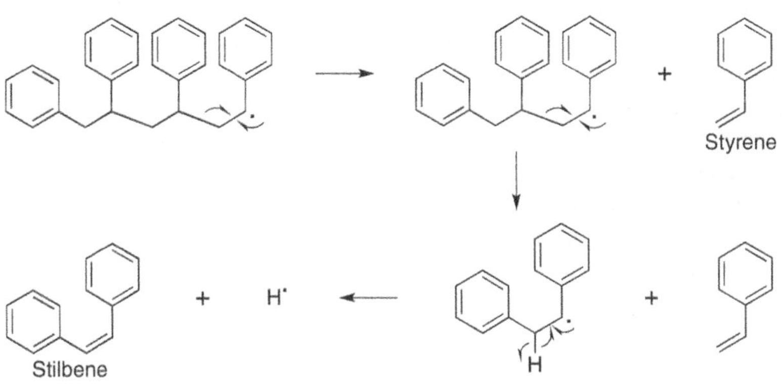

b.

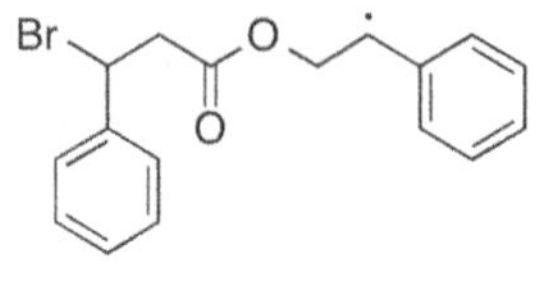

The illustrated free radical starting material is similar to that shown in Problem 5a because of the presence of a benzylic free radical. Using arrow-pushing, one can rationalize the formation of styrene as in the previous problem. On release of styrene, a carboxyl free radical is formed. On homolytic cleavage of the carbonyl–carbon bond, carbon dioxide gas is formed. Final homolytic cleavage of the carbon–bromine bond adjacent to the newly formed free radical results in the formation of a second molecule of styrene and a bromine atom.

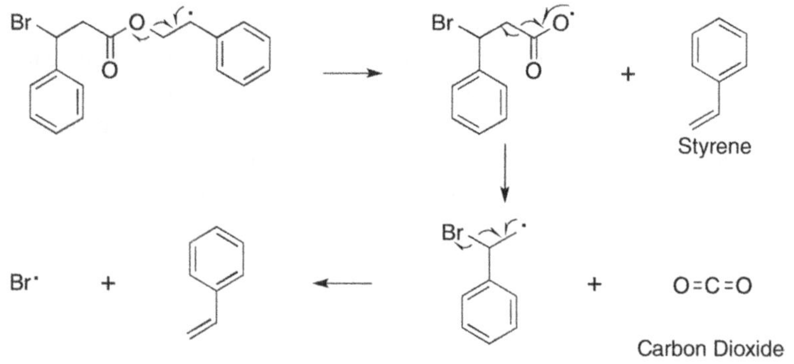

Styrene

Carbon Dioxide

6. *Sodium metal reacts with benzophenone to form a deep blue colored radical anion called a ketyl via an electron transfer process. Predict the structure of sodium benzophenone ketyl.*

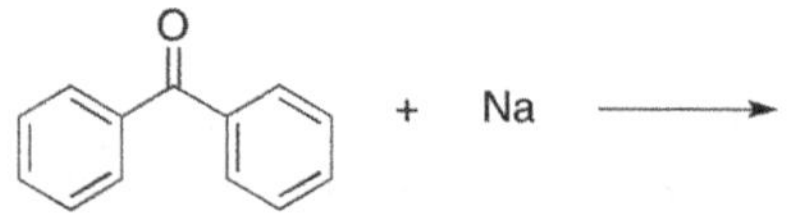

Ketones are comprised of a carbon–oxygen double bond where the oxygen electron octet comprises two lone electron pairs and one electron pair for each bond in the carbon–oxygen double bond. Similarly, the carbon octet comprises two electrons for each bond. Conceptually homolytically splitting one of the bonds from the carbon–oxygen double bond places a radical on the oxygen and another radical on the carbon. On reaction with sodium, a single electron transfers from the sodium metal generating a sodium cation. The recipient of the electron is the oxygen or the carbon of the carbon–oxygen double bond. These structures, shown in the following, are resonance forms of the same radical anion responsible for the sodium benzophenone blue color.

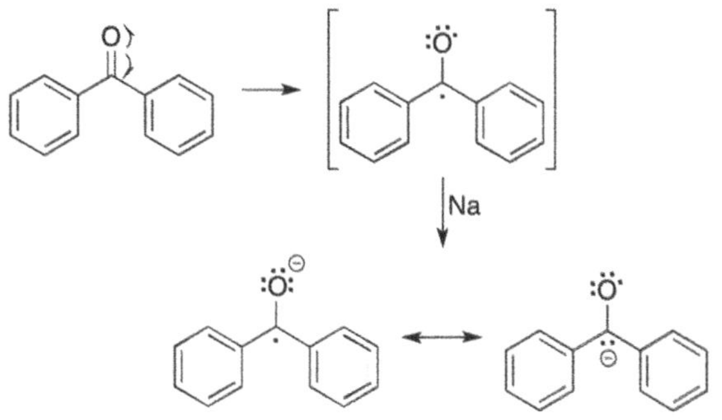

7. *Most free radicals are highly reactive species with very short half-lives. Some free radicals are stable and can be stored for long periods of time without the concern of dimerization or degradation. Explain why the (2,2,6,6-tetramethylpiperidin-1-yl)oxyl radical (TEMPO) is stable.*

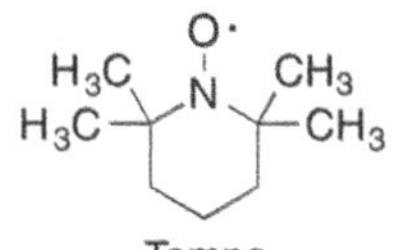

Tempo

TEMPO is stabilized by two factors. The first is understood on recognition that the free radical residing on the oxygen atom is adjacent to a lone electron pair residing on the nitrogen atom leading to stabilization by hyperconjugation. Secondly, while free radicals are prone to dimerization, TEMPO cannot dimerize due to the four methyl groups adjacent to the nitrogen–oxygen bond. Should two TEMPO molecules attempt to join, the steric interactions between methyl groups of each TEMPO molecule would create significant distance from the molecules and prevent reaction.

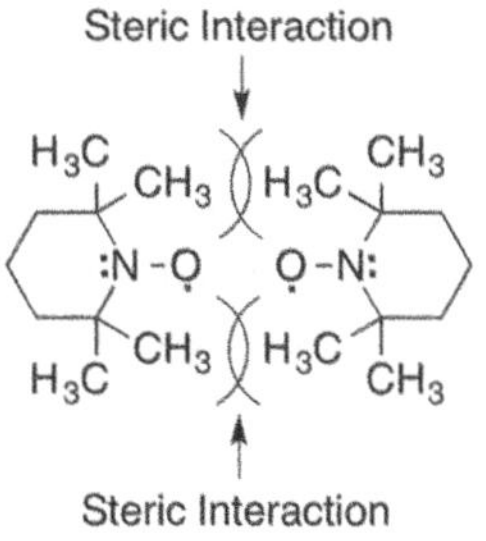

8. *Chlorofluorocarbon-mediated ozone depletion is a significant environmental concern. The sequence begins with the photochemical formation of free radicals in the upper atmosphere and continues with reaction of the free radicals with ozone. Using arrow-pushing, show the mechanisms for each step associated with ozone depletion. Account for all electrons.*

a. $Cl-CCl_2F \longrightarrow \cdot CCl_2F + Cl\cdot$

This step involves the homolytic cleavage of a carbon–chlorine bond.

$$Cl \overset{\frown}{-} CCl_2F \longrightarrow \cdot CCl_2F + Cl\cdot$$

b. $Cl\cdot + O \overset{\oplus}{\underset{\cdot}{:O}} \cdot O^{\ominus} \longrightarrow Cl-O-\overset{\oplus}{O}-\overset{\ominus}{O}$

This step involves the addition of a chlorine atom to the oxygen–oxygen double bond of ozone resulting in an unpaired electron residing on an oxygen atom.

$$Cl\cdot \overset{\curvearrowleft}{\frown} \overset{\frown}{O} \overset{\oplus}{:O} \cdot O^{\ominus} \longrightarrow Cl-O-\overset{\oplus}{O}-\overset{\ominus}{O}$$

c. $Cl-O-\overset{\oplus}{\underset{\cdot}{O}}-\overset{\ominus}{O} \longrightarrow Cl-O\cdot + O{=}O$

Counting electrons, the central oxygen radical cation has only seven electrons. Two electrons are a lone electron pair, four electrons are associated, two each, with oxygen–oxygen bonds. The final electron is a free radical. Furthermore, the anionic oxygen has a full octet of electrons with one bonded electron pair and three non-bonded lone pairs. On homolytic cleavage of the central oxygen–oxygen bond, an electron is removed from the central oxygen leaving only six electrons associated with that atom. Combined with the anionic oxygen, charges are neutralized and oxygen gas is released, leaving behind a highly reactive free radical.

$$Cl-\overset{\curvearrowleft}{O}\overset{\oplus}{\underset{\cdot}{O}}-\overset{\ominus}{O} \longrightarrow Cl-O\cdot + O{=}O$$

d. $Cl-O\cdot + O \overset{\oplus}{\underset{\cdot}{:O}} \cdot O^{\ominus} \longrightarrow Cl\cdot + 2\ O{=}O$

In a fourth stage of ozone depletion, the reactive free radical formed in step (e) reacts with ozone on addition to the oxygen–oxygen double bond. The resulting highly unstable species undergoes homolytic bond cleavage between the chlorine–oxygen bond and the central oxygen–oxygen bond, releasing a chlorine atom and two molecules of oxygen. The free chlorine atom is then able to repeat this process.

$$Cl-O\cdot \overset{\curvearrowleft}{\frown} \overset{\frown}{O} \overset{\oplus}{:O} \cdot O^{\ominus} \longrightarrow Cl\overset{\curvearrowleft}{\frown}O-\overset{\curvearrowleft}{O}\overset{\oplus}{\underset{\cdot}{O}}-\overset{\ominus}{O} \longrightarrow Cl\cdot + 2\ O{=}O$$

CHAPTER 3 SOLUTIONS

1. *Explain how the Henderson–Hasselbalch equation can be used, in conjunction with a titration curve, to determine a pK$_a$.*

   When the progression of an acid–base titration is graphed as a function of pH versus the volume of acid or base added, the curve will appear as shown in the following. If we recall, from general chemistry coursework, that the steepest point on the curve represents the equivalence point of the titration (the point where the amount of acid and base are equal), we can locate the point on the curve that represents the midpoint of the titration. This point is found at half the concentration of base added to acid (or acid added to base) to reach the equivalence point. Once we have done this, we recall the Henderson–Hasselbalch equation (Fig. 2.8)—specifically, the term dealing with the concentrations of the ionic and the neutral species. Realizing that at the midpoint of the titration, these concentrations are equal, the logarithmic term in the Henderson–Hasselbalch equation reduces to $\log(1)$, which is equal to zero. Therefore, the equation reduces to $pK_a = pH$ at the midpoint of the titration.

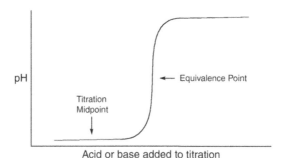

Acid or base added to titration

2. *What is the pH of a solution of acetic acid (pK$_a$ = 4.75) that has been titrated with 1/4 an equivalent of NaOH?*

   When acetic acid is titrated with 1/4 an equivalent of base, we realize that the term $\log\{[A^-]/[HA]\}$ becomes $\log(1/3)$ because one part out of four parts of acetic acid has been deprotonated. This leaves three parts acid to one part conjugate base. Filling in this value and that of the pK_a of acetic acid into the Henderson–Hasselbalch equation (Fig. 2.8), solving for pH gives us a value of 4.27 as our answer.

3. *Draw the resonance structures of the following charged molecules.*

   a.

The following represent the resonance forms of the benzyl cation:

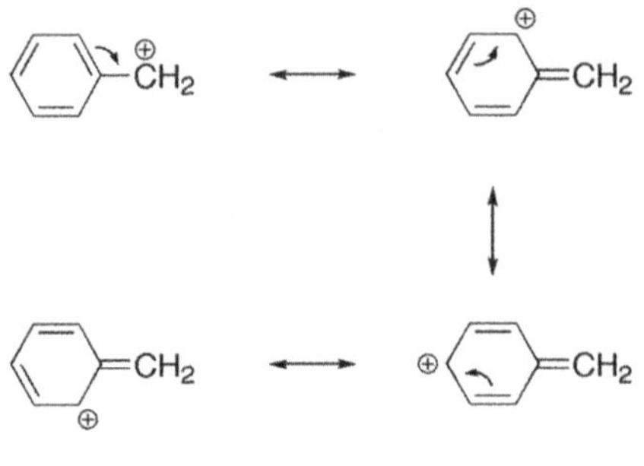

b.

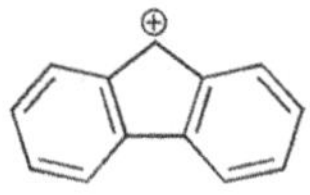

The following represent the resonance forms of the fluorenyl cation:

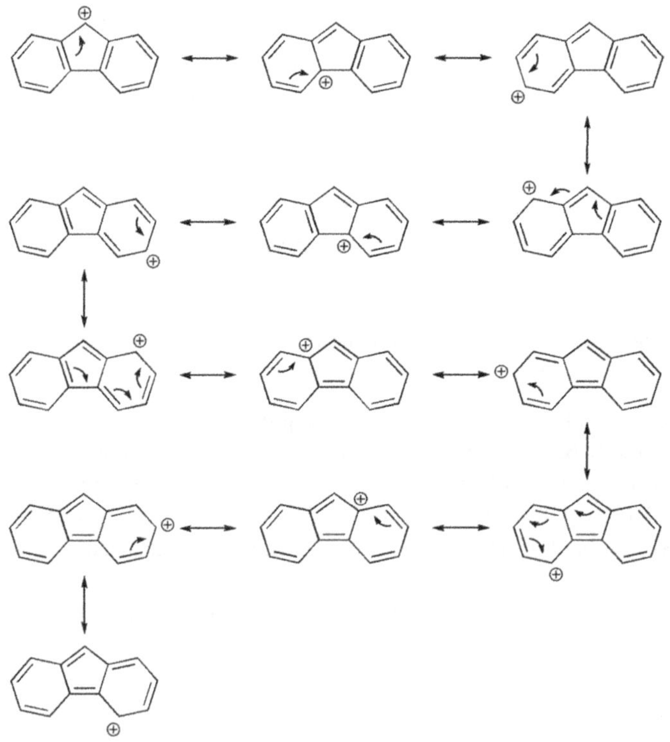

c.

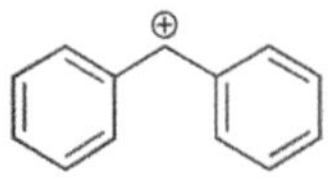

The following represent the resonance forms of the diphenylmethyl cation:

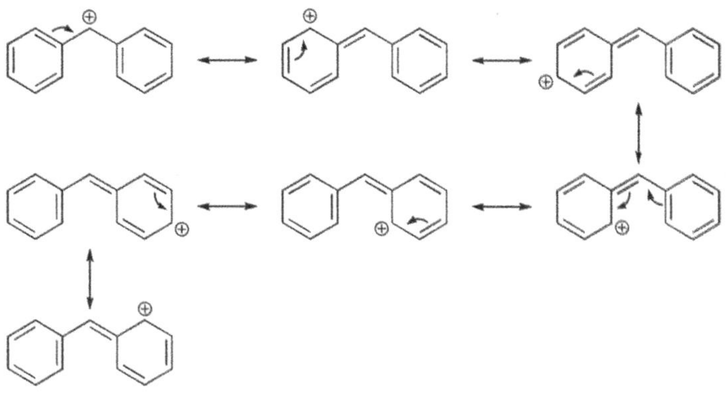

d.

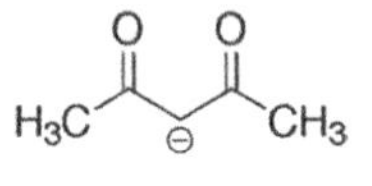

The following represent the resonance forms of the acetylacetone anion:

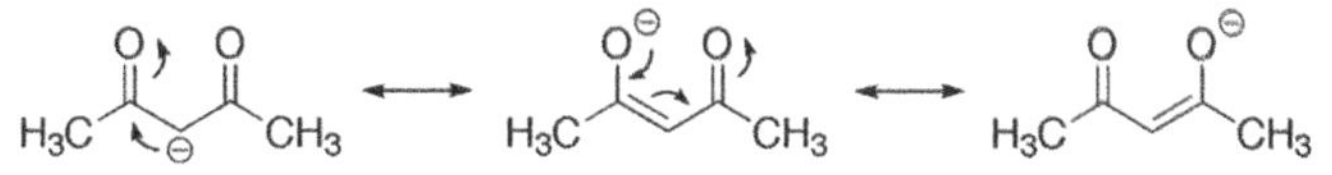

e.

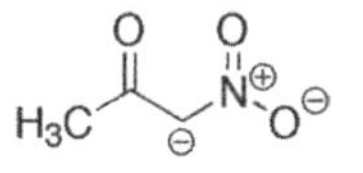

The following represent the resonance forms of the nitroacetone anion:

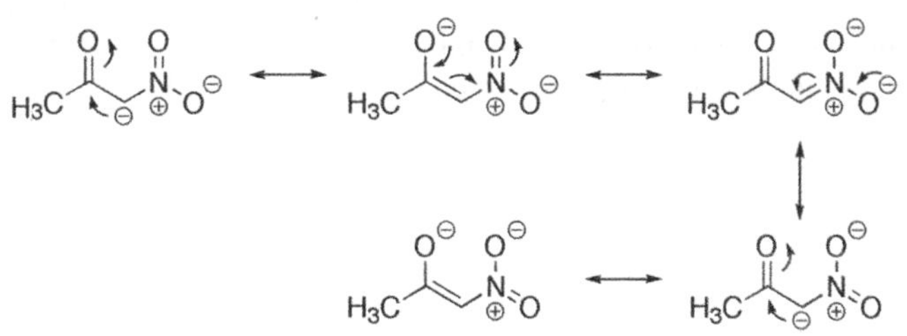

Please note that while nitro groups are so electron withdrawing that delocalization of their associated positive charge plays a minimal role in any family of resonance structures, this delocalization is technically possible. Try to identify additional resonance structures where the positive charge is delocalized.

f.

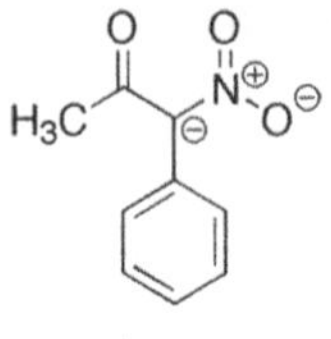

The following represent the resonance forms of the 3-nitroacetophenone anion:

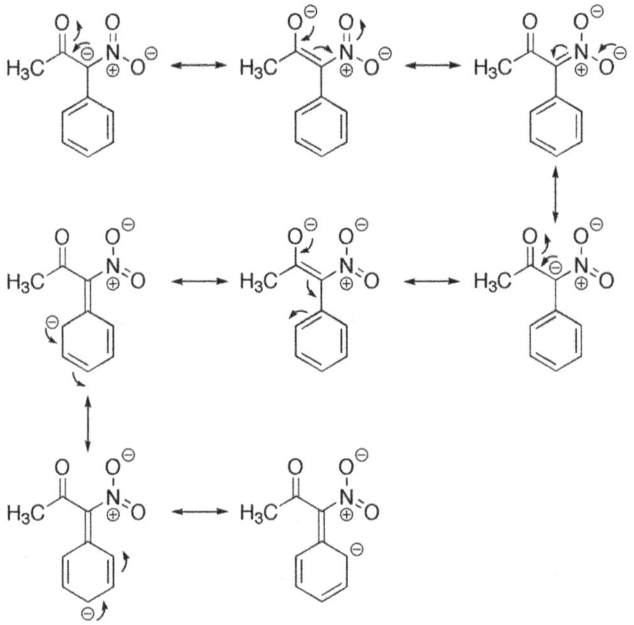

Please note that while nitro groups are so electron withdrawing that delocalization of their associated positive charge plays a minimal role in any family of resonance structures, this delocalization is technically possible. Try to identify additional resonance structures where the positive charge is delocalized.

g.

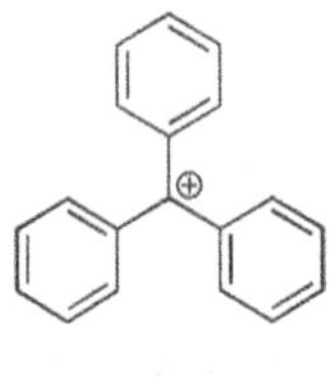

The following represent the resonance forms of the triphenylmethyl cation:

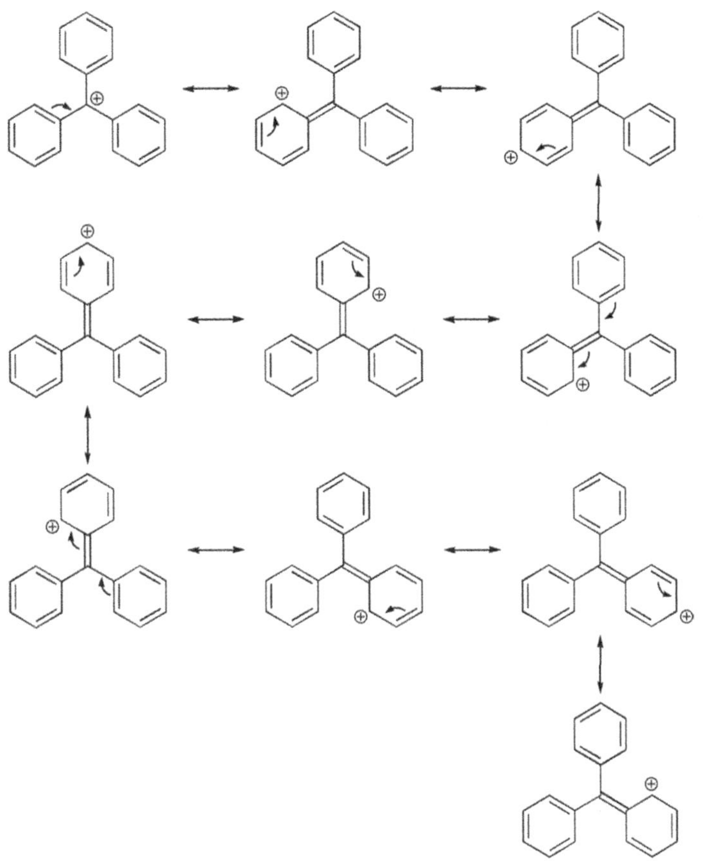

h.

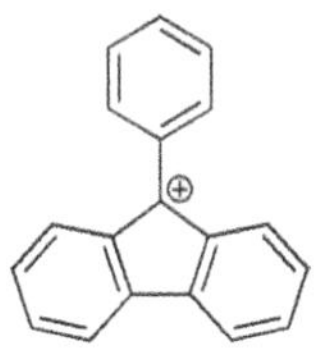

The following represent the resonance forms of the phenylfluorenyl cation:

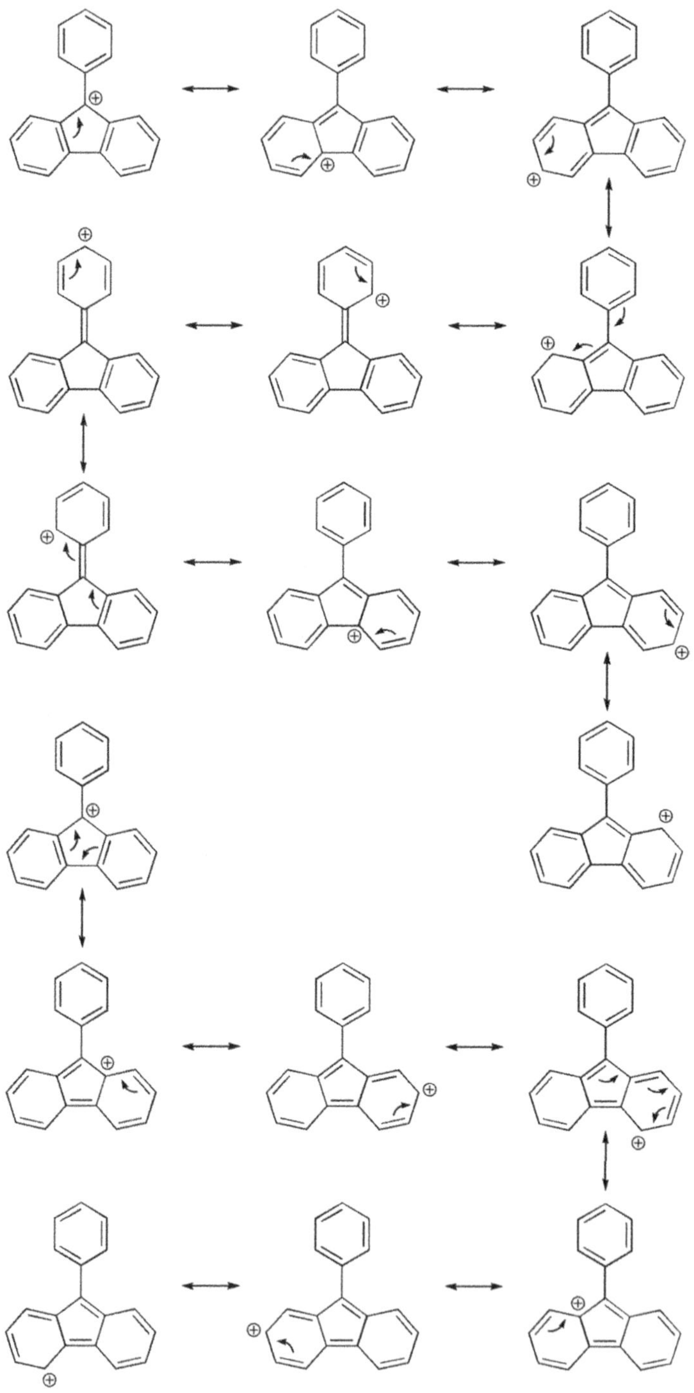

4. *Which cation from Problem 3 is more stable, 3g or 3h? Explain using partial charges.*

Of the 10 resonance forms of the triphenylmethyl cation shown in the solution for Problem 3g, no two resonance forms place a positive charge on adjacent atoms. However, when looking at the 17 resonance forms of the phenylfluorenyl cation shown in the solution for Problem 3h, there are two resonance forms (shown in the following) where the positive charge is placed on adjacent atoms. This is a disfavored electronic relationship and is destabilizing to the cation itself. Thus, through charge distribution and delocalization, because the phenylfluorenyl cation possesses partial positive charges on two adjacent atoms, the triphenylmethyl cation (3g) is more stable.

Note: There is another explanation relating to the definition of aromatic and anti-aromatic ring systems. See if you can explain the answer to this problem using these definitions.

5. *How will the following substituents affect the pK$_a$ of benzoic acid (raise, lower, or no change)? Explain using partial charges to illustrate inductive effects. Remember, o refers to ortho positions, m refers to meta positions, and p refers to the para position.* ***In addressing these problems, assume that the acidity of the carboxylic acid is influenced solely by the partial charges induced by additional ring substituents.***

Note: It is important to realize that in addition to inductive effects, there are other factors that influence acidity and pK$_a$ values. Therefore, while this problem asks for expectations regarding how inductive effects influence pK$_a$ values, in actuality, the measured values may be different than anticipated.

The pK$_a$ of benzoic acid is 4.19

Benzoic Acid

a.   *o-NO$_2$*

The structure of *o*-nitrobenzoic acid is shown in the following with partial charges assigned to the ring system. Because the electron-withdrawing nitro group is located *ortho* to the carboxylic acid, electron density is reduced adjacent to the acid functionality, effectively rendering the aromatic ring electron-withdrawing *ortho* to the nitro group. An electron-withdrawing group attached to a carboxylic acid stabilizes the anion resulting from deprotonation, thus increasing its acidity and **lowering** its pK_a. In actuality, the pK_a of o-nitrobenzoic acid is 2.16, thus supporting the conclusion of this problem.

Electron-Withdrawing Group

b. p-NO$_2$

The structure of p-nitrobenzoic acid is shown in the following with partial charges assigned to the ring system. Because the electron-withdrawing nitro group is located *para* to the carboxylic acid, electron density is reduced adjacent to the acid functionality, effectively rendering the aromatic ring electron-withdrawing *para* to the nitro group. An electron-withdrawing group attached to a carboxylic acid stabilizes the anion resulting from deprotonation, thus increasing its acidity and **lowering** its pK_a. In actuality, the pK_a of p-nitrobenzoic acid is 3.41, thus supporting the conclusion of this problem.

Electron-Withdrawing Group

c. m-NO$_2$

The structure of m-nitrobenzoic acid is shown in the following with partial charges assigned to the ring system. Because the electron-withdrawing nitro group is located *meta* to the carboxylic acid, electron density is increased adjacent to the acid functionality, effectively rendering the aromatic ring electron-donating *meta* to the nitro group. An electron-donating group attached to a carboxylic acid destabilizes the anion resulting from deprotonation, thus decreasing its acidity and **raising** its pK_a. In actuality, the pK_a of m-nitrobenzoic acid is 3.47 reflecting the electron-withdrawing nature of the nitrophenyl group. While this value does not support the conclusion of this problem, the trend, compared to Problems 5a and 5b, indicates that the m-NO$_2$ has less of an effect on acidity than o-NO$_2$ and p-NO$_2$.

Electron-Withdrawing Group
(Regardless of Partial Charges)

d. *p-OH*

The structure of *p*-hydroxybenzoic acid (salicylic acid) is shown in the following with partial charges assigned to the ring system. Because the electron-donating hydroxyl group is located *ortho* to the carboxylic acid, electron density is increased adjacent to the acid functionality, effectively rendering the aromatic ring electron-donating *para* to the hydroxyl group. An electron-donating group attached to a carboxylic acid destabilizes the anion resulting from deprotonation, thus decreasing its acidity and **raising** its pK_a. In actuality, the pK_a of *p*-hydroxybenzoic acid is 4.48, thus supporting the conclusion of this problem.

Electron-Donating Group

e. *m-OH*

The structure of *m*-hydroxybenzoic acid is shown in the following with partial charges assigned to the ring system. Because the electron-donating hydroxyl group is located *meta* to the carboxylic acid, electron density is decreased adjacent to the acid functionality, effectively rendering the aromatic ring electron-withdrawing *meta* to the hydroxyl group. An electron-withdrawing group attached to a carboxylic acid stabilizes the anion resulting from deprotonation, thus increasing its acidity and **lowering** its pK_a. In actuality, the pK_a of *m*-hydroxybenzoic acid is 4.06, thus supporting the conclusion of this problem.

Electron-Withdrawing Group

f. *p-NH₂*

The structure of *p*-aminobenzoic acid is shown in the following with partial charges assigned to the ring system. Because the electron-donating amino group is located *para* to the carboxylic acid, electron density is increased adjacent to the acid functionality, effectively rendering the aromatic ring electron-donating *para* to the amino group. An electron-donating group attached to a carboxylic acid destabilizes the anion resulting from deprotonation, thus decreasing its acidity and **raising** its pK_a. In actuality, the pK_a of *p*-aminobenzoic acid is 4.65, thus supporting the conclusion of this problem.

Electron-Withdrawing Group

g.   *m-CH$_3$*

The structure of *m*-methylbenzoic acid is shown in the following with partial charges assigned to the ring system. Because the electron-donating methyl group is located *meta* to the carboxylic acid, electron density is decreased adjacent to the acid functionality, effectively rendering the aromatic ring electron-withdrawing *meta* to the methyl group. An electron-withdrawing group attached to a carboxylic acid stabilizes the anion resulting from deprotonation, thus increasing its acidity and **lowering** its pK_a. In actuality, the pK_a of *m*-methylbenzoic acid is 4.27.

Electron-Donating Group
(Regardless of Partial Charges)

h.   *p-CH$_3$*

The structure of *p*-methylbenzoic acid is shown in the following with partial charges assigned to the ring system. Because the electron-donating methyl group is located *para* to the carboxylic acid, electron density is increased adjacent to the acid functionality, effectively rendering the aromatic ring electron-donating *para* to the methyl group. An electron-donating group attached to a carboxylic acid destabilizes the anion resulting from deprotonation, thus decreasing its acidity and **raising** its pK_a. In actuality, the pK_a of *p*-methylbenzoic acid is 4.36, thus supporting the conclusion of this problem.

Electron-Donating Group

i.   *m-CHO*

The structure of *m*-carboxybenzaldehyde (or m-formylbenzoic acid) is shown in the following with partial charges assigned to the ring system. Because the electron-withdrawing aldehyde group is located *meta* to the carboxylic acid, electron density is increased adjacent to the acid functionality, effectively rendering the aromatic ring electron-donating *meta* to the formyl (aldehyde) group. An electron-donating group attached to a carboxylic acid destabilizes the anion resulting from deprotonation, thus decreasing its acidity and **raising** its pK_a. In actuality, the pK_a of *m*-formylbenzoic acid is 3.85, reflecting the electron-withdrawing nature of the formylphenyl group.

Electron-Withdrawing Group
(Regardless of Partial Charges)

j. *p-OCH₃*

The structure of *p*-methoxybenzoic acid is shown in the following with partial charges assigned to the ring system. Because the electron-donating methoxy group is located *para* to the carboxylic acid, electron density is increased adjacent to the acid functionality, effectively rendering the aromatic ring electron-donating *para* to the methoxy group. An electron-donating group attached to a carboxylic acid destabilizes the anion resulting from deprotonation, thus decreasing its acidity and **raising** its pK_a. In actuality, the pK_a of *p*-methoxybenzoic acid is 4.47, thus supporting the conclusion of this problem.

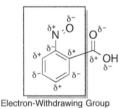

Electron-Donating Group

k. *o-NO*

The structure of *o*-nitrosobenzoic acid is shown in the following with partial charges assigned to the ring system. Because the electron-withdrawing nitroso group is located *ortho* to the carboxylic acid, electron density is reduced adjacent to the acid functionality, effectively rendering the aromatic ring electron-withdrawing *ortho* to the nitroso group. An electron-withdrawing group attached to a carboxylic acid stabilizes the anion resulting from deprotonation, thus increasing its acidity and **lowering** its pK_a. In actuality, the pK_a of *o*-nitrosobenzoic acid is less than 4, thus supporting the conclusion of this problem.

Electron-Withdrawing Group

l. *p-Cl*

In considering this problem, it is important to recognize that chlorides are electronegative and can thus attract electron density. However, chlorides have three lone electron pairs that can donate electron density into the phenyl ring. Therefore, chlorides (and halides in general) have both electron-donating and electron-withdrawing properties.

The structure of *p*-chlorobenzoic acid is shown in the following with partial charges assigned to the ring system. Acting as an electron-donating group, the chloride located *para* to the carboxylic acid increases electron density adjacent to the acid functionality, effectively rendering the aromatic ring electron-donating *para* to the chloro group. An electron-donating group attached to a carboxylic acid destabilizes the anion resulting from deprotonation, thus decreasing its acidity and **raising** its pK_a.

In actuality, the pK_a of *p*-chlorobenzoic acid is 3.98, reflecting the electron-withdrawing nature of the chlorophenyl group.

Electron-Withdrawing Group
(Regardless of Partial Charges)

m. *m-Cl*

In considering this problem, it is important to recognize that chlorides are electronegative and can thus attract electron density. However, chlorides have three lone electron pairs that can donate electron density into the phenyl ring. Therefore, chlorides (and halides in general) have both electron-donating and electron-withdrawing properties.

The structure of *m*-chlorobenzoic acid is shown in the following with partial charges assigned to the ring system. Because the electron-donating chloro group is located *meta* to the carboxylic acid, electron density is decreased adjacent to the acid functionality, effectively rendering the aromatic ring electron-withdrawing *meta* to the chloro group. An electron-withdrawing group attached to a carboxylic acid stabilizes the anion resulting from deprotonation, thus increasing its acidity and **lowering** its pK_a. In actuality, the pK_a of *m*-chlorobenzoic acid is 3.82, thus supporting the conclusion of this problem.

Electron-Withdrawing Group

6. *Arrange the following groups of molecules in order of increasing acidity. Explain your results using partial charges and inductive effects.*

Initially, when considering inductive effects, we realize that F, O, and Cl all possess partial negative charges. Therefore, we realize that all of these atoms will pull electron density from the carboxylic acid, thus stabilizing the anion resulting from deprotonation and lowering the pK_a values compared to the baseline acetic acid. The question now focuses on how strong this effect is for each atom. The answer is found in the periodic table of elements and relates to electronegativities. Of the three atoms in question, F is the most electronegative. Moving to the second row, Cl is more electronegative than O. Since the most acidic compound will have the most electronegative atoms associated with it, the order of increasing acidity is as follows:

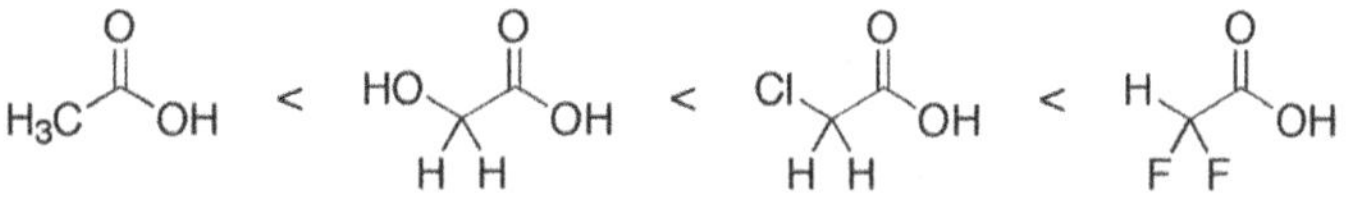

7. *Predict pK$_a$ values for the highlighted protons in following molecules. Rationalize your answers.*

When estimating the pK_a values for protons adjacent to multiple functional groups, the pK_a values can be calculated according to the following formula where n is defined as the number of relevant functional groups.

$$pK_a = \dfrac{\left\{ \dfrac{pK_a^1}{n} + \dfrac{pK_a^2}{n} + \cdots + \dfrac{pK_a^n}{n} \right\}}{n}$$

a.

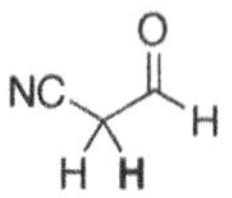

According to Appendix 1, the pK_a value for a proton adjacent to a nitrile is approximately 20–25 as is the pK_a value for a proton adjacent to an aldehyde. Recognizing that there are two relevant functional groups (an aldehyde and a nitrile), the aforementioned formula gives us a pK_a value of approximately 10–12.5.

b.

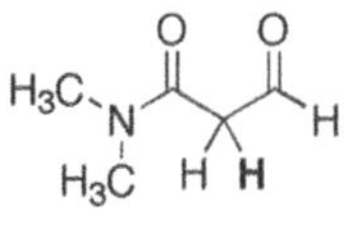

According to Appendix 1, the pK_a value for a proton adjacent to an amide is approximately 30–35, and the pK_a value for a proton adjacent to an aldehyde is approximately 20–25. Recognizing that there are two relevant functional groups (an aldehyde and an amide), the aforementioned formula gives us a pK_a value of approximately 12.5–15.

c.

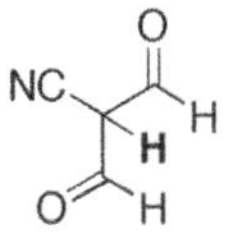

According to Appendix 1, the pK_a value for a proton adjacent to a nitrile is approximately 20–25 as is the pK_a value for a proton adjacent to an aldehyde. Recognizing that there are three relevant functional groups (two aldehydes and a nitrile), the aforementioned formula gives us a pK_a value of approximately 6.7–8.3.

d.

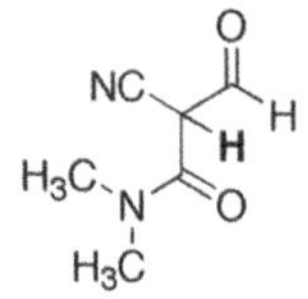

According to Appendix 1, the pK_a value for a proton adjacent to an amide is approximately 30–35, the pK_a value for a proton adjacent to an aldehyde is approximately 20–25, and the pK_a value for a proton adjacent to a nitrile is approximately 20–25. Recognizing that there are three relevant functional groups (a nitrile, an aldehyde, and an amide), the aforementioned formula gives us a pK_a value of approximately 7.8–9.4.

8. *Predict the order of deprotonation of the various protons on the following molecules. Back up your answers with appropriate* pK_a *values.*

a.

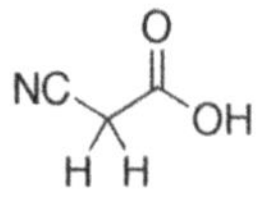

According to Appendix 1, the pK_a value for a carboxylic acid is approximately 4.75. Furthermore, if we imagine converting the carboxylic acid to an ester, we recognize that protons adjacent to esters have pK_a values of approximately 25–30. Finally, the pK_a value for a proton adjacent to a nitrile is approximately 20–25. Using the formula described in Problem 7, we calculate a pK_a value of approximately 11.25–13.75 for the protons between the two functional groups. Therefore, the order of deprotonation is as follows:

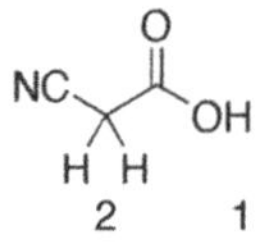

b.

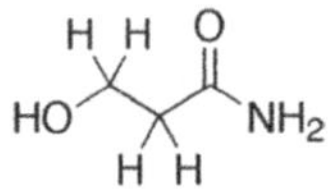

According to Appendix 1, the pK_a value for an amide is approximately 15–17, the pK_a of a primary alcohol is approximately 15–16, and the pK_a of a proton adjacent to an amide is approximately 30–35. Therefore, the order of deprotonation is as follows:

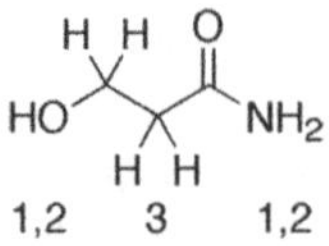

c.

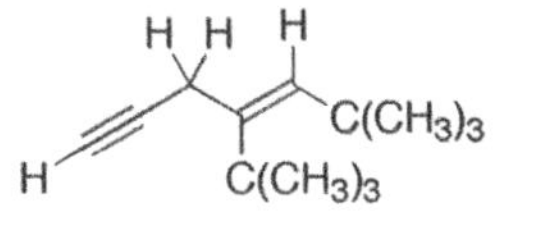

According to Appendix 1, the pK_a value for an acetylene is approximately 25, and the pK_a of a vinyl proton is approximately 35–40. While the acetylene and olefin lend delocalization effects to adjacent anions, the absence of heteroatoms incorporated in these functional groups minimizes this effect, and the pK_a at this position will resemble something between a vinyl pK_a and a hydrocarbon pK_a. Therefore, the order of deprotonation is as follows:

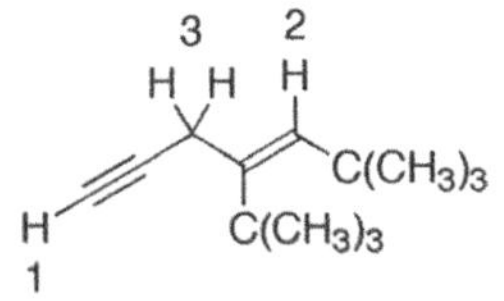

d.

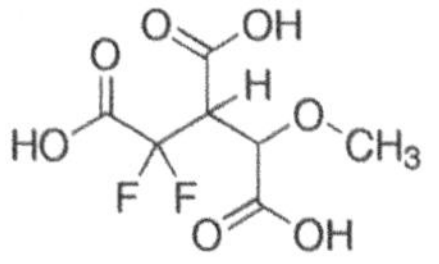

This problem relies entirely on inductive effects. Realizing that fluorine is more electronegative than oxygen, the order of deprotonation is as follows:

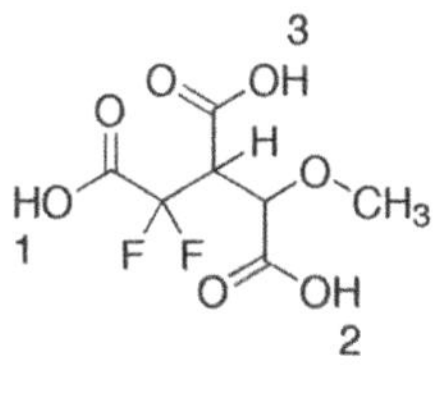

9. *Which proton is the most acidic? Rationalize your answer.*

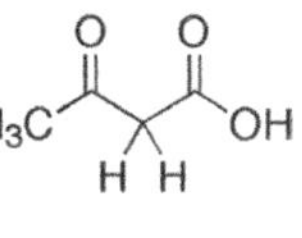

The pK_a value of a proton adjacent to a ketone carbonyl is approximately 20–25. The pK_a value of a carboxylic acid is approximately 4.75. Using the same calculations presented in the solution for Problem 8a, the pK_a value of the protons between the ketone and the carboxylic acid is approximately 11.25–13.75. Since the acidity of a proton increases as its pK_a value decreases, the most acidic proton belongs to the carboxylic acid.

10. *Using the pK$_a$ values given in Appendix 1, calculate the equilibrium constants for the following reactions.*

Recall from general chemistry that the equilibrium constant, K_{eq}, for a given reaction is defined as follows:

$$A \; + \; B \; \rightleftharpoons \; C \; + \; D$$

$$K_{eq} \; = \; \frac{[C][D]}{[A][B]}$$

Also, recall that the definition of the acid dissociation constant, K_a, is as follows:

$$AH \; \rightleftharpoons \; A^- \; + \; H^+$$

$$K_a \; = \; \frac{[A^-][H^+]}{[AH]}$$

Finally, we recognize, as shown, that an acid/base equilibrium consists of two related reactions for which K_a values can be calculated and that at equilibrium, the [H$^+$] is equivalent for each equation.

$$AH \; + \; B^- \; \rightleftharpoons \; A^- \; + \; BH$$

$$AH \; \rightleftharpoons \; A^- \; + \; H^+$$

$$H^+ \; + \; B^- \; \rightleftharpoons \; BH$$

Therefore, with K_a^1 and K_a^2 defined, K_{eq} is derived as shown.

$$K_{eq} \; = \; \frac{K_a^1}{K_a^2} \; = \; \frac{\dfrac{[A^-][H^+]}{[AH]}}{\dfrac{[B^-][H^+]}{[BH]}} \; = \; \frac{[A^-][H^+][BH]}{[B^-][H^+][AH]} \; = \; \frac{[A^-][BH]}{[B^-][AH]}$$

and K_{eq} is simply the ratio of the two relevant pK$_a$ values.

a.

From Appendix 1, we know that pK$_a^1$, the dissociation constant associated with protons adjacent to ketone carbonyls, is approximately 20–25. Furthermore, from Appendix 1, we know that pK$_a^2$, the dissociation constant associated with protonated amines, is approximately 10. Finally, remembering that pK$_a$ = $-\log K_a$, the K_{eq} for this reaction ranges from

$$\frac{10^{-20}}{10^{-10}} \;\; \text{to} \;\; \frac{10^{-25}}{10^{-10}} \;\; \text{or} \;\; 10^{-10} \text{ to } 10^{-15}$$

b.

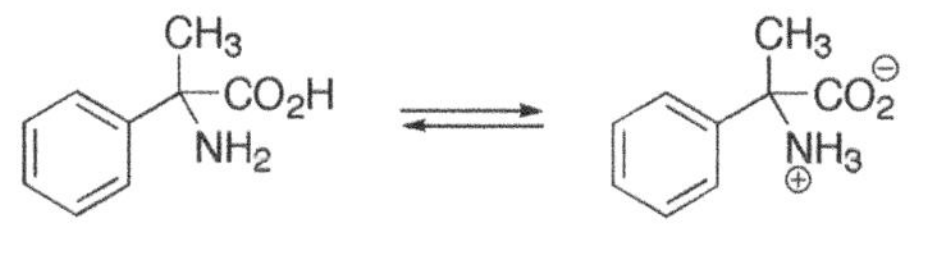

From Appendix 1, we know that pK_a^1, the dissociation constant associated with carboxylic acid protons, is approximately 4.75. Furthermore, from Appendix 1, we know that pK_a^2, the dissociation constant associated with protonated amines, is approximately 10. Finally, remembering that $pK_a = -\log K_a$, the K_{eq} for this reaction is approximately

$$\frac{10^{-4.75}}{10^{-10}} \text{ or } 10^{5.25}$$

c.   HCl + Br$^{\ominus}$ $\rightleftharpoons$ HBr + Cl$^{\ominus}$

From Appendix 1, we know that pK_a^1, the dissociation constant associated with hydrochloric acid, is approximately −2.2. Furthermore, from Appendix 1, we know that pK_a^2, the dissociation constant associated with hydrobromic acid, is approximately −4.7. Finally, remembering that $pK_a = -\log K_a$, the K_{eq} for this reaction is approximately

$$\frac{10^{-2.2}}{10^{-4.7}} \text{ or } 10^{-2.5}$$

d.

From Appendix 1, we know that pK_a^1, the dissociation constant associated with protonated amines, is approximately 10. Furthermore, from Appendix 1, we know that pK_a^2, the dissociation constant associated with protonated ketones, is approximately −7 to −9. Finally, remembering that $pK_a = -\log K_a$, the K_{eq} for this reaction ranges from

$$\frac{10^{-10}}{10^7} \text{ to } \frac{10^{-10}}{10^9} \text{ or } 10^{-17} \text{ to } 10^{-19}$$

CHAPTER 4 SOLUTIONS

1. *In each case, circle the better nucleophile. Explain your answers.*

a.

Oxygen is more electronegative than nitrogen. Therefore, the lone pair on nitrogen is not held as tightly as the lone pairs of oxygen. This greater availability of the nitrogen lone pair compared with the oxygen lone pairs makes the amine the better nucleophile.

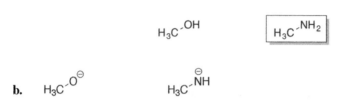

b. $H_3C-O^{\ominus}$ $H_3C-\overset{\ominus}{NH}$

Oxygen is more electronegative than nitrogen. This difference in electronegativity is reflected in the greater acidity of alcohols compared with amines. As oxygen lone pairs are held more tightly than the nitrogen lone pair, negative charges on oxygen are more stable than negative charges in nitrogen. Thus, the nitrogen anion is more available to react making it the better nucleophile.

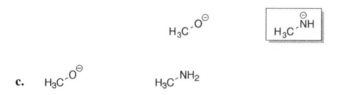

c. $H_3C-O^{\ominus}$ H_3C-NH_2

While in equivalent states, nitrogen functionalities are better nucleophiles than oxygen nucleophiles (see Problems 1a and 1b), when comparing different electronic states, the more reactive species will be the better nucleophile. Thus, the oxygen anion is a better nucleophile compared with an amine.

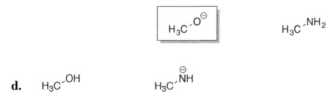

d. H_3C-OH $H_3C-\overset{\ominus}{NH}$

For all the reasons discussed under Problems 1a, 1b, and 1c, the nitrogen anion is the better nucleophile.

H_3C-OH $\boxed{H_3C-\overset{\ominus}{NH}}$

e.   Cl$^{\ominus}$ I$^{\ominus}$

The answer to this question depends on the solvent used for reaction as illustrated in Figure 3.4. Also relevant is recognition that chloride anions are hard bases and iodide anions are soft bases. Iodide is the better nucleophile in polar protic solvents, while chloride is the better nucleophile in polar aprotic solvents.

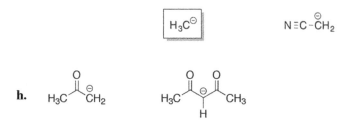

f.   N≡C$^{\ominus}$ HC≡C$^{\ominus}$

As shown in Appendix 1, the pK_a for hydrocyanic acid is approximately 9.3 and the pK_a for acetylene is approximately 25. Thus, the acetylene anion is more reactive than the cyanide anion and is therefore the better nucleophile.

N≡C$^{\ominus}$ $\boxed{\text{HC}≡\text{C}^{\ominus}}$

g.   H$_3$C$^{\ominus}$ N≡C-$\overset{\ominus}{\text{C}}H_2$

As shown in Appendix 1, the pK_a for methane is approximately 50–75 and the pK_a for acetonitrile is approximately 25. Thus, the methyl anion is more reactive than the acetonitrile anion and is therefore the better nucleophile.

$\boxed{\text{H}_3\text{C}^{\ominus}}$ N≡C-$\overset{\ominus}{\text{C}}H_2$

h.   H$_3$C—C(=O)—$\overset{\ominus}{\text{C}}H_2$ H$_3$C—C(=O)—$\overset{\ominus}{\underset{\text{H}}{\text{C}}}$—C(=O)—CH$_3$

As shown in Appendix 1, the pK_a for acetone is approximately 20. Furthermore, the pK_a for acetylacetone is approximately 10 as estimated using the formula presented in Chapter 2, Problem 7. Thus, the acetone anion is more reactive than the acetylacetone anion and is therefore the better nucleophile.

$\boxed{\text{H}_3\text{C}—\text{C}(=O)—\overset{\ominus}{\text{C}}\text{H}_2}$ H$_3$C—C(=O)—$\overset{\ominus}{\underset{\text{H}}{\text{C}}}$—C(=O)—CH$_3$

2. *Nucleophiles often participate in nucleophilic substitution reactions. The general form of these reactions may be represented by the following equation where Nu$_1^-$ and Nu$_2^-$ are nucleophiles:*

$$Nu_1^{\ominus} + \underset{R}{\overset{R}{\underset{R'}{\diagdown}}}\!\!-Nu_2 \longrightarrow Nu_1\!-\!\!\underset{R}{\overset{R}{\underset{R}{\diagdown}}}\!\!R + Nu_2^{\ominus}$$

a. *Explain what type of relationship between Nu_1^- and Nu_2^- is necessary in order for this reaction to be favored.*

In order for this reaction to proceed, Nu_1^- must be a better nucleophile than Nu_2^-.

b. *What does this say about the relative basicities of Nu_1^- and Nu_2^-?*

In general the stronger the nucleophile is, the stronger the base. Therefore, Nu_1^- is more basic than Nu_2^-.

c. *Which nucleophile has the larger pK_a?*

Remembering that a strong base is derived from a weak conjugate acid, if we consider the conjugate acids of Nu_1^- and Nu_2^-, we expect that since Nu_1^- is more basic, its conjugate acid has the larger pK_a than the conjugate acid of Nu_2^-.

d. *What generalization can be concluded about the relationship between bases and nucleophiles?*

Since nucleophiles, by definition, are species attracted to positive charges and since, by definition, protons are positively charged, nucleophiles are bases. The extent of nucleophilicity associated with a given nucleophile largely depends on the degree of its basicity. Thus, in general terms, the more nucleophilic a given nucleophile is, the more basic it is.

3. *How can pK_a values be used to describe basicity?*

By definition, pK_a values relate to the degree of acidity associated with a given acid. Referring to the Henderson–Hasselbalch equation, as acidity increases, pK_a values decrease. Conversely, as acidity decreases, pK_a values increase. Referring to the definition of a base, we realize that as acidity increases, basicity decreases. Conversely, as acidity decreases, basicity increases. Recognizing that as acidity decreases, pK_a values increase, we recognize that as pK_a values increase, basicity increases. Therefore, the higher the pK_a value, the greater the basicity, and the lower the pK_a value, the lower the basicity.

4. *As electron-donating and electron-withdrawing substituents will affect the acidity of organic molecules, so will they affect the basicity. How will the following substituents affect (raise, lower, or no change) the pK_a of aniline (aminobenzene)? Explain using partial charges to illustrate inductive effects. Remember, o refers to ortho positions, m refers to meta positions, and p refers to the para position. **In addressing these problems, assume that the acidity of the amine is influenced solely by the partial charges induced by additional ring substituents.***

Note: It is important to realize that in addition to inductive effects, there are other factors that influence acidity and pK_a values. Therefore, while this problem asks for expectations regarding how inductive effects influence pK_a values, in actuality, the measured values may be different than anticipated.

The pK_a of aniline is 4.63

Aniline

a. *o-NO$_2$*

The structure of *o*-nitroaniline is shown in the following with partial charges assigned to the ring system. Because the electron-withdrawing nitro group is located *ortho* to the amine, electron density is reduced adjacent to the amine functionality, effectively rendering the aromatic ring electron-withdrawing *ortho* to the nitro group. An electron-withdrawing group attached to an amine stabilizes the anion resulting in deprotonation, thus increasing its acidity and **lowering** its pK_a. In actuality, the pK_a of *o*-nitroaniline is −0.26, thus supporting the conclusion of this problem.

Electron-Withdrawing Group

b. *p-NO$_2$*

The structure of *p*-nitroaniline is shown in the following with partial charges assigned to the ring system. Because the electron-withdrawing nitro group is located *para* to the amine, electron density is reduced adjacent to the amine functionality, effectively rendering the aromatic ring electron-withdrawing *para* to the nitro group. An electron-withdrawing group attached to an amine stabilizes the anion resulting in deprotonation, thus increasing its acidity and **lowering** its pK_a. In actuality, the pK_a of *p*-nitroaniline is 1.0, thus supporting the conclusion of this problem.

Electron-Withdrawing Group

c. *m-NO$_2$*

The structure of *m*-nitroaniline is shown in the following with partial charges assigned to the ring system. Because the electron-withdrawing nitro group is located *meta* to the amine, electron density is increased adjacent to the amine functionality, effectively rendering the aromatic ring electron-donating *meta* to the nitro group. An electron-donating group attached to an amine destabilizes the anion resulting in deprotonation, thus decreasing its acidity and **raising** its pK_a. In actuality, the pK_a of *m*-nitroaniline is 2.47 reflecting the electron-withdrawing nature of the nitrophenyl group. Comparing this observation to the solutions for Problems 4a and 4b, the overall electron-withdrawing nature of the nitro group generally overrides any influence of partial charge distribution—regardless of positioning on the phenyl ring.

Electron-Withdrawing Group
(Regardless of Partial Charges)

d. *p-NH$_2$*

The structure of *p*-aminoaniline is shown in the following with partial charges assigned to the ring system. Because the electron-donating amino group is located *para* to the amine, electron density is increased adjacent to the amine functionality, effectively rendering the aromatic ring electron-donating *para* to the amino group. An electron-donating group attached to an amine destabilizes the anion resulting in deprotonation, thus decreasing its acidity and **raising** its pK_a. In actuality, the pK_a of *p*-phenylenediamine is 6.2, thus supporting the conclusion of this problem.

Electron-Donating Group

e. *m-CH$_3$*

The structure of *m*-methylaniline is shown in the following with partial charges assigned to the ring system. Because the electron-donating methyl group is located *meta* to the amine, electron density is decreased adjacent to the amine functionality, effectively rendering the aromatic ring electron-withdrawing *meta* to the methyl group. An electron-withdrawing group attached to an amine stabilizes the anion resulting in deprotonation, thus increasing its acidity and **lowering** its pK_a. In actuality, the pK_a of *m*-methylaniline is 4.73.

Electron-Donating Group
(Regardless of Partial Charges)

f. *p-CH$_3$*

The structure of *p*-methylaniline is shown in the following with partial charges assigned to the ring system. Because the electron-donating methyl group is located *para* to the amine, electron density is increased adjacent to the amine functionality, effectively rendering the aromatic ring electron-donating *para* to the methyl group. An electron-donating group attached to an amine destabilizes the anion resulting in deprotonation, thus decreasing its acidity and **raising** its pK_a. In actuality, the pK_a of *p*-methylaniline is 5.08, thus supporting the conclusion of this problem.

Electron-Donating Group

g. *p-OCH₃*

The structure of *p*-methoxyaniline is shown in the following with partial charges assigned to the ring system. Because the electron-donating methoxy group is located *para* to the amine, electron density is increased adjacent to the amine functionality, effectively rendering the aromatic ring electron-donating *para* to the methoxy group. An electron-donating group attached to an amine destabilizes the anion resulting in deprotonation, thus decreasing its acidity and **raising** its pK_a. In actuality, the pK_a of *p*-methoxyaniline is 5.34, thus supporting the conclusion of this problem.

Electron-Donating Group

h. *p-Cl*

The structure of *p*-chloroaniline is shown in the following with partial charges assigned to the ring system. Because the electron-donating chloro group is located *para* to the amine, electron density is increased adjacent to the amine functionality, effectively rendering the aromatic ring electron-donating *para* to the chloro group. An electron-donating group attached to an amine destabilizes the anion resulting in deprotonation, thus decreasing its acidity and **raising** its pK_a. In actuality, the pK_a of *p*-chloroaniline is 4.15, reflecting the electron-withdrawing nature of the chlorophenyl group.

Electron-Withdrawing Group
(Regardless of Partial Charges)

i. *m-Cl*

The structure of *m*-chloroaniline is shown in the following with partial charges assigned to the ring system. Because the electron-donating chloro group is located *meta* to the amine, electron density is decreased adjacent to the amine functionality, effectively rendering the aromatic ring electron-withdrawing *meta* to the chloro group. An electron-withdrawing group attached to an amine stabilizes the anion resulting in deprotonation, thus increasing its acidity and **lowering** its pK_a. In actuality, the pK_a of *m*-chloroaniline is 3.46, thus supporting the conclusion of this problem.

Electron-Withdrawing Group

5. *Arrange the following groups of molecules in order of increasing basicity. Explain your results using partial charges and inductive effects.*

Initially, when considering inductive effects, we realize that F, O, and Cl all possess partial negative charges. Therefore, we realize that all of these atoms will pull electron density from the carboxylic acid, thus stabilizing the anion resulting in deprotonation and lowering the pK_a values compared to the baseline acetic acid. The question now focuses on how strong this effect is for each atom. The answer is found in the periodic table of elements and relates to electronegativities. Of the three atoms in question, F is the most electronegative. Moving to the second row, Cl is more electronegative than O. Since the most basic compound will have the least electronegative atoms associated with it, the order of increasing basicity is as follows:

Note that this is the opposite sequence as that presented in Chapter 2, Problem 6.

6. *Predict the order of protonation of the basic sites on the following molecules. Back up your answers with* pK_a*s.*

In addressing this problem, it is important to recognize that the order of protonation depends on the basicity associated with the respective functional groups. As discussed earlier, basicity can be relayed back to the pK_a values associated with the conjugate acids of the respective sites of protonation. Thus conjugate acids with the higher pK_a values will be protonated first, while conjugate acids with lower pK_a values will be protonated last.

a.

According to Appendix 1, the pK_a value for an acetylene proton is approximately 25, and the pK_a value for a carbonyl-protonated amide is approximately 0. Therefore, the order of protonation is as follows:

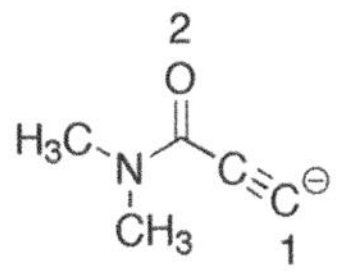

b.

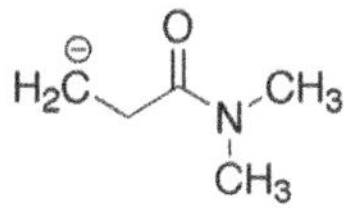

According to Appendix 1, the pK_a value for an alkane proton is approximately 50–75, and the pK_a value for a carbonyl-protonated amide is approximately 0. Therefore, the order of protonation for this hypothetical example is as follows:

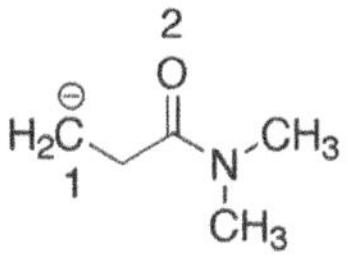

c.

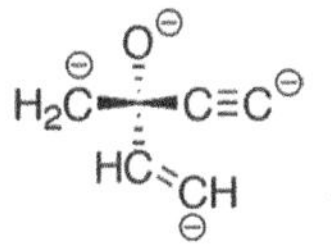

According to Appendix 1, the pK_a value for an alkane proton is approximately 50–75, the pK_a value for a vinyl proton is approximately 35–40, the pK_a value for an acetylene proton is approximately 25, and the pK_a value for an alcohol is approximately 15–19. Therefore, the order of protonation for this hypothetical example is as follows:

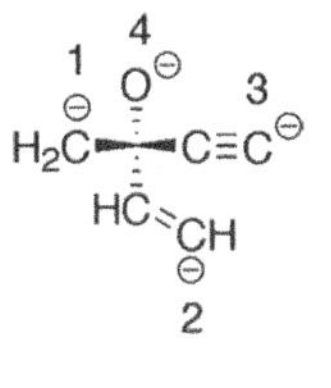

d.

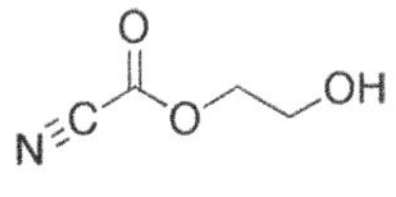

According to Appendix 1, the pK_a value for a protonated nitrile is approximately −10, the pK_a value for a carbonyl-protonated ester is approximately −6, and the pK_a value for a protonated alcohol is approximately −2. Therefore, the order of protonation is as follows:

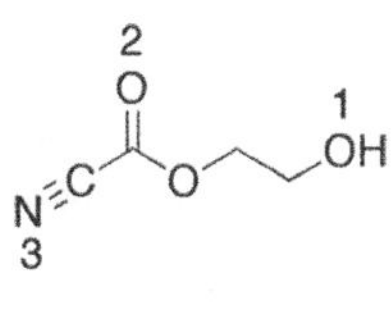

7. *Of the protons attached to the heteroatoms, which proton is the least acidic? Explain your answer.*

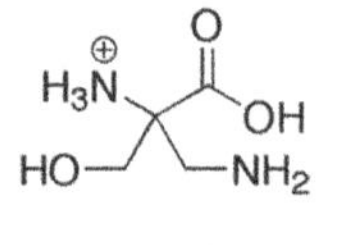

The pK_a value associated with a carboxylic acid is approximately 4.75. The pK_a value of a primary alcohol is approximately 16. The pK_a value of an amine is approximately 35. The pK_a value of a protonated amine is approximately 10. Since the highest pK_a value belongs to the amine, protons associated with the amine functionality are the least acidic.

8. *Separate the following group of bases into a group of hard bases and a group of soft bases. Rationalize your answers based on electronegativity and polarizability.*

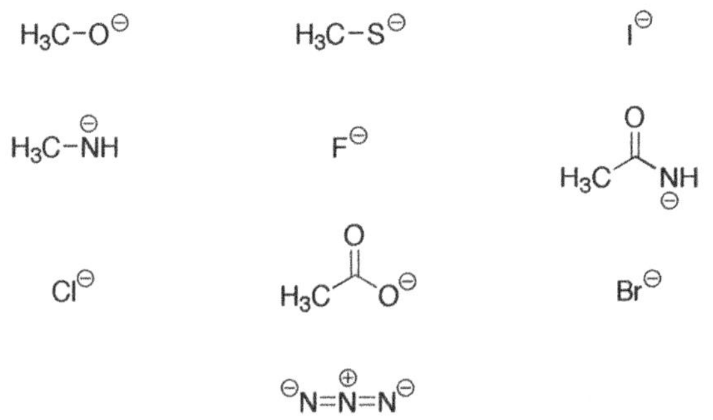

As a general rule, the basic atoms associated with soft bases have lower electronegativities and are more polarizable. Likewise, the basic atoms associated with hard bases have higher electronegativities and are less polarizable. Therefore, using the periodic table of elements, the group of bases listed earlier can be separated as illustrated.

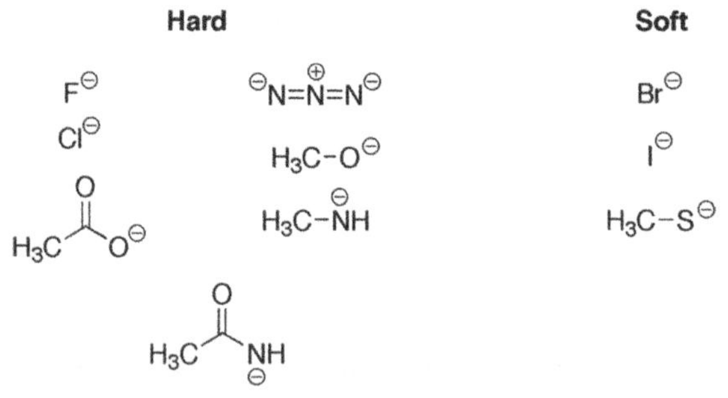

9. *Arrange the following structures in the order of increasing nucleophilicity.*

When a nucleophilic atom is surrounded by additional substituents, the degree of nucleophilicity is altered. This observation is explained because nucleophilicity depends, in part, on the ability of a given nucleophile to react with electrophiles. If the nucleophilic atom cannot approach the electrophile because of steric congestion surrounding the nucleophilic atom, then the nucleophile is rendered less effective as a nucleophile and more effective as a base.

a.

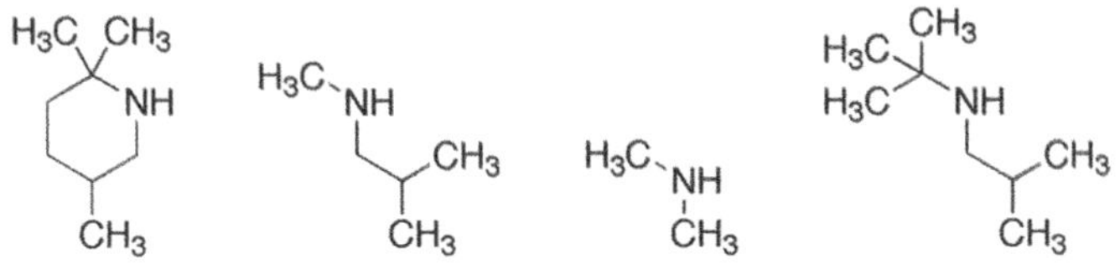

Based on the aforementioned argument, the order of increasing nucleophilicity for this group of amines is shown in the following. Regarding the two amines in the center, piperidines are typically more nucleophilic compared with acyclic amines because the alkyl groups are constrained and not able to rotate and obscure the amine lone electron pair. However, this piperidine contains two methyl groups adjacent to the nitrogen, effectively rendering the nitrogen similar to *tert*-butyl-amine. *tert*-Butylamine is not very nucleophilic because of the large alkyl group attached to the nitrogen. Through this analogy, the order of nucleophilicity is justified.

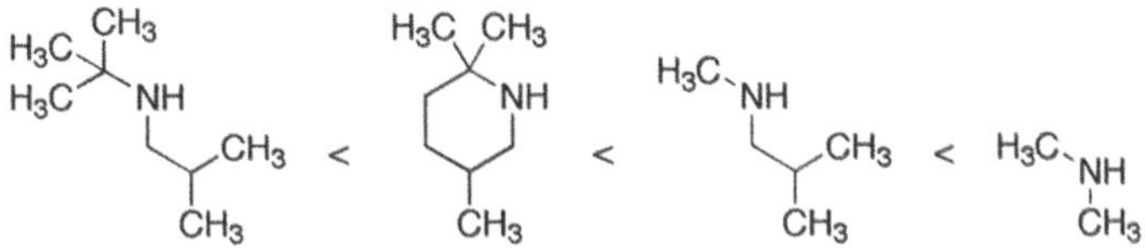

b.

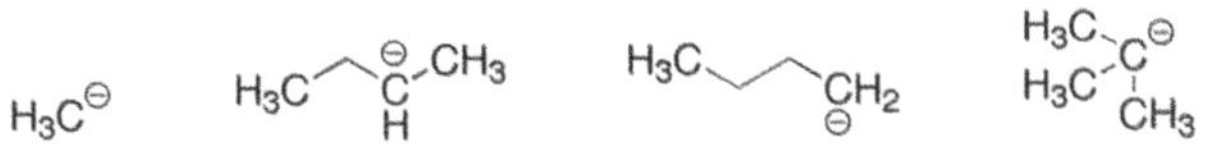

Based on the aforementioned argument, the order of increasing nucleophilicity for this group of alkyl anions is shown in the following:

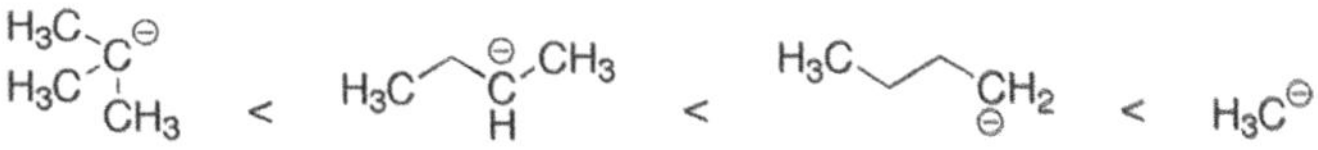

10. *For the following pairs of structures, circle the better leaving group.*

a. $Cl^{\ominus}$ $I^{\ominus}$

Compared with the chloride ion, the iodide ion is less electronegative and more polarizable. This polarizability stabilizes the anion as is reflected in the pK_a value for hydroiodic acid (−10) compared with the pK_a value for hydrochloric acid (−2.2). Therefore, iodide is the better leaving group.

$Cl^{\ominus}$ $\boxed{I^{\ominus}}$

b. $H_3C-O^{\ominus}$ $H_3C-\overset{\ominus}{N}H$

Compared with nitrogen, oxygen is more electronegative and thus holds onto its electrons more tightly. This stabilization of the oxygen anion compared with the amine anion is reflected in the pK_a values for alcohols (15–19) compared with the pK_a values for amines (35). Therefore, the alkoxide is the better leaving group.

$$\boxed{H_3C-O^{\ominus}} \qquad H_3C-\overset{\ominus}{N}H$$

c. $CH_3CH_2-O^{\ominus}$ $CF_3CH_2-O^{\ominus}$

When comparing leaving groups where the departing atoms are the same, inductive effects must be considered. Since fluorine is more electronegative than hydrogen, the presence of three fluorides pulls electron density from the alkoxide ion, thus stabilizing the anion. This is reflected in the pK_a value for trifluoroethanol (11–12) compared with the pK_a value for ethanol (15–16). Therefore, trifluoroethoxide is the better leaving group.

$$CH_3CH_2-O^{\ominus} \qquad \boxed{CF_3CH_2-O^{\ominus}}$$

d. $H_3C-S^{\ominus}$ $\qquad H_3C-O^{\ominus}$

Compared with oxygen, sulfur is less electronegative and more polarizable. This increase in polarizability stabilizes the anion as is reflected in the pK_a value for methanethiol (10.4) compared with the pK_a value for methanol (15–16). Therefore, the methylsulfide anion is the better leaving group.

$$\boxed{H_3C-S^{\ominus}} \qquad H_3C-O^{\ominus}$$

e. $F^{\ominus}$ $\qquad\qquad Br^{\ominus}$

Compared with the fluoride ion, the bromide ion is less electronegative and more polarizable. This polarizability stabilizes the anion as is reflected in the pK_a value for hydrobromic acid (−4.7) compared with the pK_a value for hydrofluoric acid (3.18). Therefore, bromide is the better leaving group.

$$F^{\ominus} \qquad\qquad \boxed{Br^{\ominus}}$$

f. $H_3C-\overset{\ominus}{N}H$ $\qquad H_3C-S^{\ominus}$

Like oxygen, sulfur is more electronegative than nitrogen. Additionally, sulfur is more polarizable. These differences stabilize the sulfur anion as reflected in the pK_a value for methanethiol (10.4) compared with the pK_a value for amines (35). Therefore, the methylsulfide anion is the better leaving group.

$$H_3C-\overset{\ominus}{N}H \qquad \boxed{H_3C-S^{\ominus}}$$

g. Br$^\ominus$ Br$-$Br$\overset{\ominus}{\text{H}}$

Bromine is more electronegative and more polarizable than oxygen. This translates to increased stability of the bromide anion compared with the oxygen anion. This stabilization is reflected in the pK_a value for hydrobromic acid (-4.7) compared with the pK_a value for methanol (15–16). Therefore, bromide is the better leaving group.

Br$^\ominus$ H$_3$C$-$O$^\ominus$

h.
$$F_3C-\overset{\overset{\displaystyle O}{\|}}{\underset{\underset{\displaystyle O}{\|}}{S}}-O^\ominus \qquad\qquad H_3C-\overset{\overset{\displaystyle O}{\|}}{\underset{\underset{\displaystyle O}{\|}}{S}}-O^\ominus$$

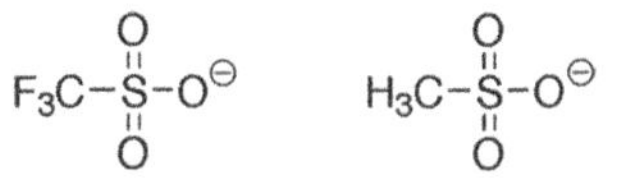

When comparing leaving groups where the departing atoms are the same, inductive effects must be considered. Since fluorine is more electronegative than hydrogen, the presence of three fluorides pulls electron density from the sulfonate ion, thus stabilizing the anion. This is the same effect noted under Problem 10c. Therefore, trifluoromethane sulfonate is the better leaving group.

$$F_3C-\overset{\overset{\displaystyle O}{\|}}{\underset{\underset{\displaystyle O}{\|}}{S}}-O^\ominus \qquad\qquad H_3C-\overset{\overset{\displaystyle O}{\|}}{\underset{\underset{\displaystyle O}{\|}}{S}}-O^\ominus$$

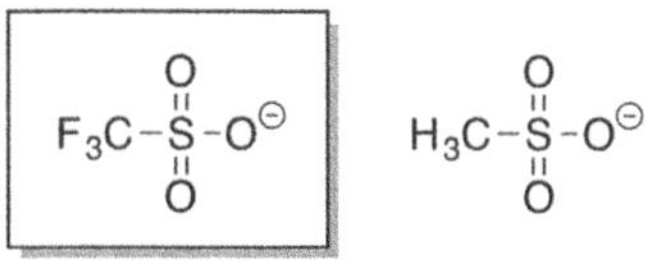

CHAPTER 5 SOLUTIONS

1. *In many S$_N$2 reactions, the nucleophile is generated by deprotonation of an organic acid. For each molecule, chose the base best suited to completely remove the labeled proton. (Consider pK$_a$ values and recognize that, in some cases, dianions should be considered.) Explain your answers.*

a.   H$_2$C(—)—C(=O)—CH$_3$ *NaOCH$_3$: (CH$_3$)$_2$NLi : CH$_3$Li*

The pK$_a$ of the highlighted proton is approximately 20. Therefore, NaOCH$_3$ (pK$_a$ of conjugate acid methanol = 16) is not a strong enough base. CH$_3$Li (pK$_a$ of conjugate acid methane = 50) will deprotonate this molecule; however, it is too nucleophilic for a base and will predominantly add to the carbonyl to produce a tertiary alcohol (see Chapter 8). (CH$_3$)$_2$NLi (pK$_a$ of conjugate acid dimethylamine = 35) is a bulkier base than CH$_3$Li and is, therefore, less nucleophilic and the best base for this case.

b.   H$_3$C—C(=O)—CH$_2$—C(=O)—CH$_3$ *NaOCH$_3$: (CH$_3$)$_2$NLi : CH$_3$Li*

The pK$_a$ of the highlighted proton is approximately 12. As described in the answer for Problem 1a, CH$_3$Li (pK$_a$ of conjugate acid methane = 50) will deprotonate this molecule; however, it is too nucleophilic for a base and will predominantly add to the carbonyl to produce a tertiary alcohol (see Chapter 8). While (CH$_3$)$_2$NLi (pK$_a$ of conjugate acid dimethylamine = 35) is a bulkier base than CH$_3$Li and is, therefore, less nucleophilic, it is also more basic than required for removal of the specified proton. NaOCH$_3$ (pK$_a$ of conjugate acid methanol = 16), on the other hand, is a milder base, and, based on pK$_a$ values, is adequate to fully deprotonate the illustrated compound.

c.   H$_3$C—C(=O)—CH$_2$—C(=O)—CH$_2$ *NaOCH$_3$: (CH$_3$)$_2$NLi : CH$_3$Li*

In this case, the most acidic proton is not the proton of interest. Therefore, it is important to remember that once the most acidic proton is removed, the resulting enolate anion renders the proton of interest even less acidic because the enolate anion is less able to stabilize the second anion. Thus, removal of the desired proton will require a comparatively stronger base. Additionally, it is important to understand (as will be explained in Chapter 8) that a negative charge next to a carbonyl makes the carbonyl much less susceptible to nucleophilic attack. Therefore, once the most acidic proton is removed with NaOCH$_3$, removal of the desired proton can subsequently be achieved using CH$_3$Li.

d.   H$_3$C—C(=O)—CH$_2$—C(=O)—OCH$_3$ *NaOH : NaOCH$_3$: NaOCH$_2$CH$_3$*

In this problem, three bases are presented, which all possess comparable pK$_a$ values and are all basic enough to remove the desired proton. In this case, however, the problem is not to recognize which base will remove the desired proton but to understand the reactivity of the target molecule in the presence of the various bases. The specific functionality of concern is the methyl ester. While the chemistry of ester groups is discussed in Chapters 7 and 8, using the principles of arrow-pushing, the answer to this problem can be derived from information already presented. Specifically, if any one of these bases is used, addition to the ester, followed by subsequent elimination of the CH_3O^- group, follows. This addition–elimination sequence produces either a carboxylic acid, an ethyl ester, or a methyl ester. Since the starting molecule possesses a methyl ester and since there is no instruction to change the nature of the ester, $NaOCH_3$ is the best base for this job.

2. *In predicting the course of S_N2 reactions, it is important to recognize groups most likely to act as nucleophiles. For each molecule, label the most nucleophilic site.*

a.

Considering the pK_a values of the respective conjugate acids, protons between two carbonyl groups have pK_a values around 12, while protons adjacent to only one carbonyl have pK_a values around 20. Therefore, the most nucleophilic site is as follows:

b.

Oxygen is more electronegative than nitrogen. As such, oxygen holds its lone pairs of electrons more tightly than nitrogen. Therefore, the most nucleophilic site is as follows:

c.

Ammonium ions, having no available electron pairs, are not nucleophilic. Carboxylate anions are nucleophilic, but the anions are stabilized through delocalization of the negative charge, thus decreasing their nucleophilicity. As mentioned in Problem 2b, nitrogen is less electronegative than oxygen. As only oxygen atoms in this compound are associated with a stable carboxylate anion, the most nucleophilic site is as follows:

d. 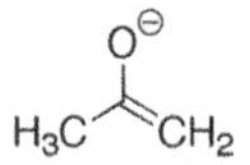 H$_3$C—C(O$^{\ominus}$)=CH$_2$ *(Hint: consider resonance)*

This structure represents an acyl anion with the negative charge delocalized to the oxygen. Since the carbon and the oxygen both possess partial negative charge characteristics, the degree of nucleophilicity depends on the relative electronegativities of carbon versus oxygen. Since oxygen is more electronegative than carbon, the most nucleophilic site is

3. *For each molecule, show the partial charges, bond polarity, and where a nucleophile is most likely to react.*

a.

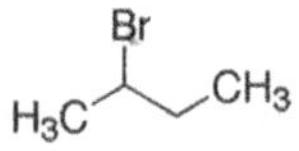

The polarity and partial charges of 2-bromobutane are dictated by the electronegativity of bromine versus the electronegativity of carbon. Therefore, the partial charges and polarity are as represented in the following, and a nucleophile is most likely to react at the carbon bearing the bromine atom.

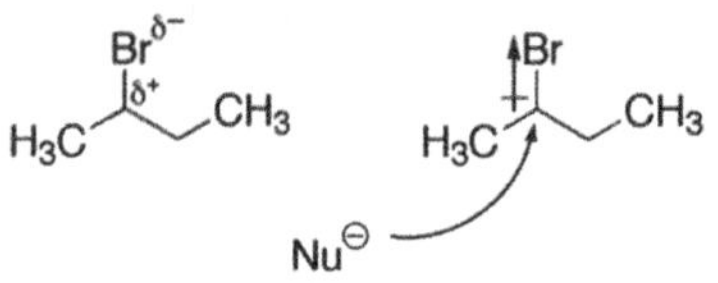

b. 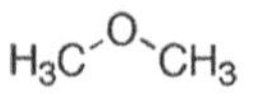   H$_3$C—O—CH$_3$

The polarity and partial charges of dimethyl ether are dictated by the electronegativity of oxygen versus the electronegativity of carbon. Therefore, the partial charges and polarity are as represented in the following, and a nucleophile is most likely to react at either of the carbon atoms.

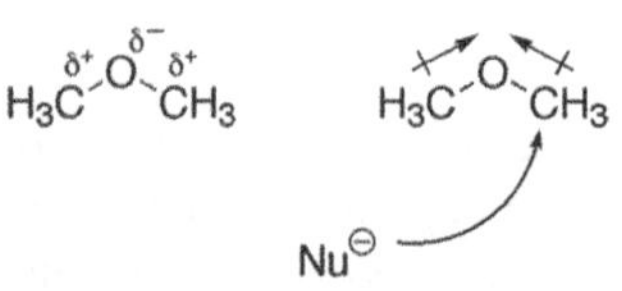

c. 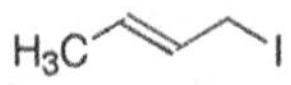   H$_3$C—CH=CH—CH$_2$—I

The polarity and partial charges of 1-iodo-2-butene are dictated by the electronegativity of oxygen versus the electronegativity of carbon. Additionally, delocalization through the double bond extends the chain of partial charges. Therefore, the partial charges and polarity are as represented in the following, and a nucleophile is most likely to react at either of the specified carbon atoms via an S$_N$2 or an S$_N$2′ mechanism.

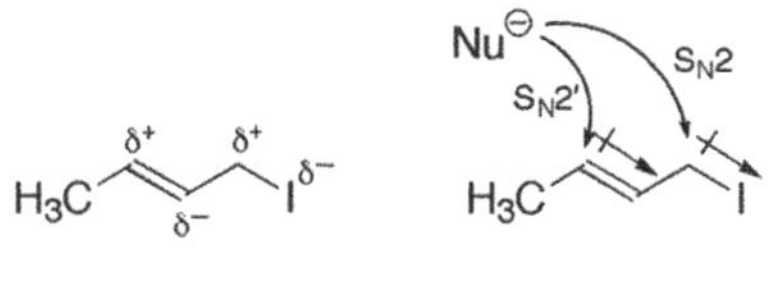

d. $H_3C \overset{\displaystyle O}{\underset{}{\|}} CH_3$

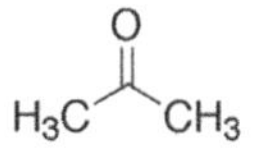

The polarity and partial charges of acetone are dictated by the electronegativity of oxygen versus the electronegativity of carbon as associated with a carbonyl. Therefore, the partial charges and polarity are as represented in the following, and a nucleophile is most likely to react at the carbonyl–carbon as will be discussed in Chapter 7.

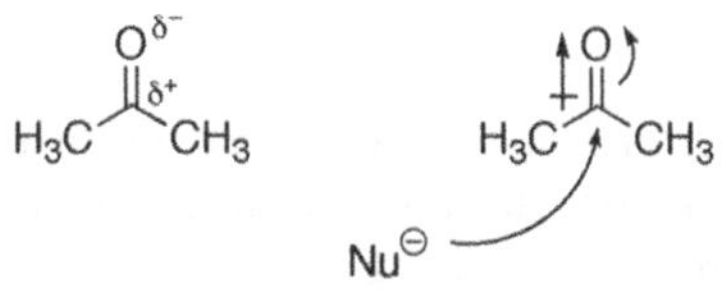

4. *For each molecule, identify the leaving group assuming that H_3C^- is the nucleophile.*

a.

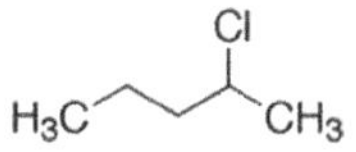

Applying partial charges based on the discussions presented in this chapter, the chlorine atom is recognized as the most electronegative. Therefore, as shown in the following, the chloride anion is the leaving group.

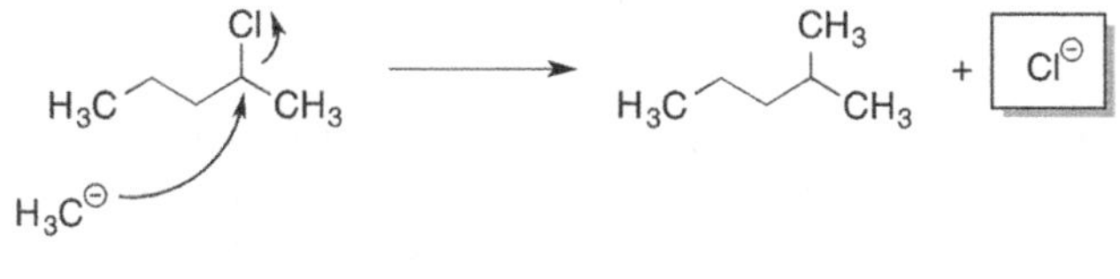

b.

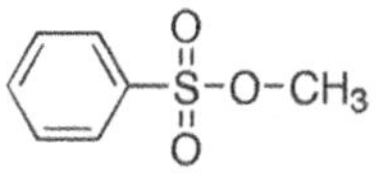

Applying partial charges based on the discussions presented in this chapter, the oxygen is recognized as more electronegative than carbon. Furthermore, an oxygen anion, derived from cleavage of a carbon–oxygen bond, is delocalized into the sulfur–oxygen double bonds and increasing its stability. Therefore, as shown in the following, the sulfonate anion is the leaving group.

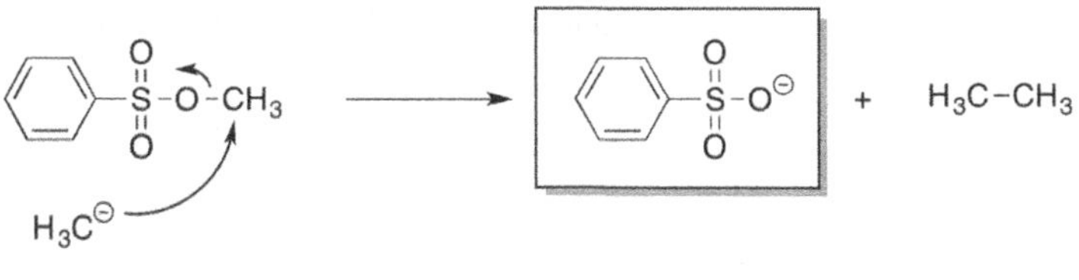

c.

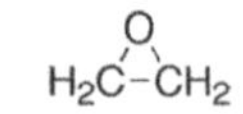

Applying partial charges based on the discussions presented in this chapter, the oxygen is recognized as more electronegative than carbon. Therefore, as shown in the following, the oxygen anion is the leaving group. Please note that in the case of an epoxide, the leaving group is attached to the reaction product.

5. *For each molecule, label the most likely leaving group. Explain your answers.*

a.

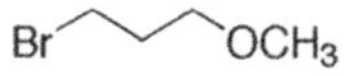

Bromine is more electronegative than oxygen. Furthermore, a bromide ion is a softer base than a methoxide ion. Because bromine can stabilize a negative charge better than oxygen, Br⁻ is the better leaving group.

b.

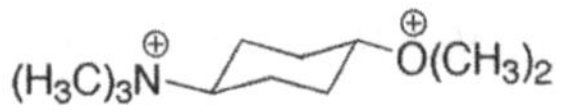

Oxygen is more electrophilic than nitrogen. Therefore, $(CH_3)_2O$ is the better leaving group.

c.

Bromide ions are softer bases than chloride ions. Therefore, bromine is more polarizable than chlorine, making Br⁻ the better leaving group.

d.

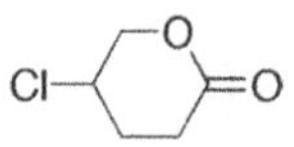

The answer to this question depends on information presented in Chapter 7. However, through an understanding of the nature of various nucleophiles coupled with application of arrow-pushing techniques, the answer can be derived from information presented thus far.

First, analyzing this structure for partial charges, we recognize the charge distribution as represented in the following:

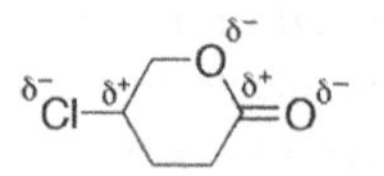

Recognizing that nucleophiles can react at two different sites, an initial thought might be direct displacement of the chloride anion in an S_N2 manner. However, as alluded to in Problem 3d, nucleophiles can add to carbonyl groups as shown in the following:

Once a nucleophile reacts with an ester carbonyl as shown earlier, the next phase of reaction depends on whether the better leaving group is an oxygen anion or the nucleophile itself. This is illustrated in the following through the ability of the newly formed oxygen anion to displace either the nucleophile or a second oxygen anion.

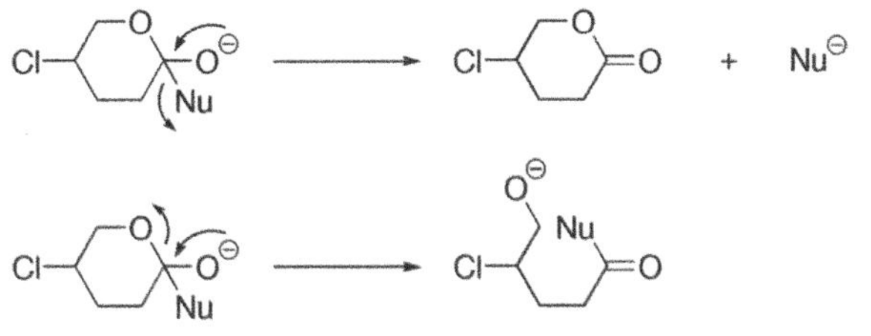

As shown earlier, if the better leaving group is the nucleophile, the result is regeneration of the starting material and the realization that displacement of Cl⁻ through an S_N2 mechanism is the most likely course of this reaction. However, if the better leaving group is the oxygen anion, then displacement of Cl⁻ generally will be a secondary reaction depending upon how much nucleophile is added to this system.

In summary, the purpose of this problem is not to solicit identification of a leaving group but rather to induce consideration of the different reaction processes that can occur. Through such an understanding, starting materials and reaction conditions can be chosen, which maximize the chances of generating a desired product with minimal side reactions.

6. *Detailed discussions focused on stereochemistry are not within the scope of this book. However, considering the products of typical S_N2 reactions, in addition to the transition state shown in Scheme 5.2, one may deduce the stereochemical course of this type of reaction. Predict the product of the following reaction and show the correct stereochemistry.*

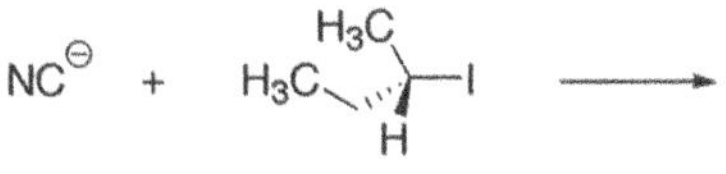

As shown in the following, initial reaction of a cyanide anion results in the formation of the transition state shown in brackets. Release of the iodide anion results in complete inversion of the stereochemistry generating the illustrated final product.

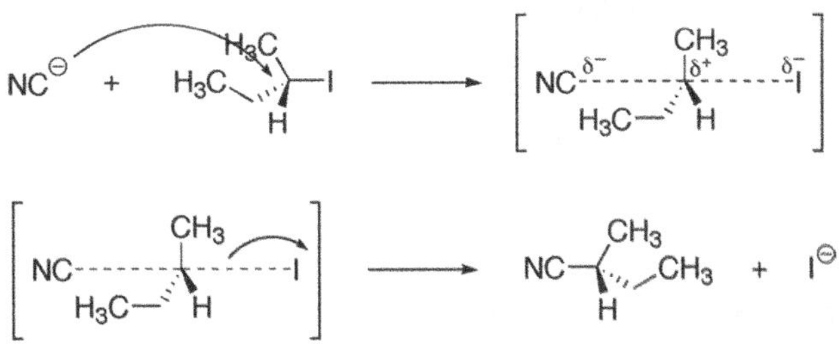

7. *Predict the products of the following reactions by pushing arrows.*

a.  I—CH$_3$ + $^{\ominus}$CN $\longrightarrow$

This is a direct S$_N$2 displacement of an iodide anion by a cyanide anion.

I⌒CH$_3$ + $^{\ominus}$CN $\longrightarrow$ H$_3$C—CN + I$^{\ominus}$

b. 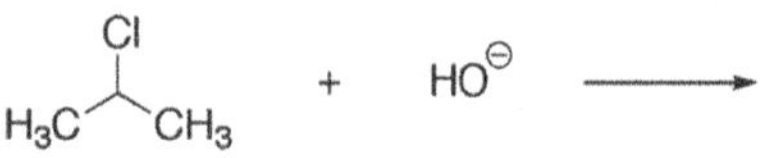 (H$_3$C)(CH$_3$)CH—Cl + HO$^{\ominus}$ $\longrightarrow$

This is a direct S$_N$2 displacement of a chloride anion by a hydroxide anion.

(H$_3$C)(CH$_3$)CH—Cl + HO$^{\ominus}$ $\longrightarrow$ (H$_3$C)(CH$_3$)CH—OH + Cl$^{\ominus}$

c. H$_3$C—C(=O)—CH$_2$$^{\ominus}$ + (H$_3$C)(CH$_3$)CH—Br $\longrightarrow$

This is a direct S$_N$2 displacement of a bromide anion by an acyl anion.

H$_3$C—C(=O)—CH$_2$$^{\ominus}$ + (H$_3$C)(CH$_3$)CH—Br $\longrightarrow$ H$_3$C—C(=O)—CH$_2$—CH(CH$_3$)—CH$_3$ + Br$^{\ominus}$

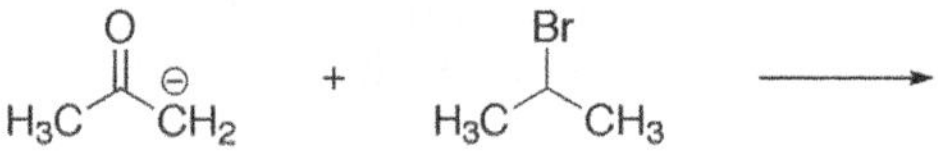

d. 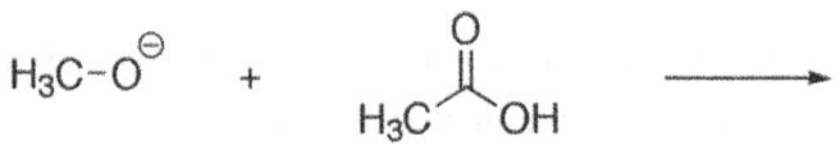 H$_3$C—O$^{\ominus}$ + H$_3$C—C(=O)—OH $\longrightarrow$

This is an acid–base proton exchange between a methoxide anion (pK_a of methanol is approximately 15) and acetic acid (pK_a = 4.75).

H$_3$C—O$^{\ominus}$ + H—O—C(=O)—CH$_3$ $\longrightarrow$ H$_3$C—OH + $^{\ominus}$O—C(=O)—CH$_3$

e. 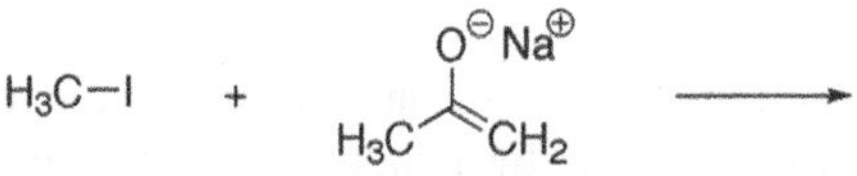 H$_3$C—I + H$_3$C—C(—O$^{\ominus}$Na$^{\oplus}$)=CH$_2$ $\longrightarrow$

This is a direct S$_N$2 displacement of an iodide anion by an acyl anion. Please note that the negative charge of the acyl anion is delocalized into the carbonyl and that the negative charge is paired with a cation.

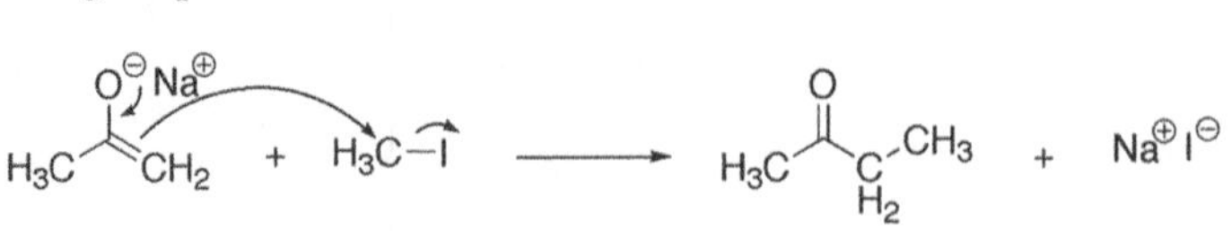

H$_3$C—C(—O$^{\ominus}$Na$^{\oplus}$)=CH$_2$ + H$_3$C—I $\longrightarrow$ H$_3$C—C(=O)—CH$_2$—CH$_3$ + Na$^{\oplus}$I$^{\ominus}$

f. 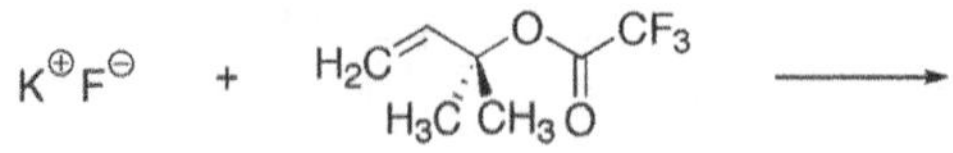 K$^{\oplus}$F$^{\ominus}$ + H$_2$C=C(CH$_3$)—O—C(CH$_3$)(O)—C(=O)—CF$_3$ $\longrightarrow$

This is an S_N2' displacement of a trifluoroacetoxy anion by a fluoride anion. The related S_N2 mechanism is not favored because of steric factors. Specifically, the trifluoroacetate resides at a tertiary center. Please note that the fluoride anion is accompanied by a potassium cation and that the final trifluoroacetoxy group is presented as its potassium salt.

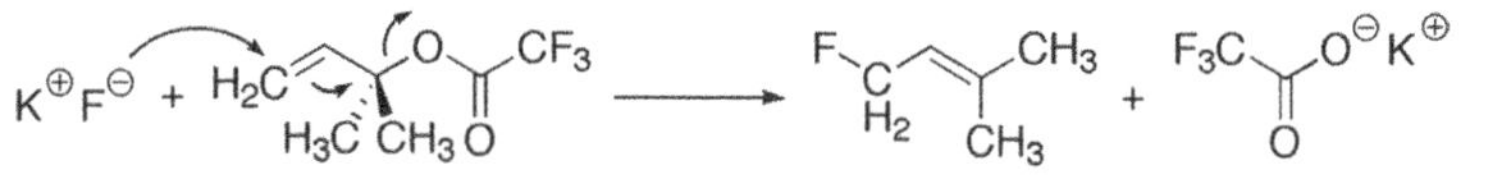

g.

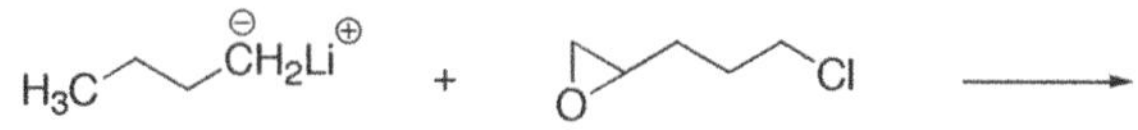

This is a two-step reaction with initial S_N2 opening of an epoxide. The opening of the epoxide is favored because of the strain associated with a three-membered ring. Subsequent S_N2 displacement of the chloride by the alkoxide resulting from epoxide opening leads to the illustrated tetrahydrofuran derivative. The purpose of this example is to illustrate that in many cases, organic reactions do not stop after an initial stage and frequently advance to generate products over several mechanistic steps.

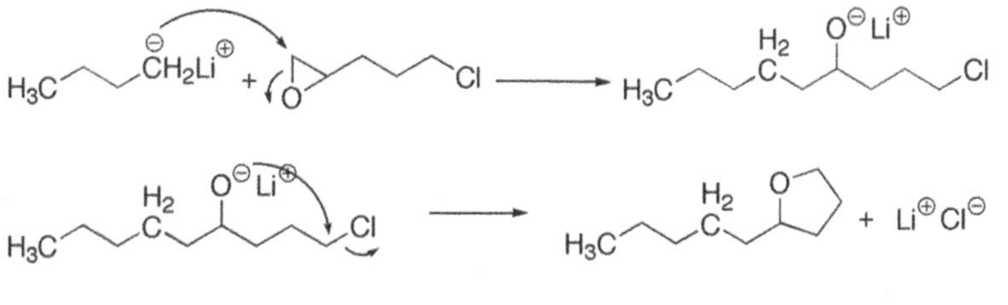

h.

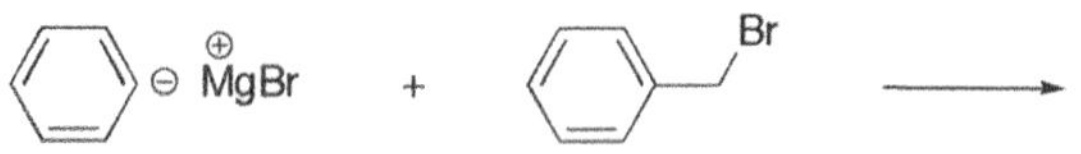

This is a direct S_N2 displacement of a bromide anion by a phenyl anion. Please note that the negative charge of the phenyl anion is accompanied by a magnesium bromide complex. This class of organic salt is known as a Grignard reagent and is characterized by the presence of magnesium and a halide such as chloride, bromide, or iodide.

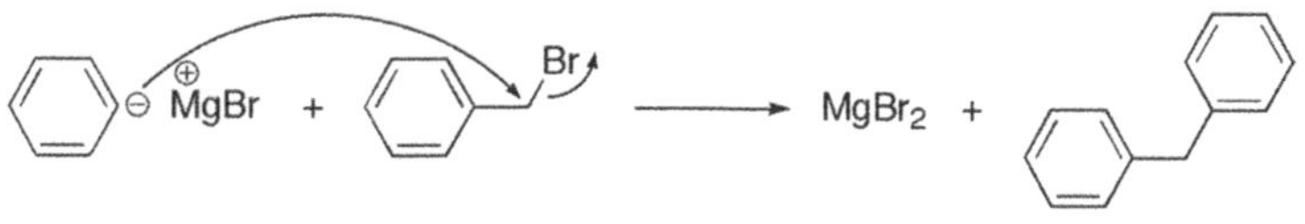

i.

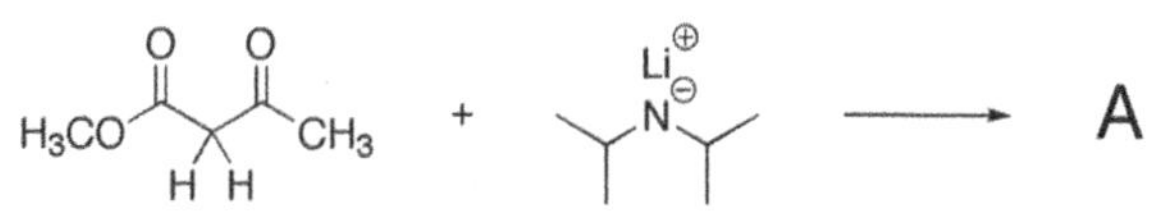

This is an acid–base proton exchange between lithium diisopropylamide (LDA) (pK_a of diisopropylamine is approximately 35) and methyl acetoacetate (pK_a is approximately 12). Please note the transfer of the lithium counterion from LDA to the deprotonated methyl acetoacetate.

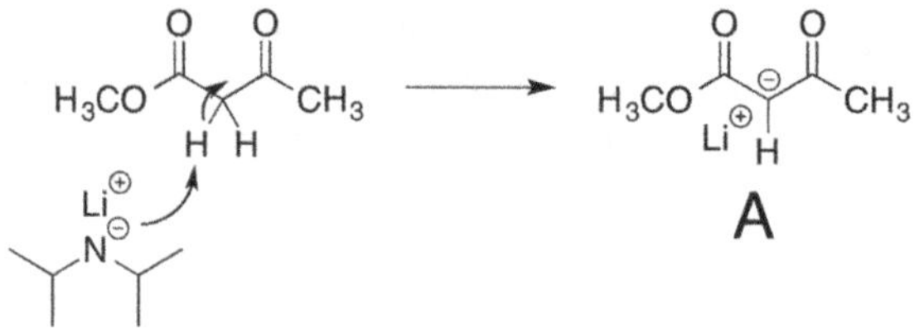

j. 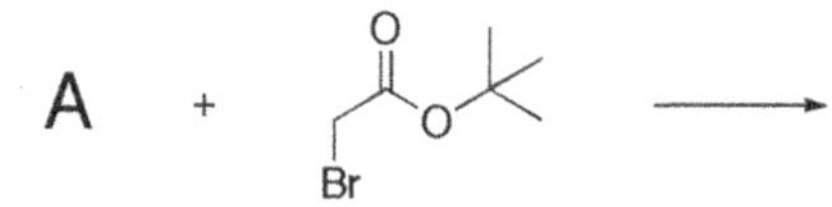   A   +   ⟶

This is a direct S_N2 displacement of the bromide anion of *tert*-butyl bromoacetate by a methyl acetoacetate anion. Lithium bromide (LiBr), the salt by-product, is not shown in the following reaction:

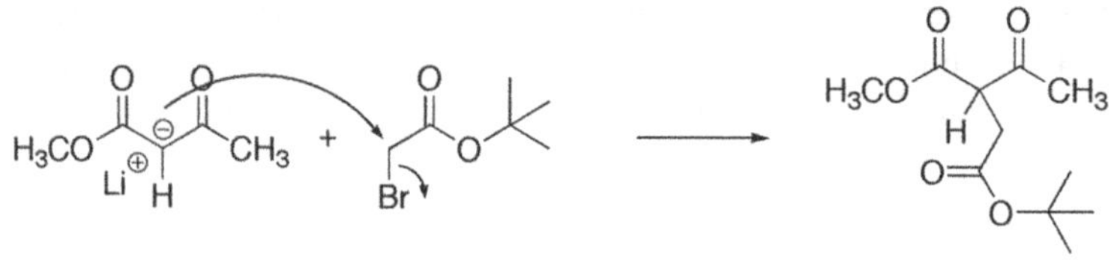

k. 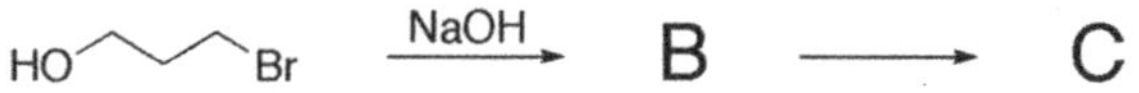 $\xrightarrow{\text{NaOH}}$ **B**   ⟶   **C**

The first step of this reaction is an acid–base proton exchange between a hydroxide anion (pK_a of water is approximately 16) and a primary alcohol (pK_a is approximately 16) forming the illustrated alkoxide, **B**. Formation of water, the by-product, is not shown in this step.

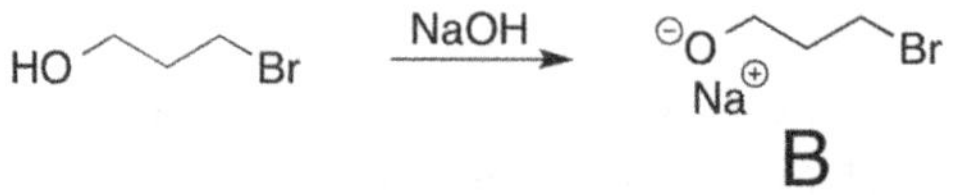

The second step of this reaction is a direct S_N2 displacement of a bromide anion by the alkoxide anion present in the same molecule. This step leads to the formation of oxetane, **C**. Formation of sodium bromide (NaBr), the salt by-product of this step, is not shown.

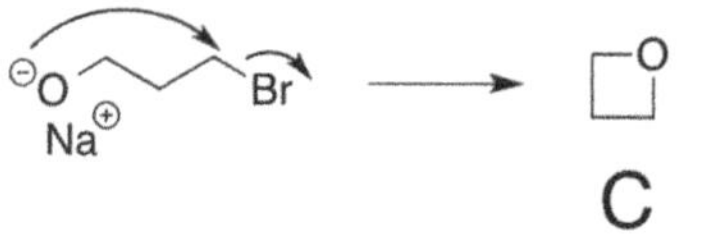

l. 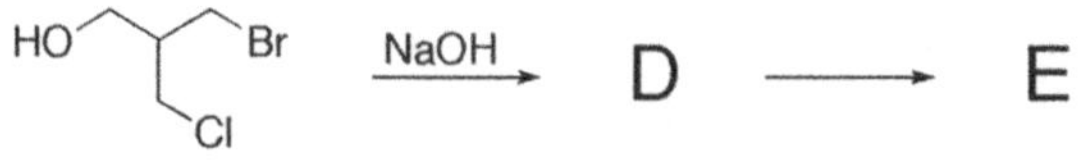 $\xrightarrow{\text{NaOH}}$ **D**   ⟶   **E**

The first step of this reaction is an acid–base proton exchange between a hydroxide anion (pK_a of water is approximately 16) and a primary alcohol (pK_a is approximately 16) forming the illustrated alkoxide, **D**. Formation of water, the by-product, is not shown in this step.

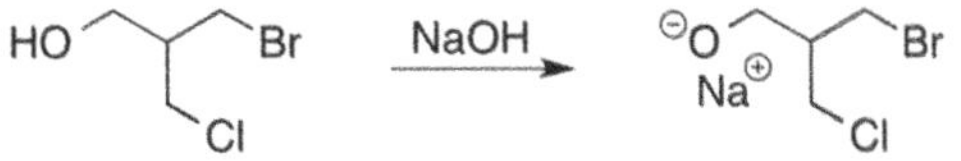

The second step of this reaction is a direct S_N2 displacement of a bromide anion by the alkoxide anion present in the same molecule. This step leads to the formation of the oxetane, **E**. Please note that displacement of the bromide is preferred over displacement of the chloride because bromide is a better leaving group than chloride. Formation of sodium bromide (NaBr), the salt by-product of this step, is not shown.

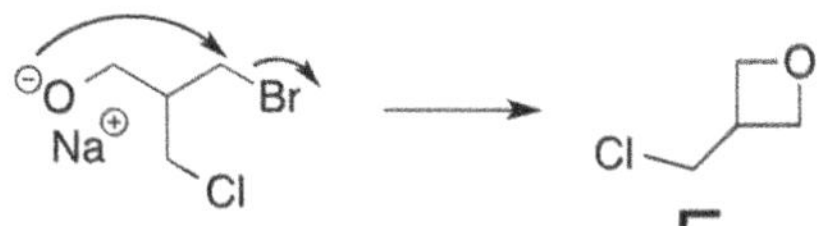

m.

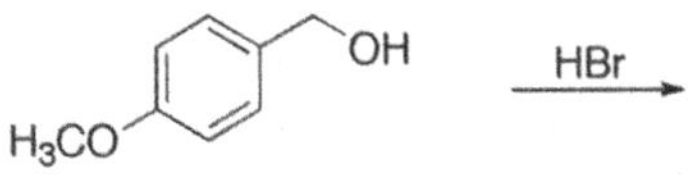

This is a solvolysis reaction that proceeds in two steps. The first step involved protonation of the hydroxyl group of *p*-methoxybenzyl alcohol. Once protonated, a bromide ion displaces water generating the illustrated product. The reaction shown in the following demonstrates this reaction through an S_N2 mechanism; however, this reaction can also be represented through an S_N1 reaction involving initial dissociation of water followed by reaction of the resulting cation with a bromide anion.

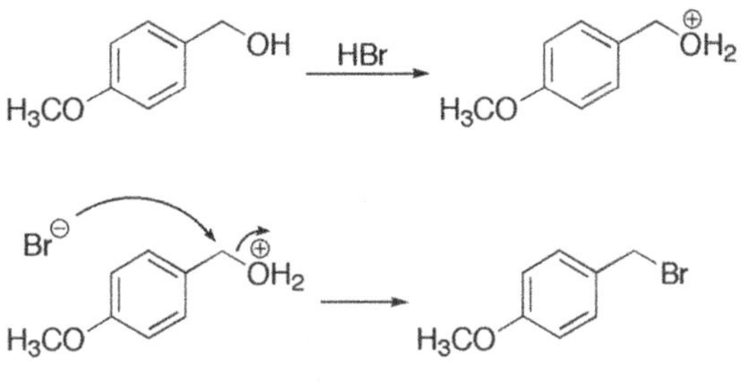

n.

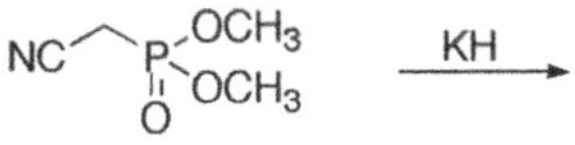

Potassium hydride (KH) is a reactive base possessing a potassium cation and a hydrogen anion (hydride ion). The hydride ion reacts as any other base mentioned thus far and extracts acidic protons generating hydrogen gas and leaving behind anions with associated potassium cations. In this case, the dimethyl cyanomethylphosphonate anion, **F**, is formed.

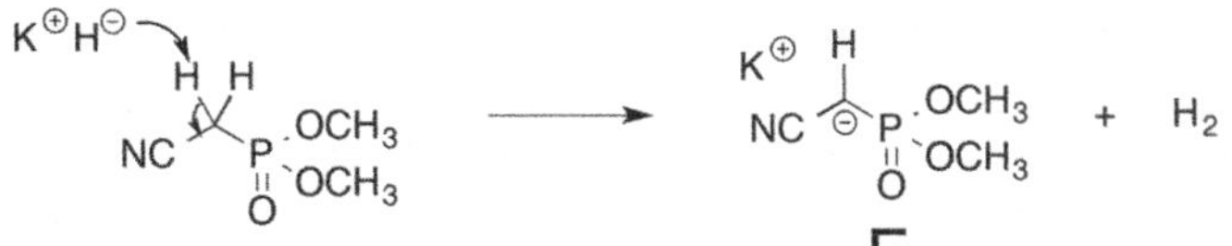

o. 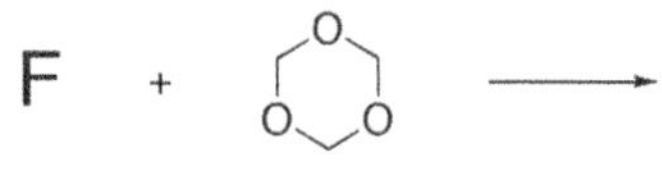   F   +

At first glance, this reaction appears simple with the phosphonate anion illustrated in Problem 7n displacing an alkoxide anion from trioxane as illustrated in the following.

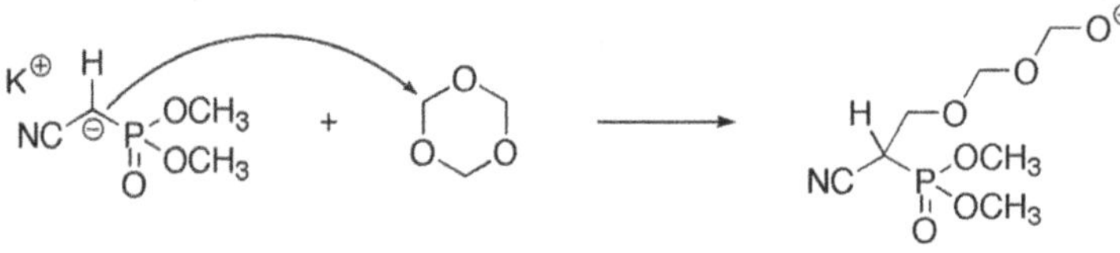

However, as continually alluded to, anions, once formed, can participate in further reactions. Trioxane is essentially a trimer of formaldehyde ($H_2C=O$) and is more stable and easier to handle. When an anion opens the trioxane ring, the resulting anion degrades, as shown in the following, with release of two equivalents of formaldehyde. The resulting species is essentially that resulting from reaction of the initial phosphonate anion with formaldehyde itself. Please note the net incorporation of only one carbon atom and only one oxygen atom. Additionally, the potassium cation is omitted from the remainder of the illustrations for clarity.

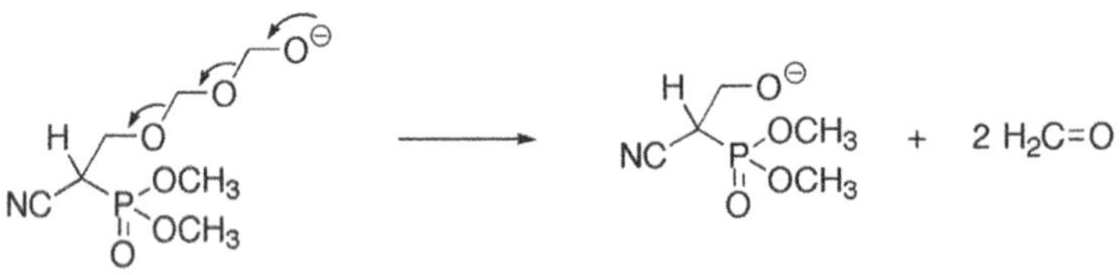

Again, referring to the ability of anions to undergo further transformations, we must recognize that phosphorus is a unique element with a strong affinity for oxygen. Furthermore, the phosphorus–oxygen double bond bears much of the same reactivity of a carbon–oxygen double bond and will accept addition of a nucleophile into the system as shown in the following. The illustrated four-membered species is known as a phosphetane.

As phosphorus exhibits a strong affinity for oxygen, phosphetane rings are known to undergo further reactions. As illustrated in the following, the negative charge on the oxygen is capable of breaking the adjacent carbon–phosphorus bond and transferring the negative charge to the carbon atom. Carrying this cycle forward, the negatively charged carbon atom participates in an E2 elimination (Chapter 7) with formation of a new double bond and cleavage of the adjacent carbon–oxygen bond. The resulting two species are an olefin and a phosphate anion.

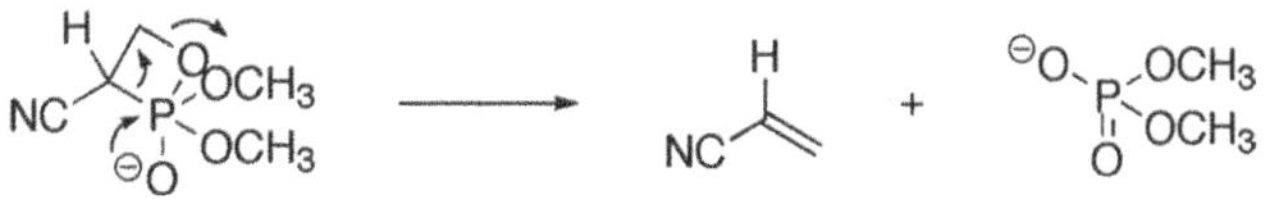

This reaction, known as a Horner–Emmons olefination, was presented to illustrate that through consideration of the electronic nature of a given starting material and the transient species involved in reactions with this material, products of more complex reactions may be identified. However, it is important to note that while this sequence appears complex, each step involved utilizes principles of arrow-pushing easily applied from material presented in this book.

8. *Addition reactions and conjugate addition reactions, to be discussed in Chapter 8, are related to S_N2 and S_N2' reactions, respectively. We can make these comparisons if we recognize that the carbonyl double bond contains a leaving group. Specifically, if a nucleophile adds to the carbon of a carbonyl, the carbonyl double bond becomes a carbon–oxygen single bond with a negative charge residing on the oxygen. Additionally, the trigonal planar geometry of the carbonyl–carbon is converted to tetrahedral geometry. With these points in mind, predict the products of the following reactions and explain your answers. For Problem 8b, the nucleophile is a methyl anion associated with the illustrated cuprate.*

a.

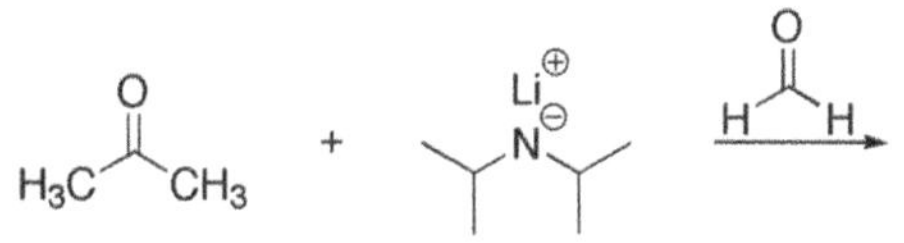

The first stage of this reaction is deprotonation of acetone by LDA in a manner analogous to that demonstrated in Problem 7i.

The second stage of the reaction is addition of the acetone anion to formaldehyde as shown in the following:

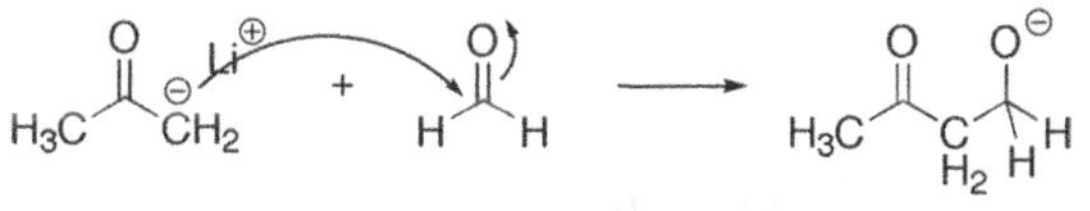

Protonation of the resulting alkoxide anion leads to the alcohol illustrated in the following. This reaction is known as an aldol condensation.

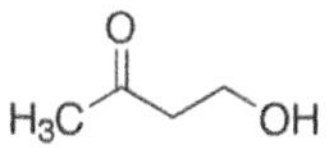

b.

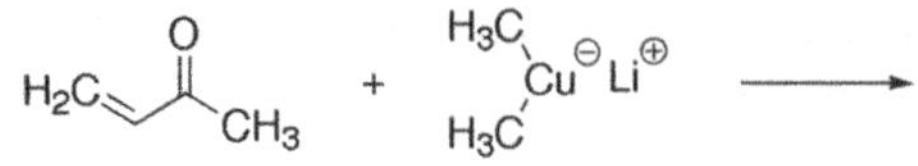

The copper-based reagent shown in the aforementioned reaction is known as a cuprate. This specific compound is dimethyllithiocuprate and is an excellent carrier of methyl anions. Cuprates are unique in their ability to preferentially deliver nucleophiles to carbonyl groups through adjacent double bonds and in manners analogous to S_N2' mechanisms. Thus, as illustrated in the following, arrow-pushing demonstrates how cuprates add nucleophiles to unsaturated carbonyl systems.

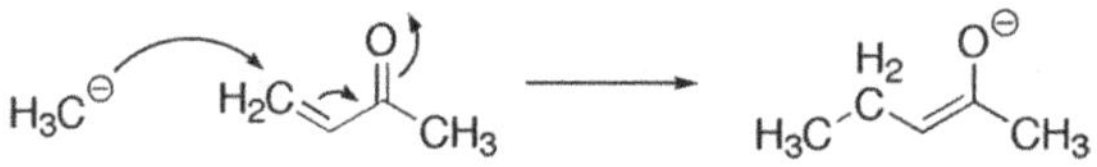

When the illustrated anion is treated with acid, proton transfer generates the final product as shown in the following:

9. *Propose a reasonable mechanism for each of the following reactions. Explain your answers by pushing arrows.*

a.

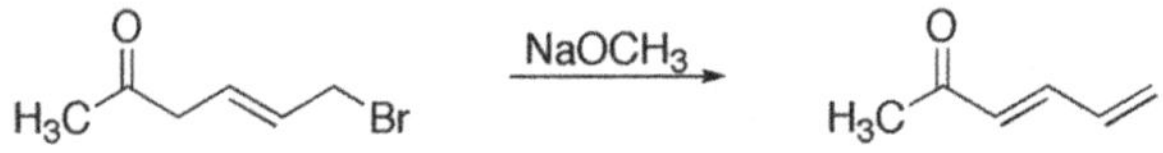

This reaction proceeds through initial deprotonation adjacent to the ketone followed by an S_N2'-type movement of electrons through the double bond and elimination of a bromide ion.

b.

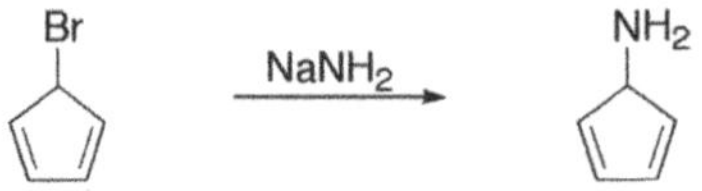

This reaction is an S_N2' displacement of a bromide anion.

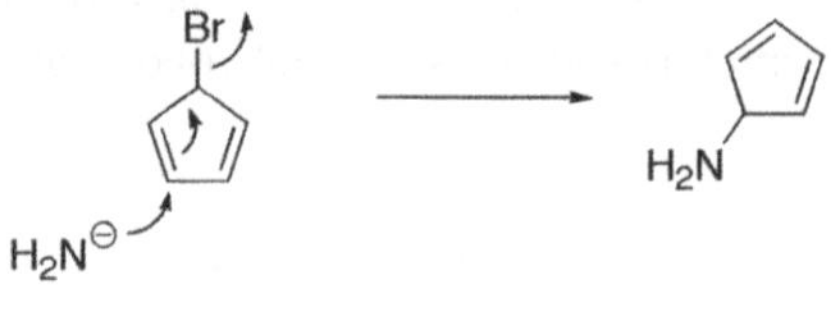

c.

The first step of this reaction is deprotonation of the alcohol with sodium hydride.

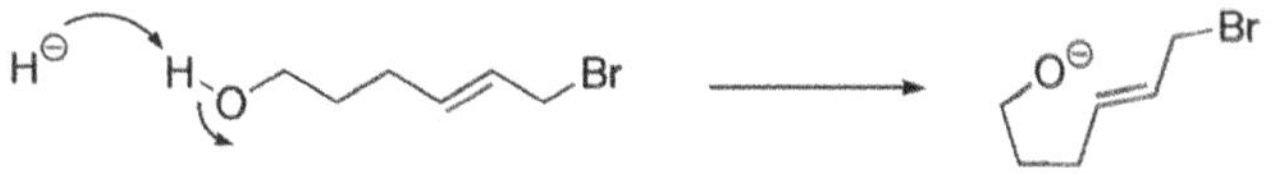

The second step of this reaction is an intramolecular S_N2' reaction with the alkoxide anion displacing the bromide anion through the double bond.

d.

This reaction is an S_N2' displacement of an alkoxide anion through the double bond.

10. *α,β-Unsaturated carbonyls are readily formed from the corresponding β-hydroxy ketones. Explain the product of the following reaction.*

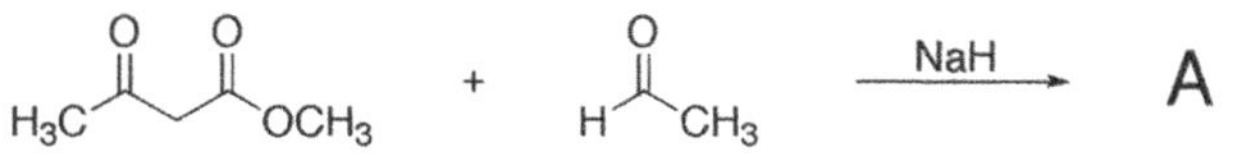

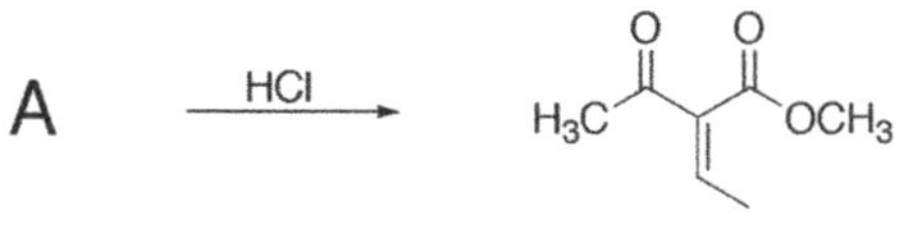

Upon examining the reaction, the initial phase of this sequence can be defined as an aldol condensation (see Problem 8a). Under the specified conditions, hydride is used to deprotonate methyl acetoacetate, and the resulting anion adds to the acetaldehyde carbonyl giving the aldol product, **A**.

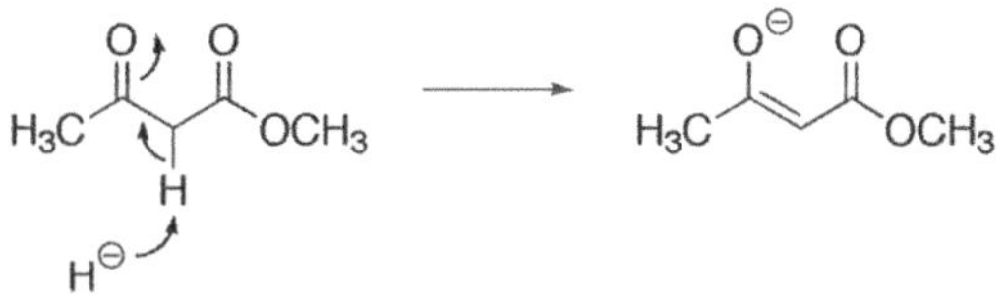

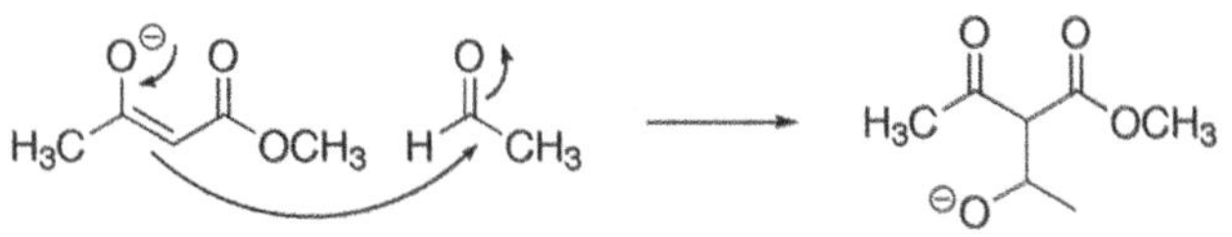

Treating the aldol adduct, **A**, with hydrochloric acid protonates the alkoxide anion and then protonates the resulting alcohol as part of a solvolysis reaction. Water then leaves generating a carbocation. The carbocation then undergoes an E1 elimination (see Chapter 7) giving the illustrated product.

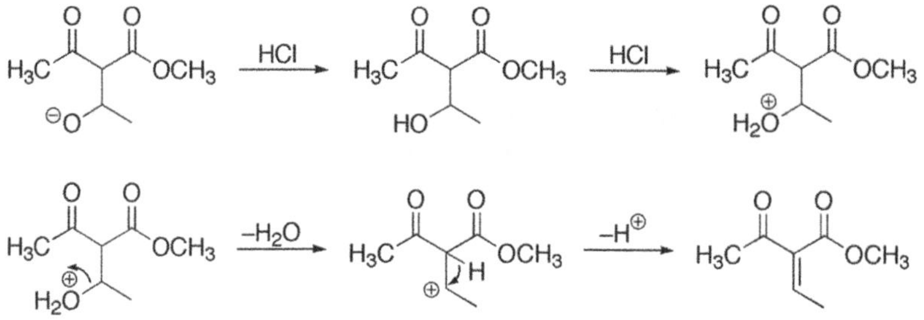

CHAPTER 6 SOLUTIONS

1. *For the following molecules, state the hybridization (sp, sp², sp³) of the orbitals associated with the highlighted bond. Also, state the geometry of the bound atomic centers (linear, bent, trigonal planar, tetrahedral).*

a.   N≡C–CH₃

The highlighted bond joins a nitrile carbon atom to a methyl carbon atom. The nitrile carbon atom, being joined to a nitrogen atom via a carbon–nitrogen triple bond, can only be joined to one additional atom. Therefore, this atom is *sp*-hybridized. However, the methyl carbon is joined to the nitrile carbon and three hydrogen atoms. Therefore, because the methyl carbon is bound to four separate atoms, this carbon is *sp³*-hybridized. Based on the atomic hybridizations, the nitrile carbon is connected to its bound atoms in a linear geometry, and the methyl carbon is connected to its bound atoms in a tetrahedral geometry.

N≡C–CH₃
sp³-Hybridized, Tetrahedral
sp-Hybridized, Linear

b.   N≡C–C(H)=CH₂

The highlighted bond joins a nitrile carbon atom to a vinyl carbon atom. The nitrile carbon atom, being joined to a nitrogen atom via a carbon–nitrogen triple bond, can only be joined to one additional atom. Therefore, this atom is *sp*-hybridized. However, the vinyl carbon is joined to the nitrile carbon, a hydrogen atom, and a second vinyl carbon atom. Because the two vinyl carbon atoms are joined by a double bond, there can be no more than three atoms bound to the highlighted vinyl carbon. Therefore, because the vinyl carbon is bound to three separate atoms, this carbon is *sp²*-hybridized. Based on the atomic hybridizations, the nitrile carbon is connected to its bound atoms in a linear geometry, and the vinyl carbon is connected to its bound atoms in a trigonal planar geometry.

sp²-Hybridized, Trigonal planar

N≡C–C(H)=CH₂

sp-Hybridized, Linear

c.   N≡C–C≡CH

The highlighted bond joins a nitrile carbon atom to an alkyne carbon atom. The nitrile carbon atom, being joined to a nitrogen atom via a carbon–nitrogen triple bond, can only be joined to one additional atom. Therefore, this atom is *sp*-hybridized. Additionally, the alkyne carbon is joined to the nitrile carbon and a second alkyne carbon atom. Because the two alkyne carbon atoms are joined by a triple bond, there can be no more than two atoms bound to the highlighted alkyne carbon. Therefore, because the alkyne carbon is bound to two separate atoms, this carbon is *sp*-hybridized. Based on the atomic hybridizations, the nitrile carbon is connected to its bound atoms in a linear geometry, and the alkyne carbon is connected to its bound atoms in a linear geometry.

$$N{\equiv}C{-}C{\equiv}CH$$

sp-Hybridized, Linear
sp-Hybridized, Linear

d. $H_3C{-}CH_3$

The highlighted bond joins two methyl carbon atoms. Each methyl carbon is joined to a methyl carbon and three hydrogen atoms. Therefore, because each methyl carbon is bound to four separate atoms, they are *sp³*-hybridized. Based on the atomic hybridizations, each methyl carbon is connected to its bound atoms in tetrahedral geometries.

$$H_3C{-}CH_3$$

sp³-Hybridized, Tetrahedral
sp³-Hybridized, Tetrahedral

e. $N{\equiv}C{-}NH_2$

The highlighted bond joins a nitrile carbon atom to an amine nitrogen atom. The nitrile carbon atom, being joined to a nitrogen atom via a carbon–nitrogen triple bond, can only be joined to one additional atom. Therefore, this atom is *sp*-hybridized. However, the amine nitrogen is joined to the nitrile carbon and two hydrogen atoms. Additionally, the amine nitrogen possesses one lone electron pair. Therefore, because the amine nitrogen is bound to three separate atoms and possesses one lone electron pair, this nitrogen is *sp³*-hybridized. Based on the atomic hybridizations, the nitrile carbon is connected to its bound atoms in a linear geometry, and the amine nitrogen is connected to its bound atoms and lone electron pair in a tetrahedral geometry. Please note that because the nitrogen is only bound to three atoms, the tetrahedral relationship between the bound atoms and lone electron pair can also be referred to as trigonal pyramidal (not considering the contributions of the lone electron pair to the geometry) because the geometry represents a three-sided pyramid.

$$N{\equiv}C{-}NH_2$$

sp³-Hybridized, Tetrahedral or Trigonal Pyramidal
sp³-Hybridized, Linear

f. $H_3C{-}N{\equiv}CH_2$

The highlighted bond joins a methyl carbon atom to an imine nitrogen atom. The methyl carbon atom, being joined to three hydrogen atoms and an imine nitrogen atom, is bound to four separate atoms and is, therefore, sp^3-hybridized. The imine nitrogen atom is bound to a methyl carbon atom through a single bond and to a second carbon atom through a double bond. Additionally, the imine nitrogen possesses one lone electron pair. Therefore, because the imine nitrogen is bound to two separate atoms and possesses one lone electron pair, this nitrogen is sp^2-hybridized. Based on the atomic hybridizations, the methyl carbon is connected to its bound atoms in a tetrahedral geometry, and the imine nitrogen is connected to its bound atoms and lone electron pair in a trigonal planar geometry. Please note that because the nitrogen is only bound to two atoms, the trigonal planar relationship between the bound atoms and lone electron pair can also be referred to as bent (not considering the contributions of the lone electron pair to the geometry).

sp^2-Hybridized, Trigonal Planar or Bent

H_3C•N⟍CH_2

sp^3-Hybridized, Tetrahedral

g. $N{\equiv}C$•^{OH}

The highlighted bond joins a nitrile carbon atom to a hydroxy oxygen atom. The nitrile carbon atom, being joined to a nitrogen atom via a carbon–nitrogen triple bond, can only be joined to one additional atom. Therefore, this atom is sp-hybridized. However, the hydroxy oxygen is joined to the nitrile carbon and one hydrogen atom. Additionally, the hydroxy oxygen possesses two lone electron pairs. Therefore, because the hydroxy oxygen is bound to two separate atoms and possesses two lone electron pairs, this oxygen is sp^3-hybridized. Based on the atomic hybridizations, the nitrile carbon is connected to its bound atoms in a linear geometry, and the hydroxy oxygen is connected to its bound atoms and lone electron pairs in a tetrahedral geometry. Please note that because the oxygen is only bound to two atoms, the tetrahedral relationship between the bound atoms and lone electron pair can also be referred to as bent (not considering the contributions of the lone electron pairs to the geometry).

$N{\equiv}C$•^{OH}

sp^3-Hybridized, Tetrahedral or Bent

sp-Hybridized, Linear

h. $H_2C{=}C{=}O$ *(Answer for both double bonds.)*

For this compound, the CH_2 carbon is bound to the central carbon through a double bond. Furthermore, this carbon atom is bound to two hydrogen atoms. Because this carbon atom is bound to only three atoms, it is sp^2-hybridized. However, the central carbon atom, being bound to the CH_2 carbon atom through a double bond, is bound to an oxygen atom through a double bond. Thus, the central carbon atom is bound to only two atoms and is sp-hybridized. Finally, the oxygen atom is bound to the central atom through a double bond. Additionally, the oxygen atom possesses two lone electron pairs. Because the oxygen atom is bound to only one atom and possesses two lone electron pairs, it is sp^2-hybridized. Regarding geometry, the CH_2 carbon, being bound to three atoms, is trigonal planar. Furthermore, the central carbon, being bound to two atoms, is linear. Lastly, the oxygen atom, being bound to only one atom and possessing two lone electron pairs, is trigonal planar or linear (not considering the contributions of the lone electron pairs to the geometry).

$$H_2C=C=O$$

sp^2-Hybridized, Trigonal Planar or Linear
sp-Hybridized, Linear
sp^2-Hybridized, Trigonal Planar

i.

$$H_3C \overset{\oplus}{\underset{H}{C}} \overset{H}{\underset{}{C}} = CH_2$$

For this compound, the positively charged CH carbon is bound to a vinyl carbon, a methyl carbon, and a hydrogen through single bonds. Because this carbon atom is bound to only three atoms, it is sp^2-hybridized. Additionally, the vinyl carbon atom is bound to the positively charged carbon atom, a hydrogen, and a second vinyl carbon. Because the two vinyl carbon atoms are joined by a double bond, there can be no more than three atoms bound to the highlighted vinyl carbon. Therefore, because the vinyl carbon is bound to three separate atoms, this carbon is sp^2-hybridized. Based on the atomic hybridizations, the positively charged carbon is connected to its bound atoms in a trigonal planar geometry. Likewise, the vinyl carbon is connected to its bound atoms in a trigonal planar geometry.

sp^2-Hybridized, Trigonal Planar

$$H_3C \overset{\oplus}{\underset{H}{C}} \overset{H}{\underset{}{C}} = CH_2$$

sp^2-Hybridized, Trigonal Planar

j.

$$H_3C \overset{\oplus}{\underset{CH_3}{C}} C \equiv CH$$

For this compound, the positively charged carbon is bound to an alkyne carbon and two methyl carbons through single bonds. Because this carbon atom is bound to only three atoms, it is sp^2-hybridized. Additionally, the alkyne carbon atom is bound to the positively charged carbon atom and a second alkyne carbon atom. Because the two alkyne carbon atoms are joined by a triple bond, there can be no more than two atoms bound to the highlighted alkyne carbon. Therefore, because the alkyne carbon is bound to two separate atoms, this carbon is sp-hybridized. Based on the atomic hybridizations, the positively charged carbon is connected to its bound atoms in a trigonal planar geometry, and the alkyne carbon is connected to its bound atoms in a linear geometry.

$$H_3C \overset{\oplus}{\underset{CH_3}{C}} C \equiv CH$$

sp-Hybridized, Linear
sp^2-Hybridized, Trigonal Planar

2. *Predict all of the products of the following reactions.*

a.

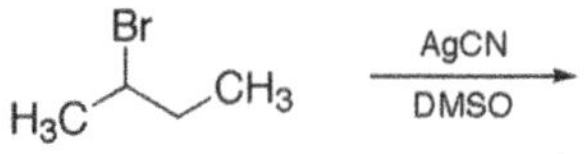

Silver is very efficient at removing halides resulting in generation of carbocations. Applied to 2-bromobutane, once a carbocation is formed, a 1,2-hydride shift applied to the illustrated secondary carbocation can only generate a less stable primary carbocation or an identical secondary carbocation. Therefore, addition of the cyanide anion to the cation results in the formation of a single product.

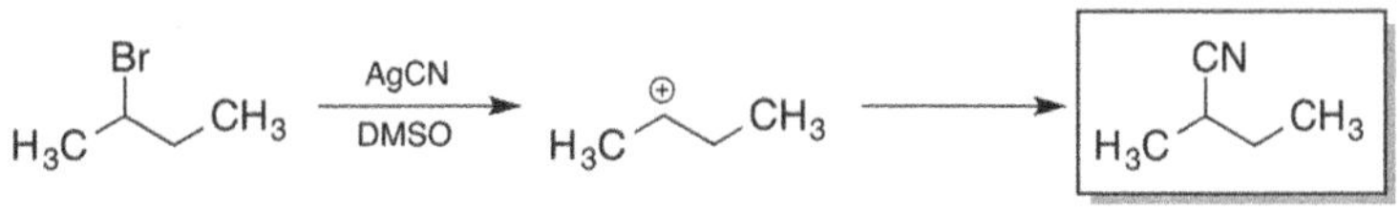

b.

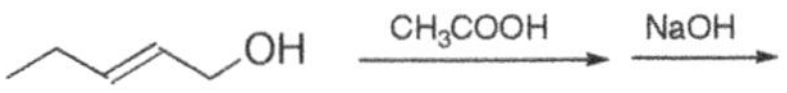

This is a solvolysis reaction where the alcohol is protonated and water leaves generating a carbocation. The resulting carbocation then joins with acetic acid or migrates through the double bond (note the arrow-pushing). The migrated carbocation then joins with acetic acid. In both cases, the resulting acetates are cleaved with sodium hydroxide generating a mixture of two alcohols—regenerated starting material and 3-hydroxy-1-pentene.

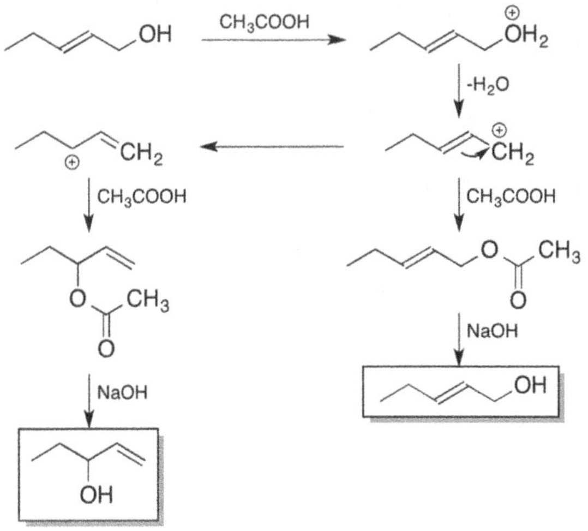

c.

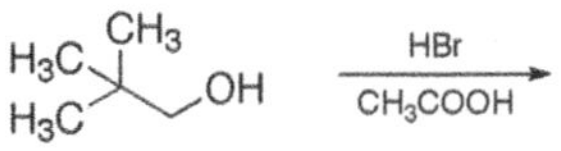

Like the previous example, this is a solvolysis reaction. Initial protonation of the alcohol followed by water leaving generates a primary carbocation. The bromide can then add to this carbocation generating neopentyl bromide. Since, for this carbocation, 1,2-hydride shifts cannot occur, a 1,2-alkyl shift generates a more stable tertiary carbocation. This new carbocation is not subject to possible 1,2-hydride shifts because any such transformation would generate either a less stable secondary carbocation or a less stable primary carbocation. When bromide adds to the tertiary carbocation, a second alkyl bromide is formed.

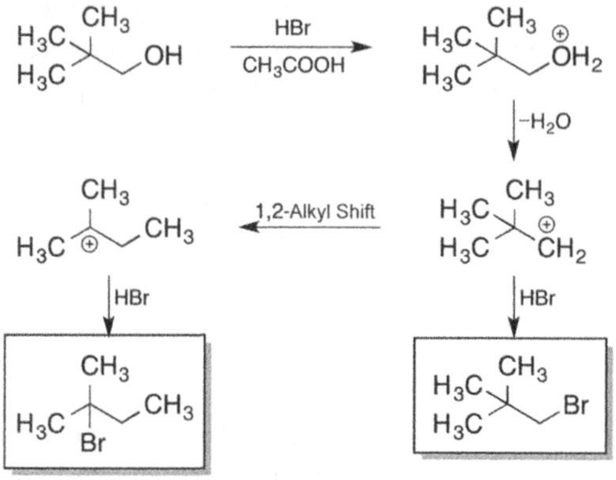

d.

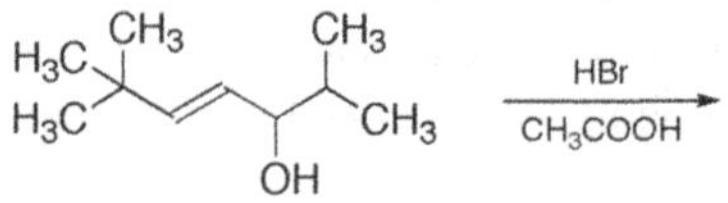

Like Problems 2b and 2c, this is also a solvolysis reaction. However, due to the increased complexity of the starting compound, the potential product mixture is more complex. Specifically, if we consider the initial solvolysis step and elimination of water, we notice that an allyl carbocation is formed that is adjacent to a migratable hydrogen atom. While reaction of this carbocation with bromide generates a secondary allyl bromide, a 1,2-hydride shift followed by reaction with bromide generates a tertiary bromide. Alternatively, if the positive charge migrates through the double bond (see arrow-pushing), an allylic carbocation adjacent to a *tert*-butyl group results. Reaction of this carbocation with bromide generates a new allyl bromide. However, if a 1,2-alkyl shift occurs, the resulting tertiary carbocation can react with bromide to form a new tertiary bromide.

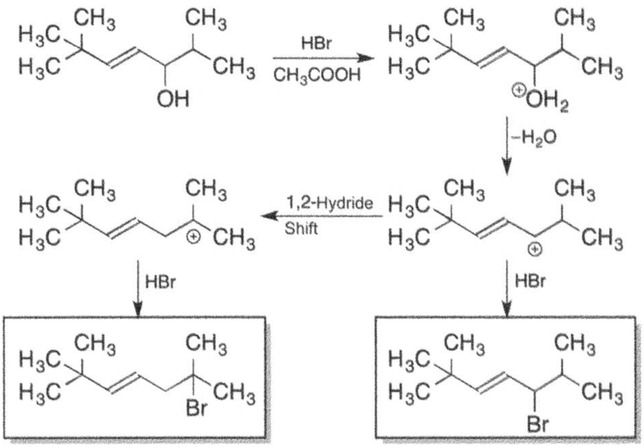

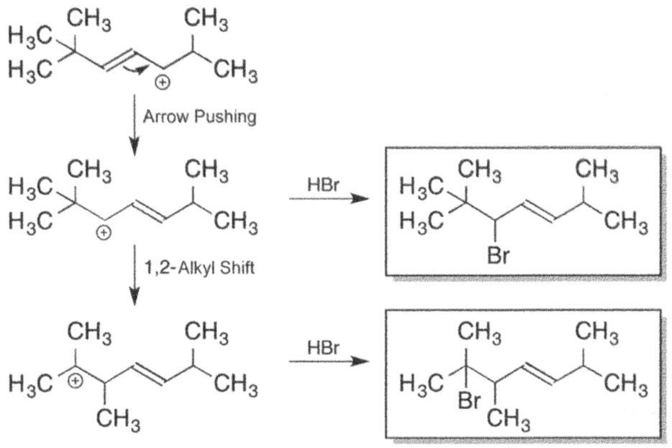

3. *For each of the following reactions, determine which will proceed via an S_N1 or an S_N2 mechanism. In cases where both may be applicable, list appropriate reaction conditions (e.g., solvents, reagents) that would favor S_N1 over S_N2 and vice versa. Explain your answers.*

a.

Because tertiary centers are not susceptible to S_N2 reactions, this reaction will proceed via an S_N1 mechanism.

b.

This reaction will show competition between S_N1 and S_N2 mechanisms due to the fact that this center is less hindered than a tertiary center but more hindered than a primary center. An S_N1 mechanism will be favored using highly polar aprotic solvents to stabilize the forming carbocation. An S_N2 mechanism will be favored when non-polar solvents are used.

c.

This reaction will proceed through an S_N2 mechanism. In general, primary centers are not sterically encumbered enough to inhibit S_N2 reactions. Additionally, recall that primary carbocations are much less stable than tertiary carbocations, making an S_N1 mechanism highly unlikely for this transformation.

4. *In studying 1,2 alkyl and hydride shifts, we explored the observation that shifts will not occur unless the newly formed carbocation is more stable than the starting carbocation. Additionally, as illustrated in Figure 6.12, these shifts were explained using hyperconjugation, thus requiring that the orbital containing the positive charge and the bond containing the shifting group lie within the same plane. This is necessary in order to allow sufficient orbital overlap for the shift to take place.*

   In addition to 1,2-shifts, which occur between adjacent bonds, other shifts are possible where the migrating group apparently moves across space. As with 1,2-shifts, these additional shifts can only occur when the positively charged empty p-orbital lies within the same plane as the bond containing the migrating group, thus allowing sufficient orbital overlap. With this in mind, explain the following 1,5-hydride shift. (Hint: Consider different structural conformations. You may want to use models.) Asterisk (*) marks enrichment with ^{13}C.

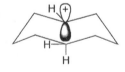

If the eight-membered ring is drawn as illustrated in the following, a planar relationship can be found between the empty p-orbital and a carbon–hydrogen bond on the opposite side of the ring. If the hydrogen atom is located on a carbon atom designated **1**, by numbering the carbon atoms around the ring, the positive charge is localized on carbon atom **5**. Thus, the established relationship between a hydrogen atom on carbon **1** and a positive charge on carbon **5** allows recognition that a 1,5-hydride shift can occur and is required to explain the described transformation.

CHAPTER 7 SOLUTIONS

1. *E2 eliminations do not require acidic protons in order to proceed. Explain how this can occur.*

   The orientation of any proton in a *trans*-periplanar relationship to a given leaving group is usually enough to allow elimination to occur under basic conditions even when, in the absence of an electron-withdrawing group, the proton is not acidic enough to be removed.

2. *When $CH_3OCH_2CH_2CH_2Br$ is treated with magnesium, we get the Grignard reagent $CH_3OCH_2CH_2CH_2MgBr$. However, when $CH_3OCH_2CH_2Br$ is treated with magnesium, the product isolated is $H_2C=CH_2$. Explain this result.*

   Grignard reagents are carbanions stabilized by a MgBr cation. As with all anionic species, if a leaving group is situated on an adjacent center, the structure is subject to an E2 elimination process. Furthermore, CH_3O^- is a sufficient leaving group when it is located adjacent to an anionic center. Therefore, in the case of bromomethoxyethane, E2 elimination leads to the formation of ethylene when the negative charge adopts a *trans*-periplanar relationship to the methoxy group.

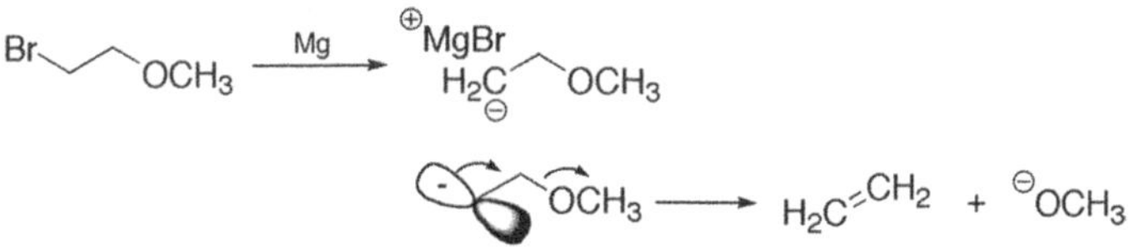

3. *With an understanding of E1 mechanisms, one may realize that under S_N1 reaction conditions, multiple products may form. In addition to the products predicted in Chapter 6 for the following molecules, predict plausible elimination products.*

   a.

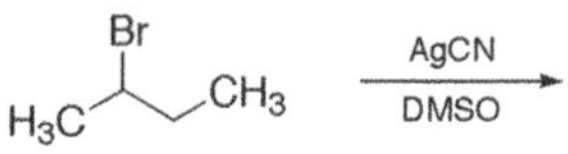

   Silver is very efficient at removing halides resulting in generation of carbocations. Because protons adjacent to carbocations are acidic and, therefore, participate in E1 elimination reactions, several potential products can be identified. These are illustrated in the following using arrow-pushing.

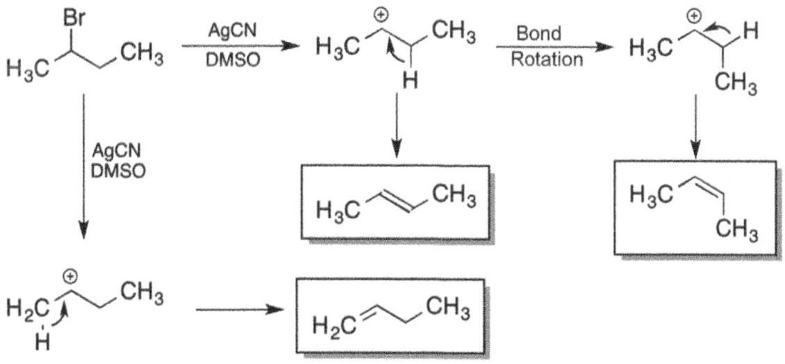

b.

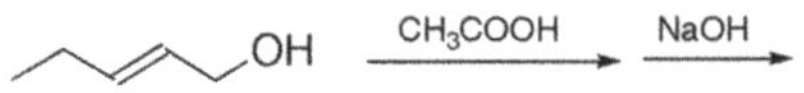

This is a solvolysis reaction where the alcohol is protonated and water leaves generating a carbocation. Because protons adjacent to carbocations are acidic and, therefore, participate in E1 elimination reactions, several potential products can be identified. These are illustrated in the following using arrow-pushing.

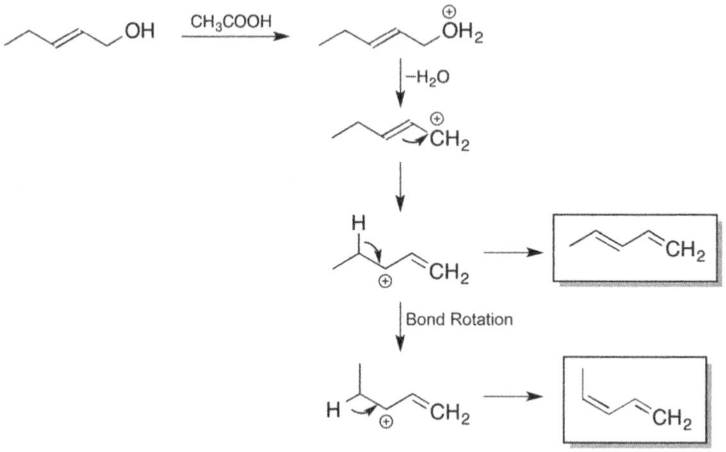

c.

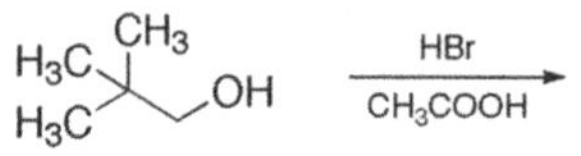

Like the previous example, this is a solvolysis reaction. Initial protonation of the alcohol followed by water leaving generates a primary carbocation. Since, for this carbocation, there are no protons adjacent to the carbocation, no direct E1 elimination products can form. However, if a 1,2-alkyl shift occurs, the resulting tertiary carbocation can participate in such reactions. Potential E1 elimination products are illustrated in the following using arrow-pushing.

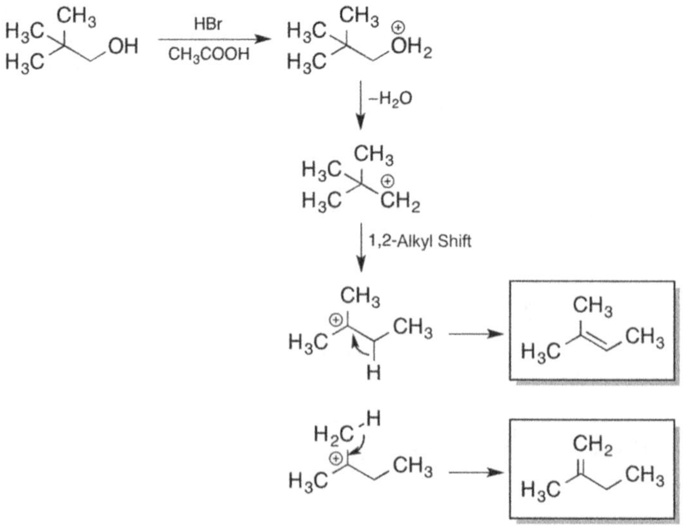

d.

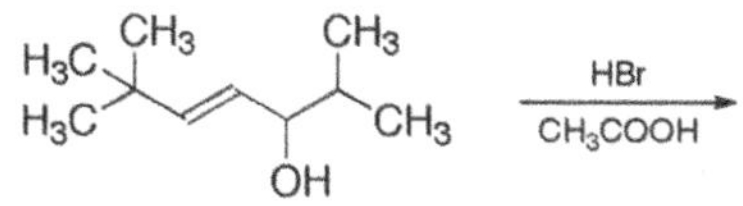

Like Problems 3b and 3c, this is a solvolysis reaction. However, due to the increased complexity of the starting compound, the potential product mixture is more complex. Specifically, if we consider the initial solvolysis step and elimination of water, we notice that an allyl carbocation is formed that is adjacent to a migratable hydrogen atom. While this carbocation can undergo an E1 elimination reaction, a 1,2-hydride shift generates a new carbocation that is also capable of forming E1 elimination products. Furthermore, if the positive charge migrates through the double bond (see arrow-pushing), an allylic carbocation adjacent to a *tert*-butyl group results. While this new carbocation bears no adjacent hydrogen atoms, a 1,2-alkyl shift generates a new carbocation that does possess adjacent protons. This new carbocation can liberate E1 elimination products. All potential E1 elimination products are illustrated in the following using arrow-pushing.

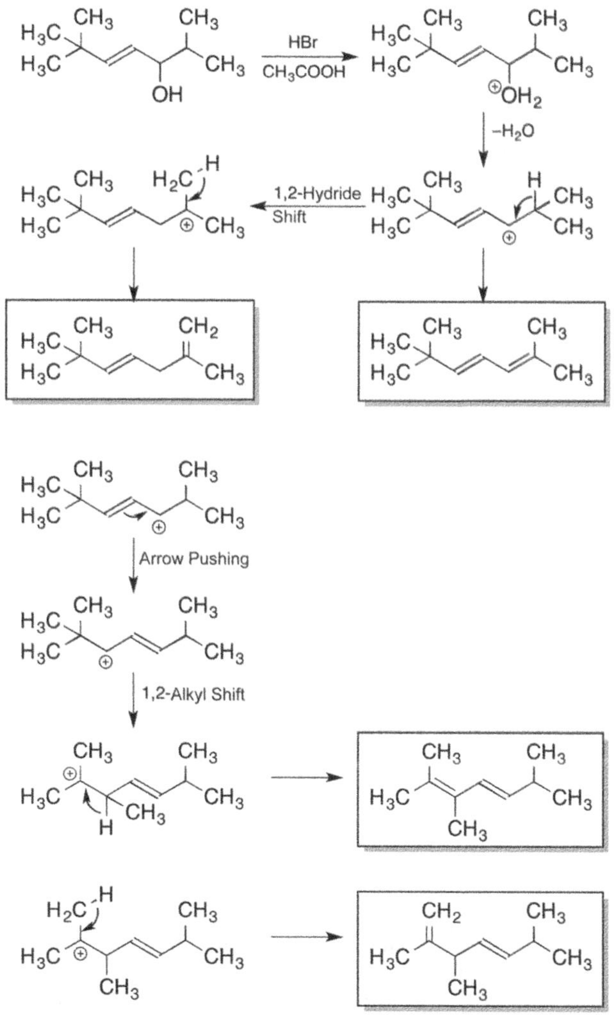

Please note: The most stable products possess conjugated double bonds.

4. *Presently, several different organic reaction mechanisms have been presented. Keeping all of these in mind, predict all of the possible products of the following reactions, and list the mechanistic type or types that these products result from.*

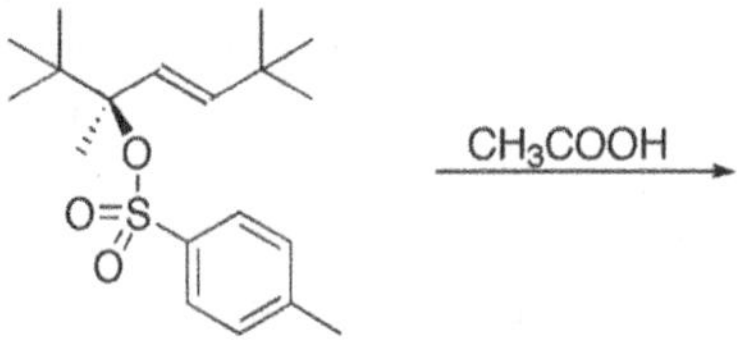

a.

Following initial solvolysis of the tosylate, addition of acetic acid to the carbocation generates an S_N1 product. Please note that there is no preservation of the stereochemical configuration in this reaction.

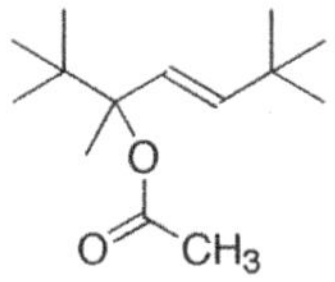

E1 elimination applied to the carbocation formed during solvolysis liberates an olefin.

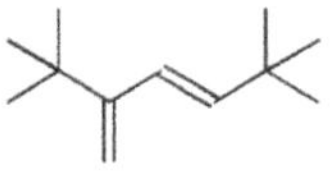

Please note: Since the carbocation formed during solvolysis is both tertiary and allylic, it is very stable, and migration reactions are not likely to occur.

b.

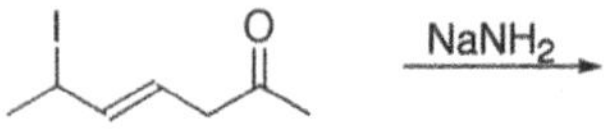

S_N2 displacement of the iodide generates an allylic amine.

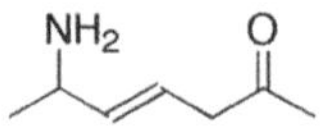

S_N2' displacement of the iodide generates a mixture of two allylic amines with the amine placed on the opposite sides of the double bond.

E2 elimination resulting from removal of a proton adjacent to the carbonyl liberates a diene. Please note that, depending upon the spatial relationship between the carbonyl and the double bond, additional illustrated dienes can form.

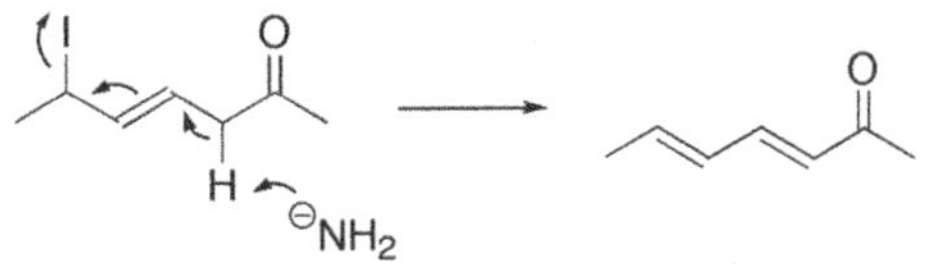

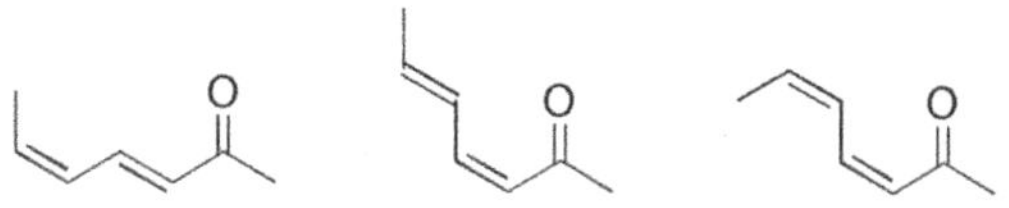

c.

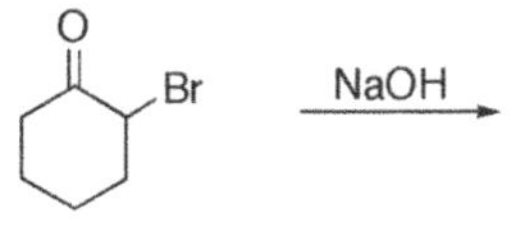

Direct S_N2 displacement of the bromide would be expected to liberate the illustrated hydroxyketone.

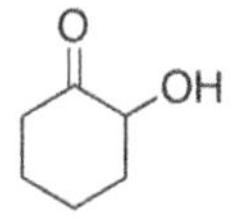

However, this is a special case reaction known as the Favorskii rearrangement. As illustrated in the following using arrow-pushing, sodium hydroxide extracts a proton adjacent to the ketone, and the resulting anion displaces the bromide ion generating a new three-membered ring.

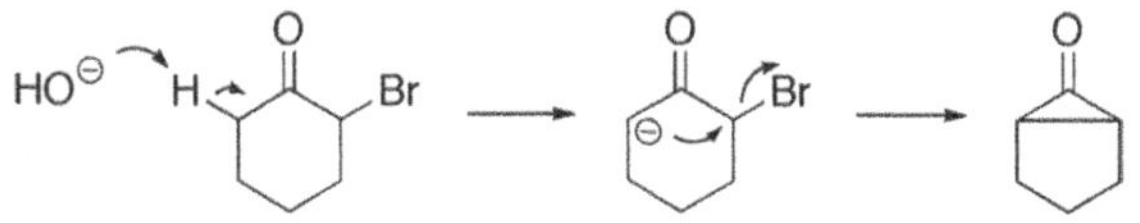

As alluded to in previous discussions, carbonyl groups are polarized with a partial positive charge residing on the carbon atom and a partial negative charge residing on the oxygen atom. This polarization has been used in discussions of charge delocalization. As will be addressed in Chapter 8, the polarized nature of carbonyls render them good electrophiles and, as such, capable of accepting nucleophiles at the partial positive center. As illustrated in the following, a hydroxide anion can now add to the carbonyl, placing a negative charge on the original carbonyl oxygen. That negative charge can then return to the original carbonyl–carbon atom and open the three-membered ring, thus forming a cyclopentane carboxylic acid.

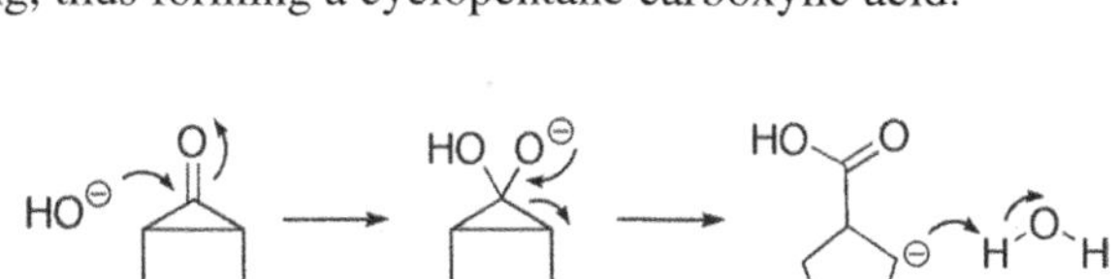

d.

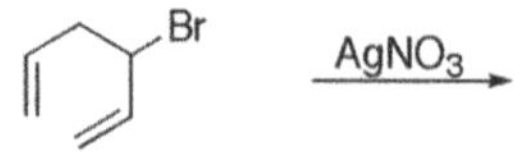

As previously mentioned (see Problem 3a), silver cations are very efficient at removing halide anions. Therefore, under these conditions, liberation of an allylic cation is favorable. This cation can then generate the E1 elimination products shown. Please note that these products are dependent upon the relationship between the two terminal double bonds.

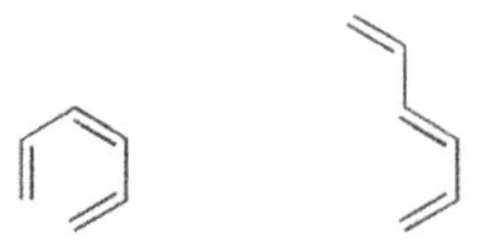

Additionally, as mentioned in Chapter 1, concerted reaction mechanisms can be described using arrow-pushing. As illustrated in the following, both the starting bromide and one of the trienes can undergo Cope rearrangements to form new products. The Cope rearrangement, and other pericyclic reactions, are discussed in Chapter 10.

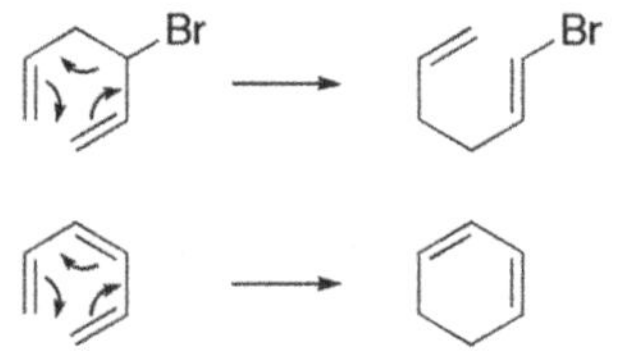

e.

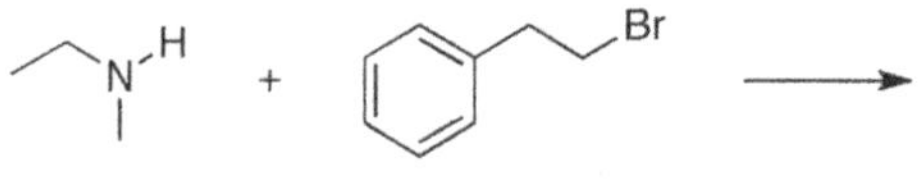

As discussed in Chapter 4, amines are both basic and nucleophilic. Typically when combining a nucleophile with an electrophile such as an alkyl bromide, alkylation of the amine is expected. However, in the case of 2-phenethyl bromide, the protons adjacent to the phenyl ring are weakly acidic. When a weak acid is combined with a base, deprotonation can occur. If an anion forms adjacent to the phenyl ring, elimination of the bromide via an E1cB mechanism is expected with styrene being the observed product.

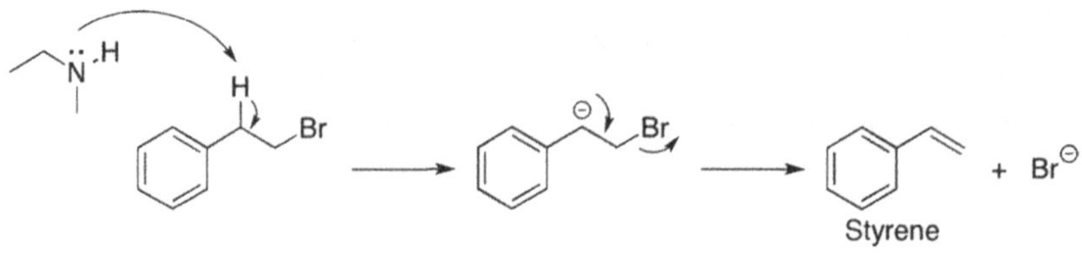

Styrene

f.

Unlike the reaction outcome illustrated in the previous problem, benzyl bromide is a strong electrophile largely due to the partial negative charge residing on the bromine atom. This partial negative charge reduces the basicity of the proton adjacent to the phenyl ring. Therefore, benzyl bromide reacts with ethyl methylamine through an S_N2 mechanism forming ethyl methyl benzylamine.

5. *As mentioned earlier, stereochemistry is not of great concern in the chapters of this book. However, certain mechanistic types will show specific stereochemical consequences when acting on chiral molecules. With this in mind, predict the product resulting from the E2 elimination of HBr when the illustrated isomer of 4-bromo-3-methyl-2-pentanone is treated with sodamide. Show all stereochemistry and explain your answer.*

In order to approach this problem, we must first identify the structure of the starting compound when the acidic proton is oriented *trans*-periplanar to the bromide. The relevant configuration is illustrated in the following and can be visualized using molecular models.

Realizing that the two methyl groups are oriented as projecting out of the same side of the molecule, E2 elimination of HBr can only form a product with the methyl group *cis* to one another. The formation of this product is illustrated in the following using arrow-pushing.

6. *Based on the answer to Problem 5, predict the product of the following reactions and show all stereochemistry.*

a.

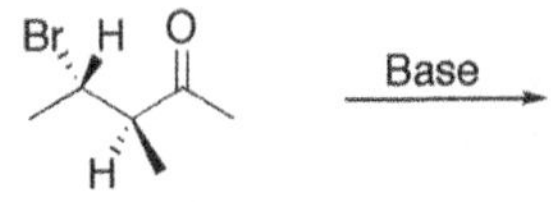

Aligning the acidic proton with the bromide in a *trans*-periplanar orientation allows formation of the illustrated product as shown using arrow-pushing.

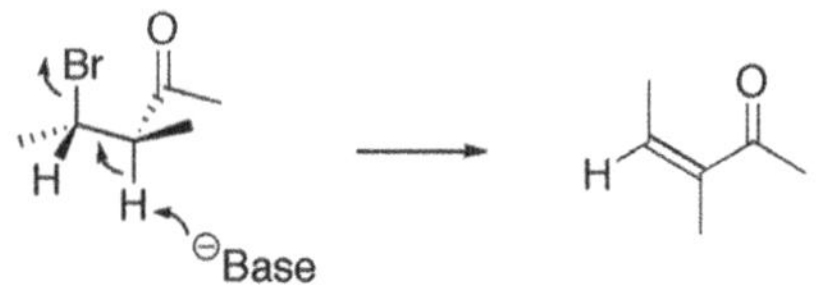

b.

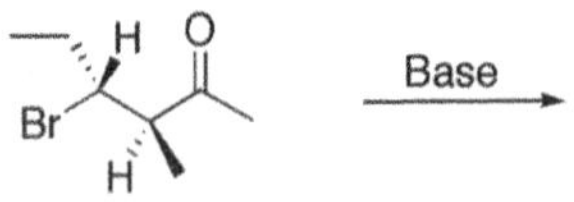

Aligning the acidic proton with the bromide in a *trans*-periplanar orientation allows formation of the illustrated product as shown using arrow-pushing.

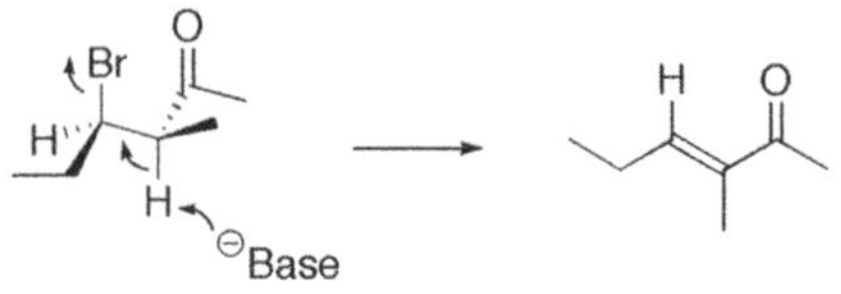

c.

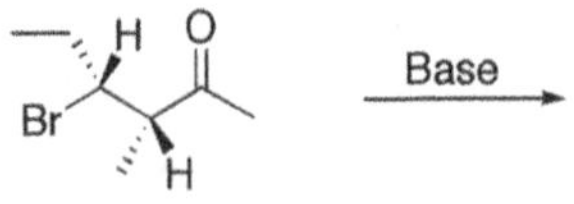

Aligning the acidic proton with the bromide in a *trans*-periplanar orientation allows formation of the illustrated product as shown using arrow-pushing.

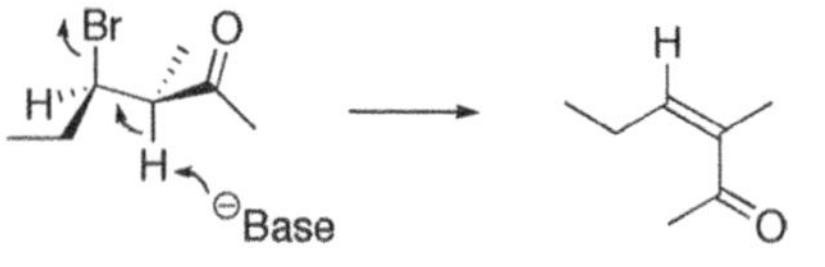

d.

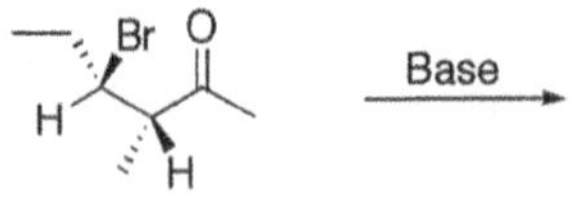

Aligning the acidic proton with the bromide in a *trans*-periplanar orientation allows formation of the illustrated product as shown using arrow-pushing.

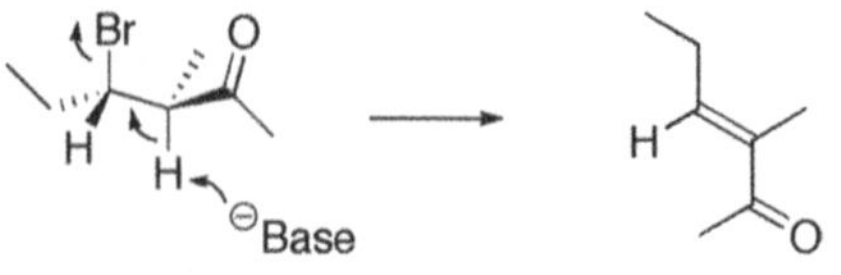

7. *Explain the results of the following experiment.*

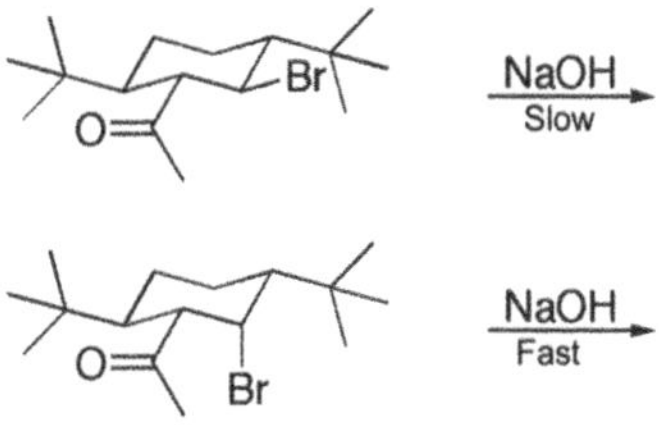

The product for both of these reactions is the cyclohexene as shown in the following:

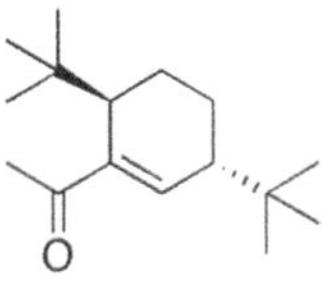

This product forms via an E2 elimination mechanism. Consequently, the elimination reaction is only favored if a *trans*-periplanar relationship exists between the acidic proton and the bromide. In the case of the starting material used in the "fast" reaction, this is the case. However, looking at the starting material used in the "slow" reaction, no *trans*-periplanar relationship exists between the acidic proton and the bromide.

Because the "slow" reaction does, in fact, form the same product as that formed in the "fast" reaction, transformations allowing a *trans*-periplanar arrangement to form must take place. As illustrated in the following, these transformations begin with initial deprotonation adjacent to the carbonyl group. The resulting anion then inverts through reversible delocalization of the negative charge into the carbonyl. Next, the chair form of the ring inverts, allowing placement of the bromide and anion into axial positions. At this point, elimination to the cyclohexene occurs.

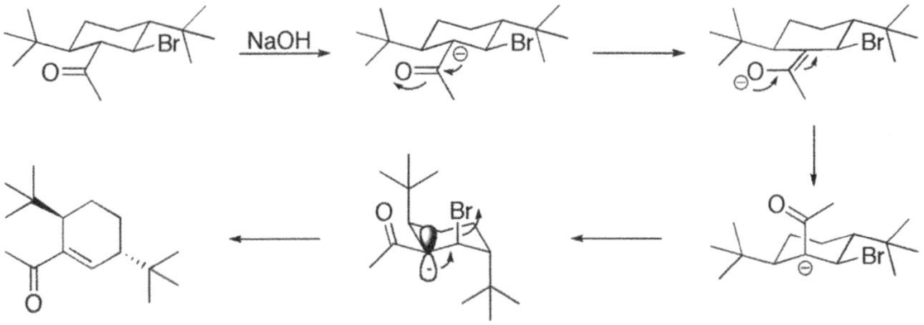

8. *Reaction (I) proceeds through the E2 elimination mechanism and reaction (II) proceeds through the E1cB mechanism. Using arrow-pushing, explain these observations.*

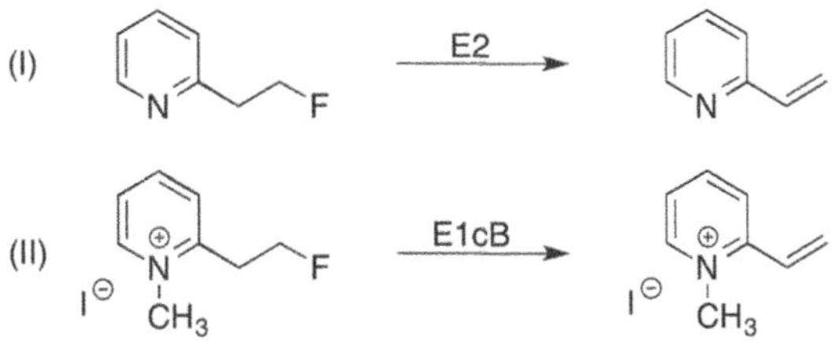

In reaction (I), the proton adjacent to the pyridyl ring is not very acidic. This is rationalized by recognizing that a partial negative charge resides on the pyridyl ring nitrogen. As illustrated, removal of the α-proton produces a negative charge that, when delocalized into the pyridyl ring, increases the negative charge density on the ring nitrogen. Due to the low acidity of this system, elimination occurs through the E2 mechanism.

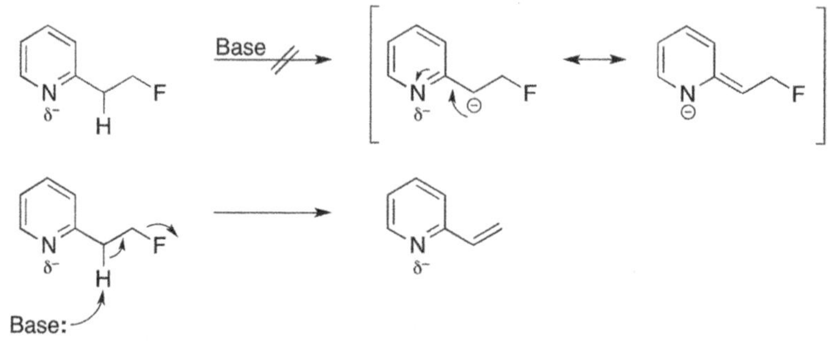

In reaction (II), the proton adjacent to the pyridyl ring is acidic. This is rationalized by recognizing that a positive charge resides on the methylated pyridyl ring nitrogen. As illustrated, removal of the α-proton produces a negative charge that, when delocalized into the pyridyl ring, neutralizes all charges at the expense of a the aromatic ring. Because aromaticity is a stabilizing factor, the pyridyl ring remains fully conjugated with anionic character remaining associated with the adjacent carbon atom. This conjugate base then undergoes elimination through the E1cB mechanism.

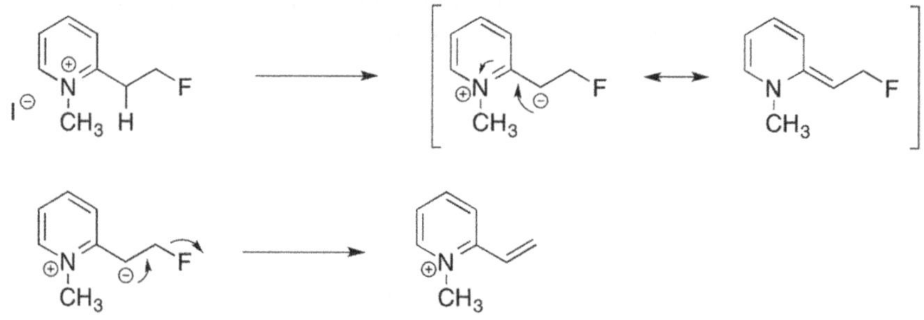

CHAPTER 8 SOLUTIONS

1. *Predict the products of the following reactions and then answer the following questions. Consider stereochemistry.*

I. 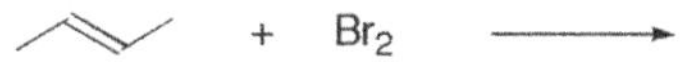   +   Br$_2$ ⟶

The addition of bromine across a double bond proceeds with attachment of each bromine atom to opposite faces of the starting olefin. In the case of the present example, the products are illustrated in the following. Please note that the two illustrated products are enantiomers of one another.

II.    +   Br$_2$ ⟶

The addition of bromine across a double bond proceeds with attachment of each bromine atom to opposite faces of the starting olefin. In the case of the present example, the products are illustrated in the following. Please note that the two illustrated products are enantiomers of one another.

III.    +   Br$_2$ ⟶

The addition of bromine across a double bond proceeds with attachment of each bromine atom to opposite faces of the starting olefin. In the case of the present example, the products are illustrated in the following. Please note that the two illustrated products are enantiomers of one another.

IV. 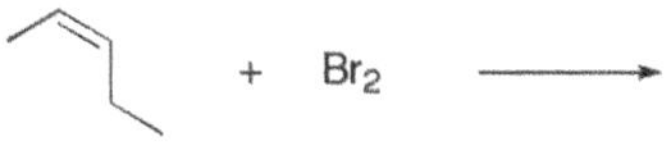   +   Br$_2$ ⟶

The addition of bromine across a double bond proceeds with attachment of each bromine atom to opposite faces of the starting olefin. In the case of the present example, the products are illustrated in the following. Please note that the two illustrated products are enantiomers of one another.

a.   *Are the products of reactions (I) and (II) the same or are they different? Explain your answer.*

The products of the first two reactions (I) and (II) are different. While the mechanistic delivery of a bromine to opposite faces of an olefin is the same for both reactions, the products of reaction (I) are diastereomers relative to the products of reaction (II). The difference in stereochemistry is the result of the olefin of reaction (I) being *trans*, while the olefin of reaction (II) is *cis*.

b.   *How do you account for the products of reactions (I) and (II)?*

When the initial addition of bromine to the double bond occurs, the addition takes place on only one side of the molecule forming what is known as a bridged bromonium ion. Therefore, the resulting three-membered intermediate retains the geometry of the starting olefin. Nucleophilic attack then occurs from the opposite side of the molecule, thus inverting the stereochemistry at one of the two centers. Since the substrates are symmetrical, only enantiomers are formed in reactions (I) and (II).

c.   *Are the products of reactions (III) and (IV) the same or are they different? Explain your answer.*

The products of the first two reactions (III) and (IV) are different. While the mechanistic delivery of a bromine to opposite faces of an olefin is the same for both reactions, the products of reaction (III) are diastereomers relative to the products of reaction (IV). The difference in stereochemistry is the result of the olefin of reaction (III) being *trans*, while the olefin of reaction (IV) is *cis*.

2. *Predict all of the products of the following reactions.*

a. 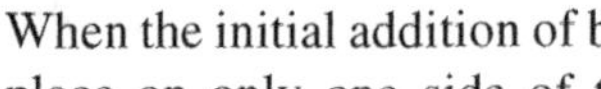   ⫽   +   HBr   ⟶

Since the starting olefin is symmetrical, there can be only one product as illustrated in the following:

⫽   +   HBr   ⟶   ⌒Br

b.    ⟋⟍   +   HBr   ⟶

Since the starting olefin is asymmetrical, there are two possible products as illustrated in the following:

⟋⟍   +   HBr   ⟶   Br⟍⟋⟍   +   Br↓⟍

Considering Markovnikov's rule, 2-bromopropane is expected to form in greater quantity compared with 1-bromopropane.

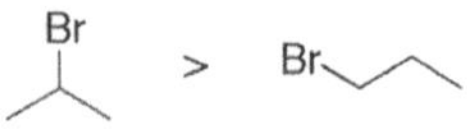

Br↓⟍   >   Br⟍⟋⟍

c.

Since the starting olefin is asymmetrical, there are initially only two products to consider as illustrated in the following:

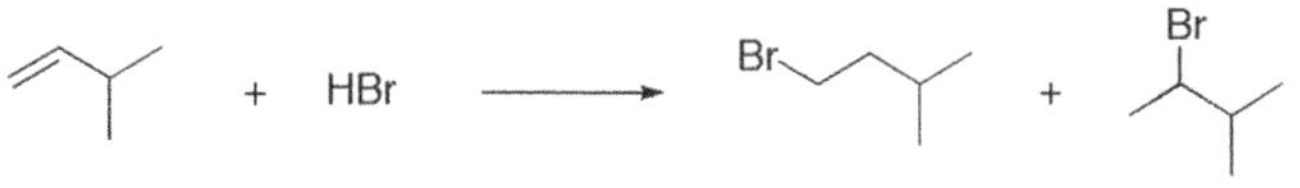

However, recognizing that the initial protonation of the olefin generates positive charges on two adjacent carbon atoms (Scheme 7.6) and that the positive charge at the secondary center is capable of receiving a 1,2-hydride shift (Chapter 5), generation of a tertiary carbocation is possible as illustrated in the following:

1,2-Hydride Shift

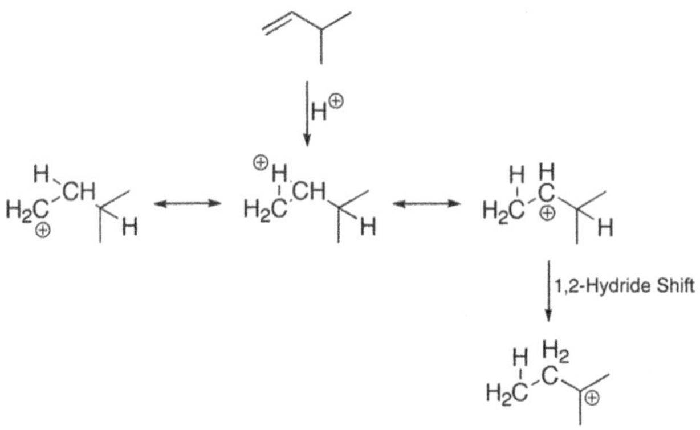

Thus, there are three potential products from this reaction.

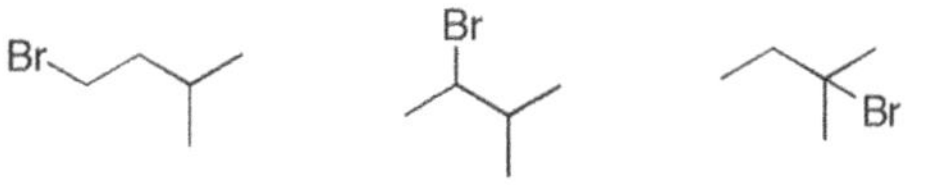

3. *Explain the results of the following reactions. Use arrow-pushing and specify mechanistic types.*

a.

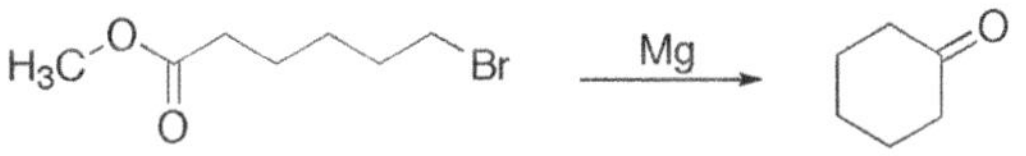

Magnesium reacts with alkyl halides to form alkylmagnesium bromide salts known as Grignard reagents. As mentioned throughout this book, these species bear nucleophilic carbon atoms. As illustrated using arrow-pushing, the alkyl anion adds to the carbonyl and subsequently eliminates methoxide. This addition–elimination process leads to the formation of cyclohexanone.

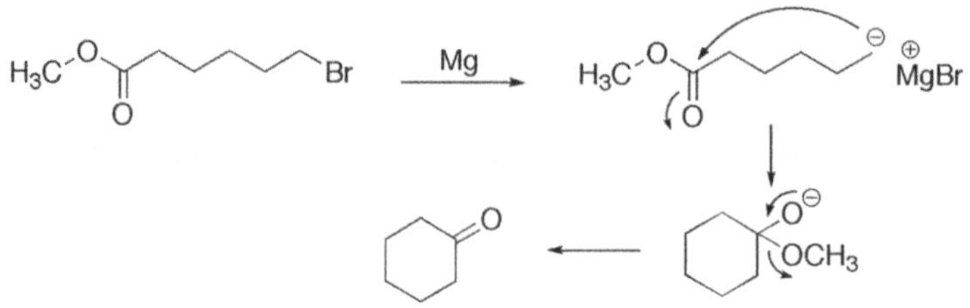

b.

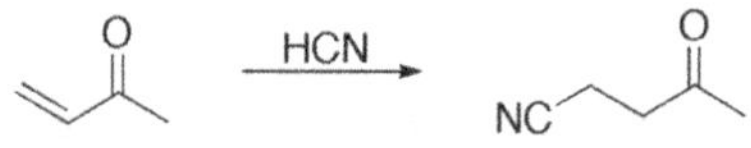

As illustrated using arrow-pushing, the cyanide anion adds to the unsaturated ketone via a 1,4-addition.

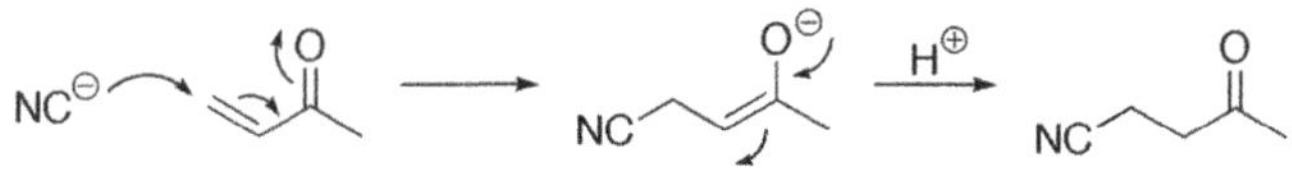

c.

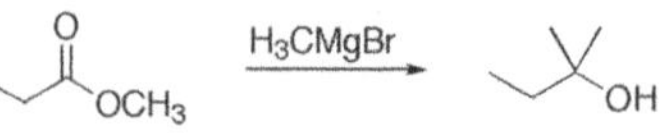

As illustrated using arrow-pushing, the first methyl anion drives an addition–elimination reaction forming a ketone. The second methyl anion then adds to the carbonyl in a 1,2-addition, generating the final alcohol.

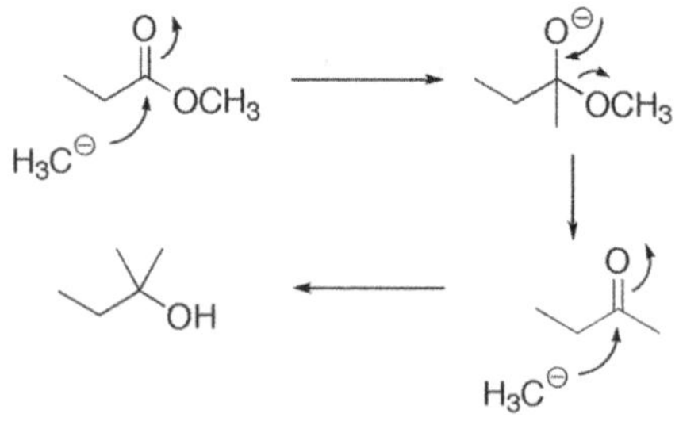

d.

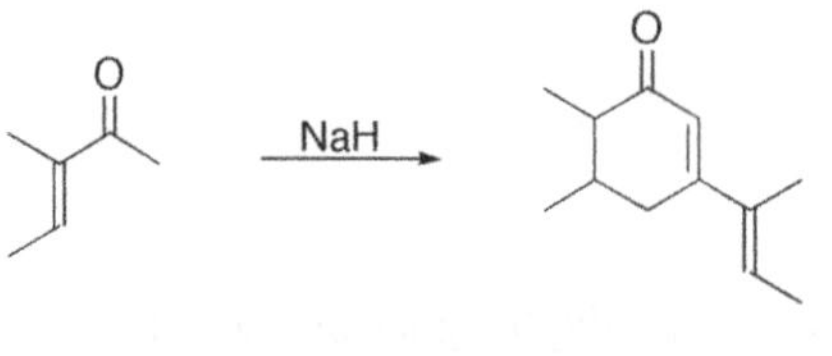

The illustrated product results from a dimerization of the starting material through a multistep process. As illustrated in the following, initial deprotonation of the starting material with sodium hydride generates an acyl anion that adds, through a 1,4-addition, to the carbonyl of a second starting material molecule. Subsequent proton transfer sets up the intermediate species for an aldol condensation.

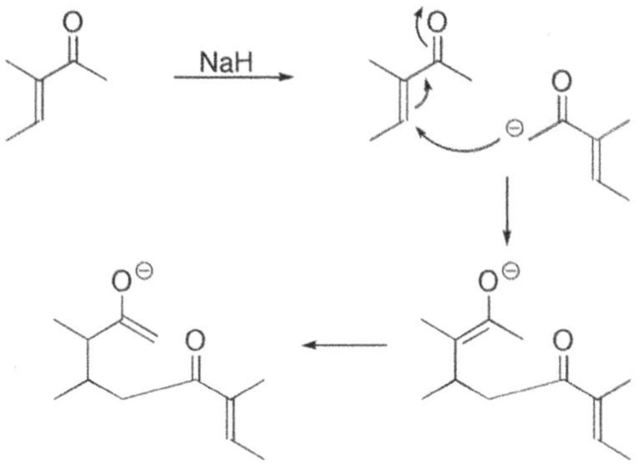

In the second phase of this transformation, illustrated in the following, a six-membered ring is formed through an intramolecular 1,2-addition. Subsequent protonation of the alkoxide anion and elimination of water generates the final product.

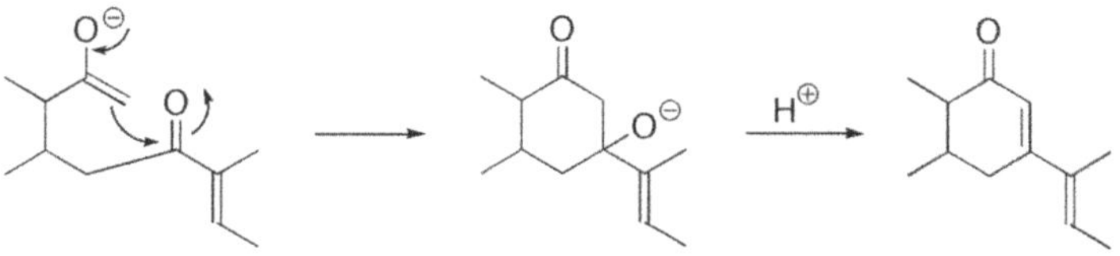

This sequence of steps is known as the Robinson annulation.

4. *Explain the following reactions in mechanistic terms. Show arrow-pushing.*

a.

As presented in this chapter, olefins can become protonated under acidic conditions, leading to the formation of electrophilic and cationic carbon atoms. Furthermore, because olefins have nucleophilic character, they can add to sites of positive charge. The cascading of this mechanism, illustrated in the following, generates polycyclic systems through the cation–π cyclization.

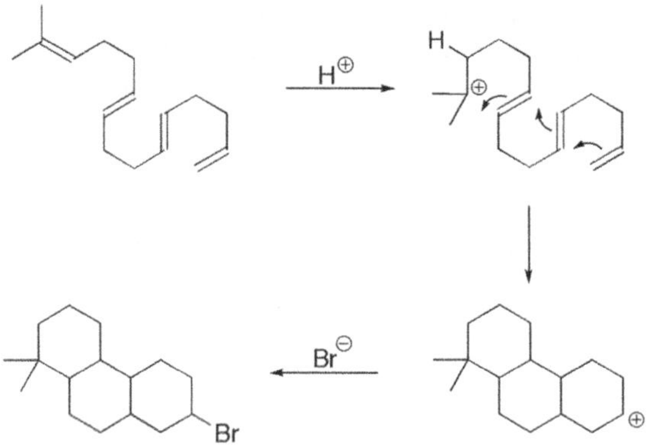

b.

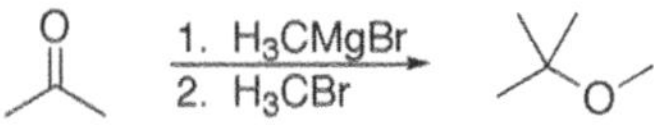

The first step in this sequence is a 1,2-addition of methylmagnesium bromide to acetone. The second step is an S_N2 displacement of bromide with the alkoxide formed in the first step. This two-step process is illustrated in the following using arrow-pushing.

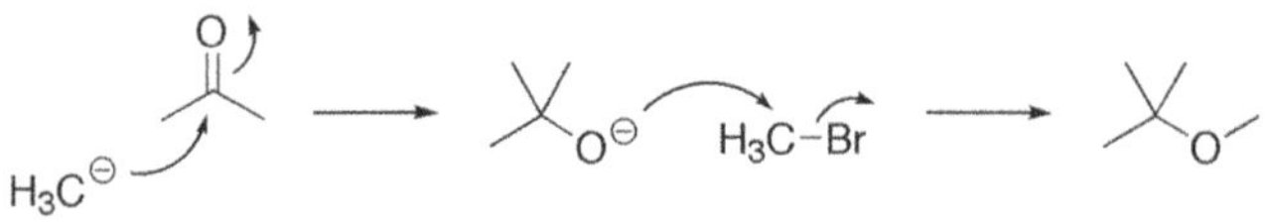

c.

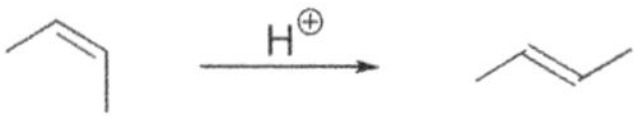

The product of this reaction is the result of a sequence of equilibrium processes. As illustrated in the following, initial protonation of the *cis*-olefin allows transient formation of single bond character. This single bond character then allows for rotation around the central carbon–carbon bond. Final deprotonation liberates the *trans*-olefin. The overall process is driven by the reduced steric interactions present in the *trans*-olefin compared with the *cis*. Specifically, the *cis*-olefin possesses methyl–methyl interactions that are not present in the *trans*.

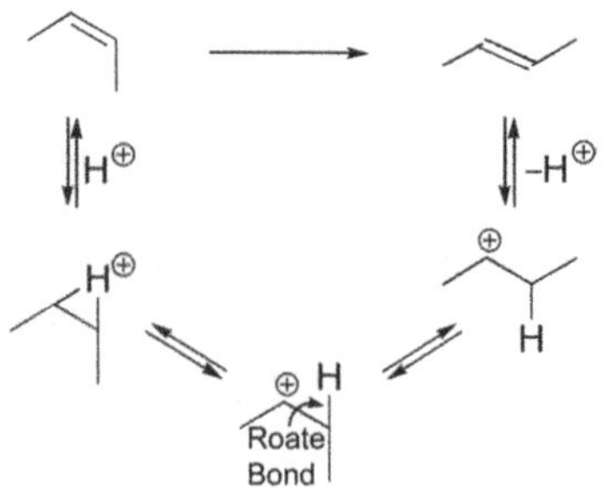

d.

This reaction is a trimerization of acetaldehyde. The mechanism is based on the nucleophilicity of the carbonyl oxygen coupled with the electrophilicity of the carbonyl–carbon. The mechanism is illustrated in the following using arrow-pushing.

5. *Explain the following products resulting from the reaction of amines with carbonyls. Use arrow-pushing and specify mechanistic types.*

a.

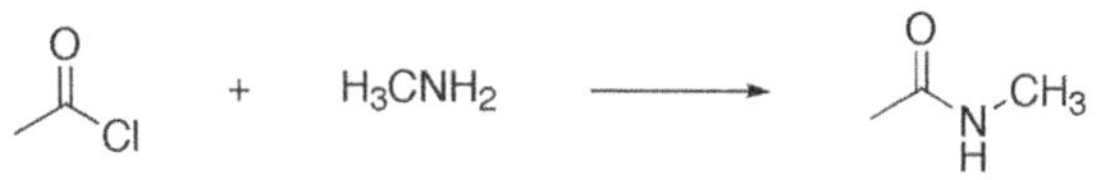

This is an addition–elimination reaction involving addition of methylamine to the acid chloride and elimination of hydrochloric acid. The mechanism is illustrated in the following using arrow-pushing.

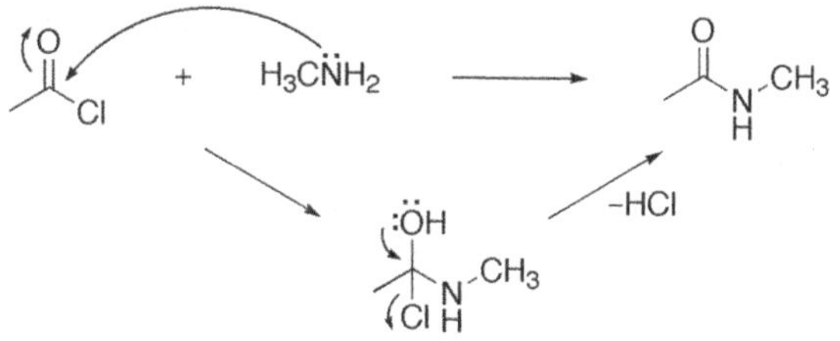

b.

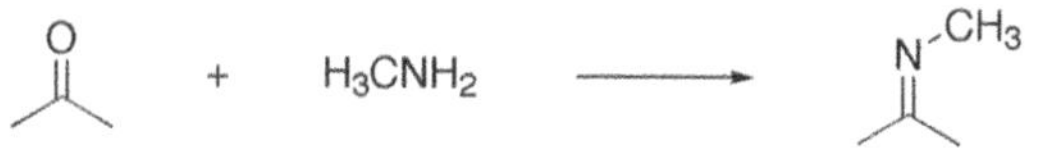

The product of this reaction is an imine resulting from 1,2-addition of methylamine to the carbonyl followed by dehydration. Please note that in the dehydration step, the amine contributes a hydrogen to match the leaving hydroxide group. The mechanism is illustrated in the following using arrow-pushing.

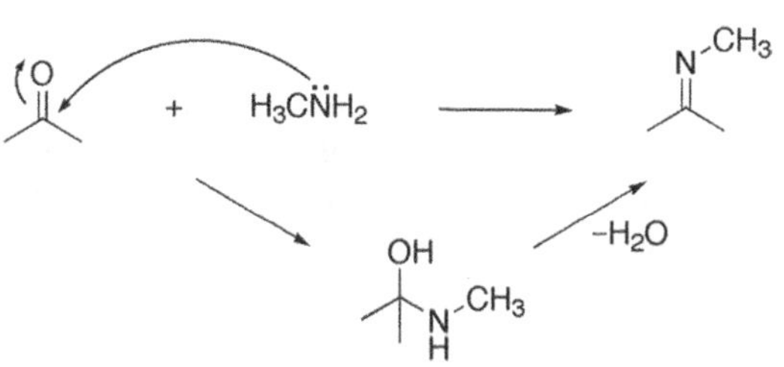

c.

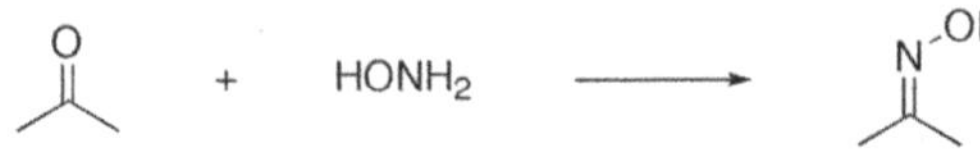

The product of this reaction is an oxime, resulting from 1,2-addition of hydroxylamine to the carbonyl followed by dehydration. Please note that in the dehydration step, the amine contributes a hydrogen to match the leaving hydroxide group. The mechanism is illustrated in the following using arrow-pushing.

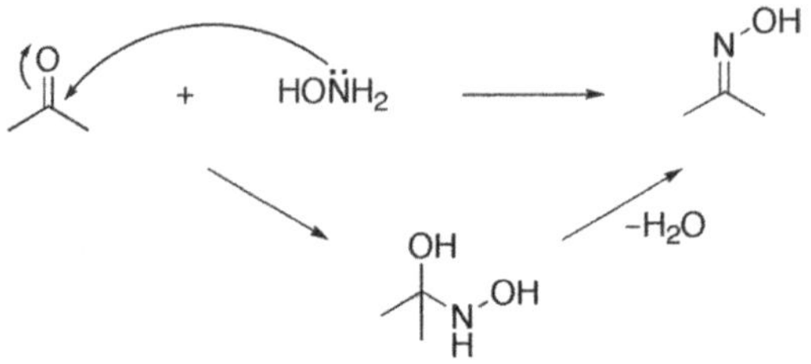

d.

The product of this reaction is an enamine, resulting from 1,2-addition of dimethylamine to the carbonyl followed by dehydration. Please note that in the dehydration step, an adjacent methyl group contributes a hydrogen to pair with the leaving hydroxide group. The mechanism is illustrated in the following using arrow-pushing.

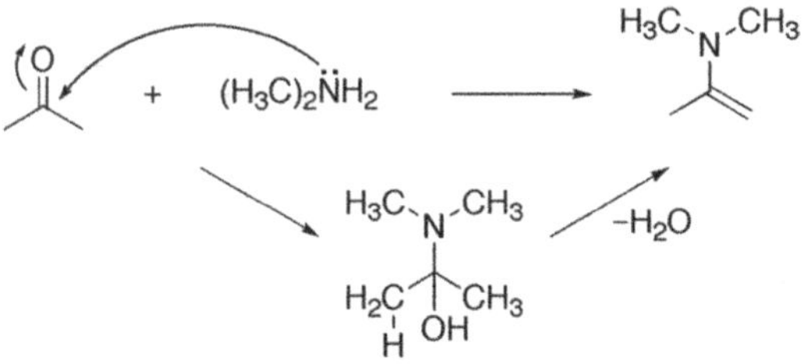

6. *Provide mechanisms for the following reactions. Show arrow-pushing.*

a.

This is an addition–elimination reaction of methanol with acetyl chloride forming methyl acetate. As illustrated in the following using arrow-pushing, methanol is being added while hydrochloric acid is being eliminated. The driving force behind this reaction lies with the relative electronegativities of chlorine and oxygen. Chlorine being more electronegative than oxygen translates to a chlorine anion (chloride) being a better leaving group than an oxygen anion (alkoxide).

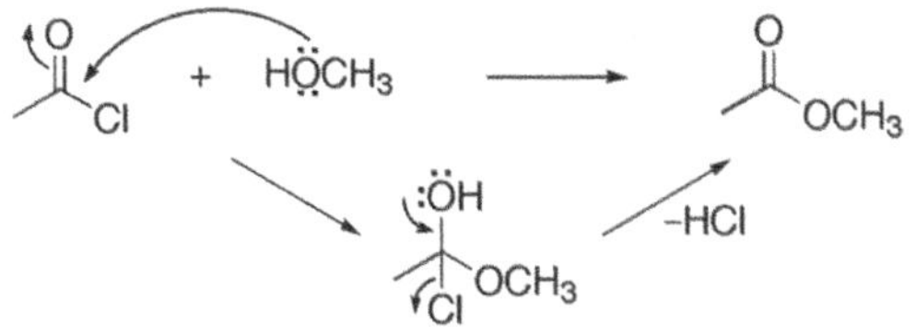

b.

This is an addition–elimination reaction of methylamine with acetyl chloride form-ing methyl acetamide. As illustrated in the following using arrow-pushing, methyl-amine is being added while hydrochloric acid is being eliminated. The driving force behind this reaction lies with the relative electronegativities of chlorine and nitrogen. Chlorine being more electronegative than nitrogen translates to a chlorine anion (chloride) being a better leaving group than a nitrogen anion (amide).

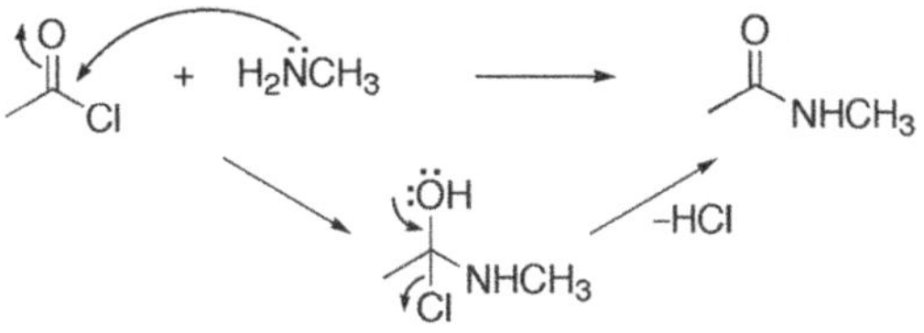

c.

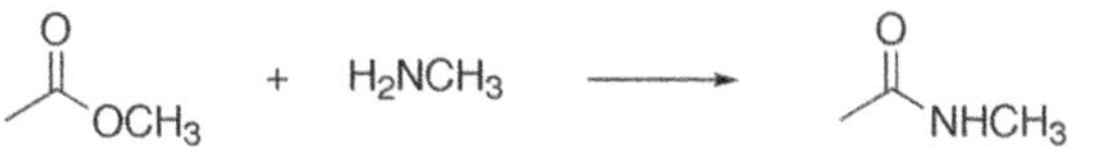

This is an addition–elimination reaction of methylamine with methyl acetate forming methyl acetamide. As illustrated in the following using arrow-pushing, methylamine is being added while methanol is being eliminated. The driving force behind this reaction lies with the relative electronegativities of oxygen and nitrogen. Oxygen being more electronegative than nitrogen translates to an oxygen anion (alkoxide) being a better leaving group than a nitrogen anion (amide).

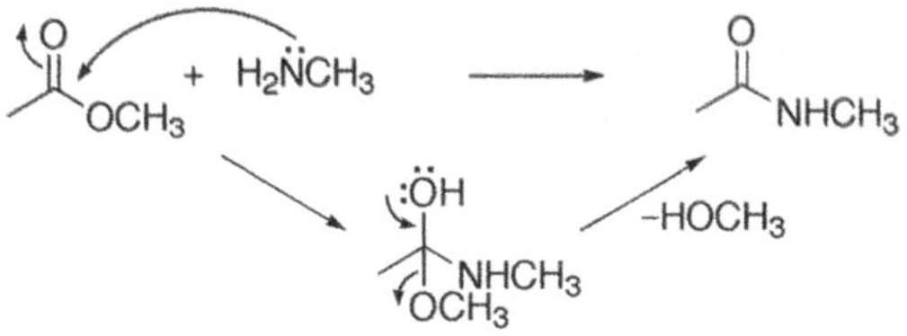

d.

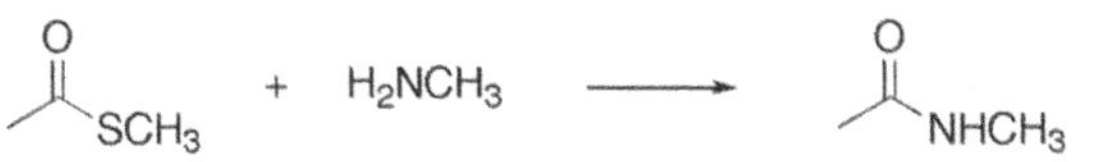

This is an addition–elimination reaction of methylamine with methyl thioacetate forming methyl acetamide. As illustrated in the following using arrow-pushing, methylamine is being added while methanethiol is being eliminated. The driving force behind this reaction lies with the relative polarizabilities of sulfur and nitrogen.

Sulfur being more polarizable than nitrogen translates to a sulfur anion (sulfide) being a better leaving group than a nitrogen anion (amide).

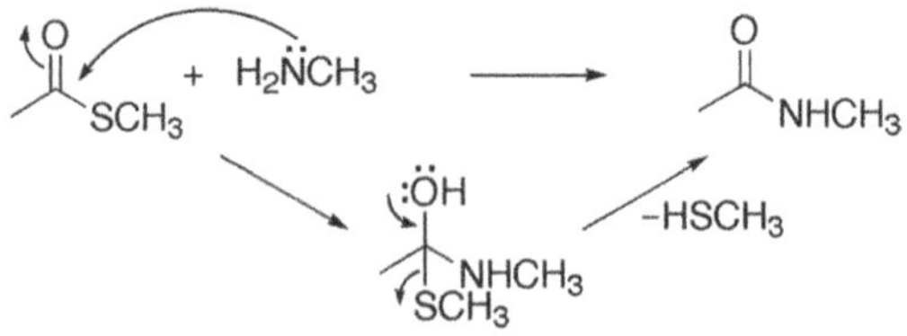

e.

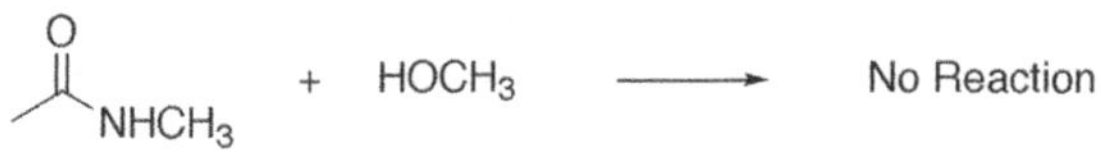

The failure of this attempted addition–elimination reaction is driven by the relative electronegativities of oxygen and nitrogen. Oxygen being more electronegative than nitrogen translates to an oxygen anion (alkoxide) being a better leaving group than a nitrogen anion (amide). Thus, while methanol may add to the amide, methanol will be the only group eliminated, and there will be no net reaction.

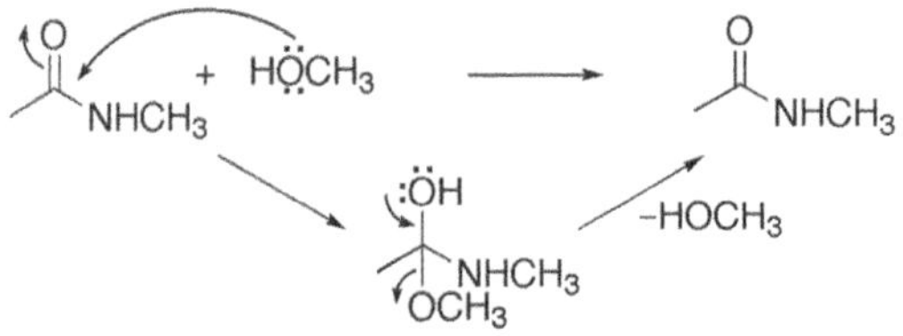

f.

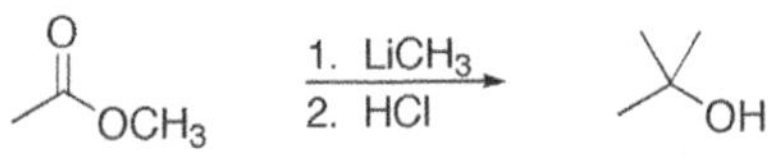

This is a two-step transformation. The first step is an addition–elimination reaction of methyllithium with methyl acetate transiently forming acetone. The second step is a 1,2-addition of methyllithium to acetone forming the final *tert*-butyl alcohol. Hydrochloric acid is present only to quench the formed anions and liberate a neutral product. The steps of this transformation are illustrated in the following using arrow-pushing. Please note that, for simplicity, association of the lithium cations with the anions of the illustrated mechanistic pathway is not shown.

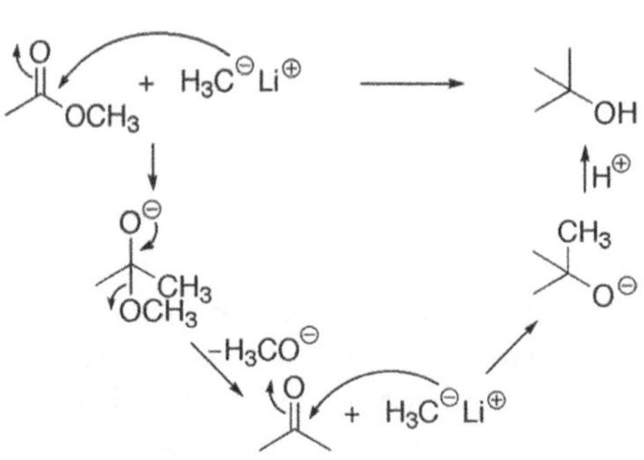

g.

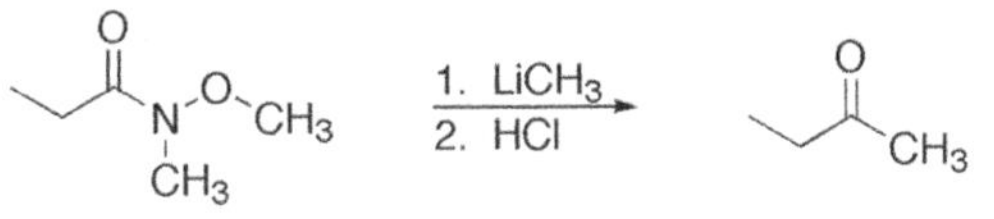

This is an addition–elimination reaction of methyllithium with *N*-methyl-*N*-methoxypropionamide forming 2-butanone. As illustrated in the following using arrow-pushing, methyllithium initially adds to the amide. Unlike the process illustrated in Problem 6f, a second methyllithium does not add and an alcohol is not formed. This is explained by the ability of lithium to coordinate between the two present oxygen atoms. The first is the oxygen of the former carbonyl, and the second is the oxygen associated with the methoxy component of the illustrated amide. Due to the stability of this type of five-membered interaction, initial collapse of the anionic intermediate with loss of *N*-methyl-*N*-methoxyamine is prevented. In Problem 6f, collapse of the anionic intermediate led to regeneration of a carbonyl capable of reacting with a second methyllithium. In this example, this does not happen, and quenching with hydrochloric acid allows exclusive formation of the ketone shown. This process is illustrated in the following using arrow-pushing.

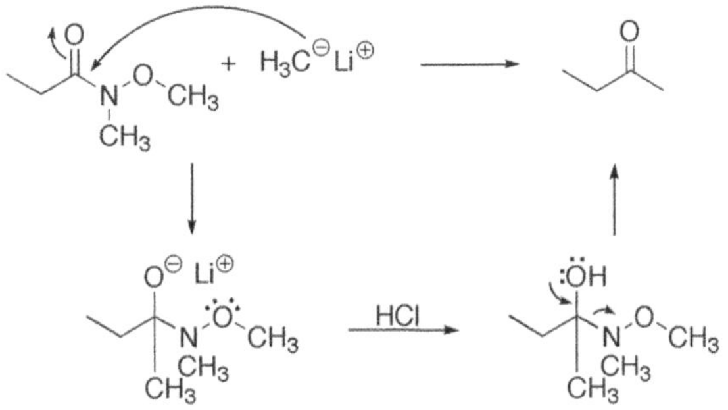

h.

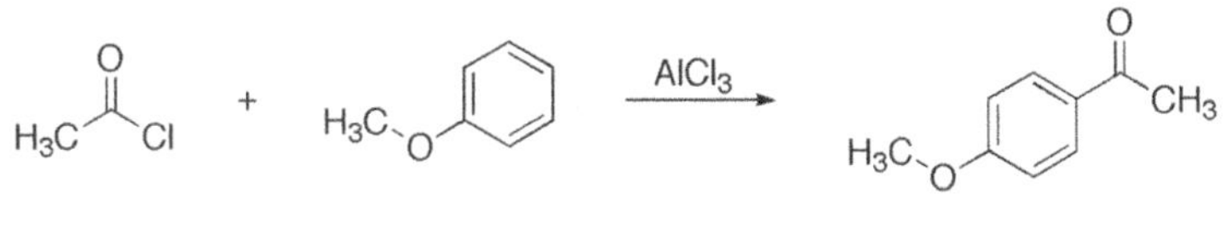

Just as double bonds possess nucleophilic characteristics, so do aromatic rings. By analyzing the charge distribution around an aromatic ring, sites of partial positive charge and sites of partial negative charge can be identified. The sites of partial positive charge are electrophilic in nature, and the sites of partial positive charge are nucleophilic in nature. The partial charge distribution for methoxybenzene was the subject of Problem 2h from Chapter 1 and is shown in the following:

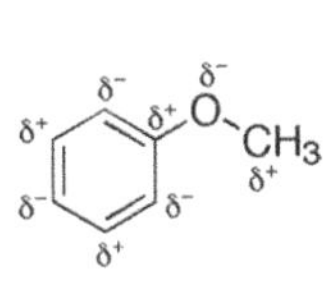

Having identified the nucleophilic sites, this mechanism now becomes an addition–elimination reaction between methoxybenzene and acetyl chloride where methoxybenzene is being added and chloride is being eliminated. As shown in the following, using arrow-pushing, electron movement starts with the methoxy oxygen and moves through the aromatic ring. The addition–elimination steps occur as shown in Problem 6a. Finally, due to the conjugated and charged system, the proton present on the reactive carbon atom of the phenyl ring becomes acidic. Loss of this proton allows rearomatization and neutralization of the cationic intermediate, thus allowing conversion to the final product.

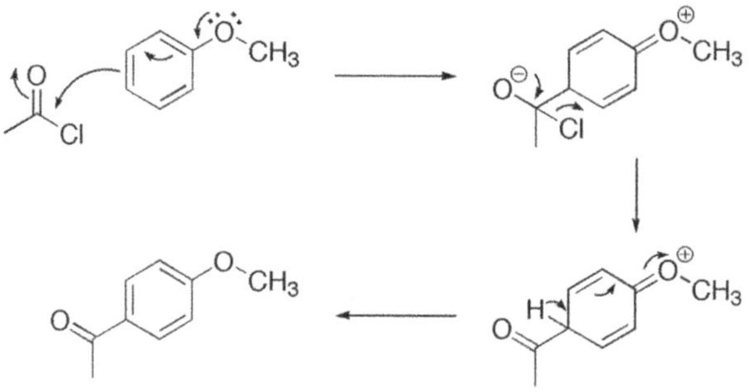

The reaction presented in this problem is known as a Friedel–Crafts acylation. Technically, this example belongs to a class of reactions referred to as electrophilic aromatic substitutions. Furthermore, the actual mechanism associated with this reaction, utilizing Lewis acid reagents as catalysts, proceeds through initial formation of an electrophilic acyl cation followed by reaction with an aromatic ring acting as a nucleophile. This mechanism, shown in the following, reflects distinct parallels to standard addition–elimination reaction mechanisms warranting introduction at this time.

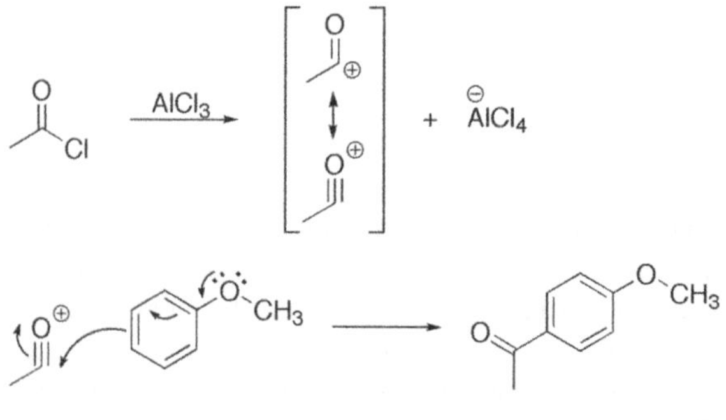

7. *Explain the following amide-forming reactions using arrow-pushing. Specify the structures of A, B, and C, and show all relevant mechanistic steps.*

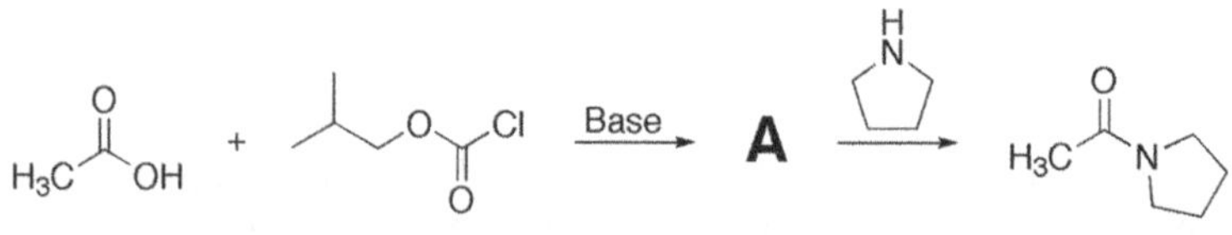

The first step of this sequence is deprotonation of the carboxylic acid by an amine base.

Next, the carboxylate anion participates in an addition–elimination reaction with isobutyl chloroformate. Elimination of a chloride anion results in formation of intermediate **A**. Such reactions are generally facilitated by the introduction of an amine base such as triethylamine (not shown in this problem). The mechanism is illustrated in the following using arrow-pushing, and the illustrated product belongs to a class of compounds known as mixed carbonic anhydrides.

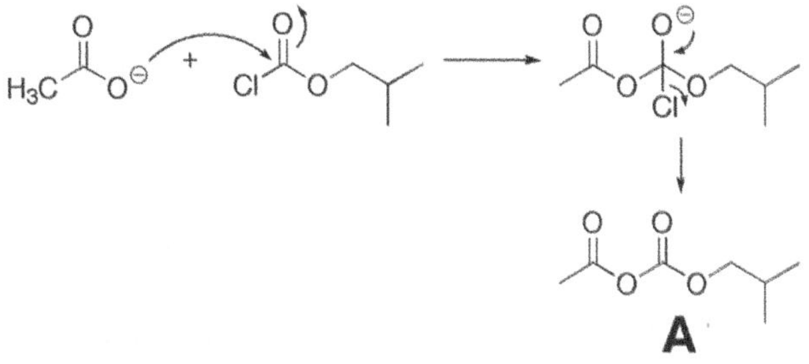

A

Mixed carbonic anhydrides are a form of activated esters that can react with amines to form amides. The addition–elimination mechanism, illustrated in the following using arrow-pushing, involves addition of an amine followed by an elimination step driven by the release of carbon dioxide.

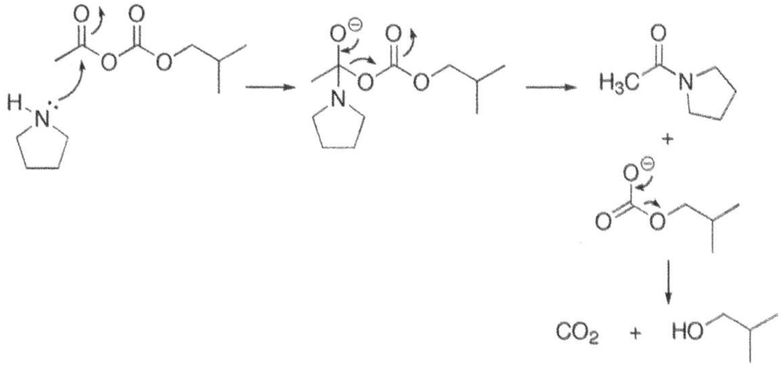

CO_2 + HO⟋⟍

b.

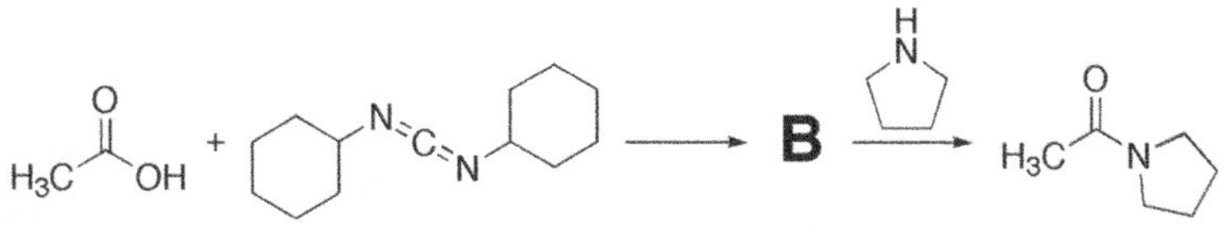

As with Problem 7a, the first step in this reaction is a proton transfer. In this case, the base is a nitrogen atom present on dicyclohexylcarbodiimide.

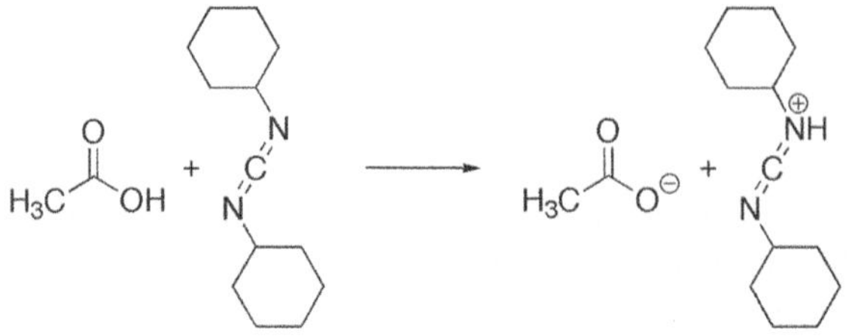

Following proton transfer, the resulting carboxylate anion adds to the protonated dicyclohexylcarbodiimide.

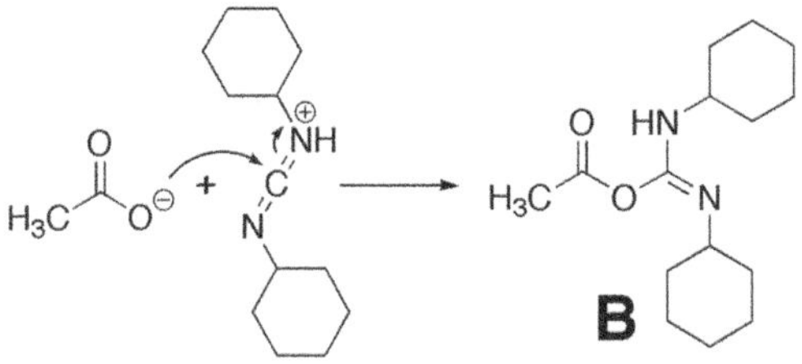

Like the mixed carbonic anhydride (intermediate **A** from Problem 7a), intermediate **B** is an activated ester that can react with amines to form amides. The addition–elimination mechanism, illustrated in the following using arrow-pushing, involves addition of an amine followed by elimination of dicyclohexylurea.

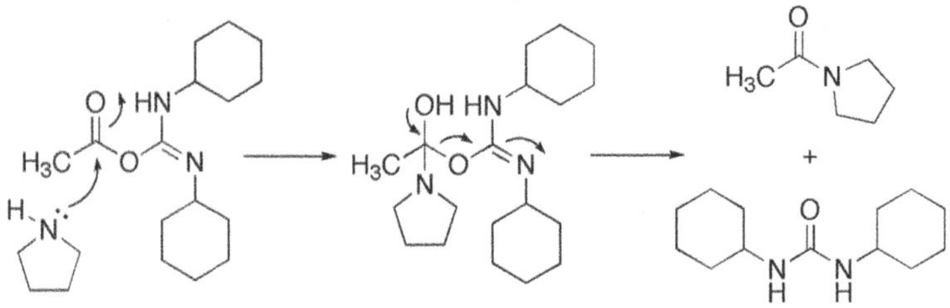

c.

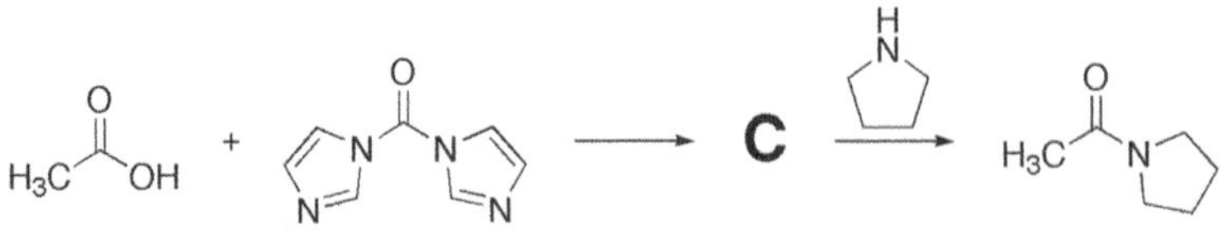

Like Problems 7a and 7b, the first step of this reaction is a proton transfer. In this case, the basic nitrogen is a nitrogen atom present on carbonyl diimidazole.

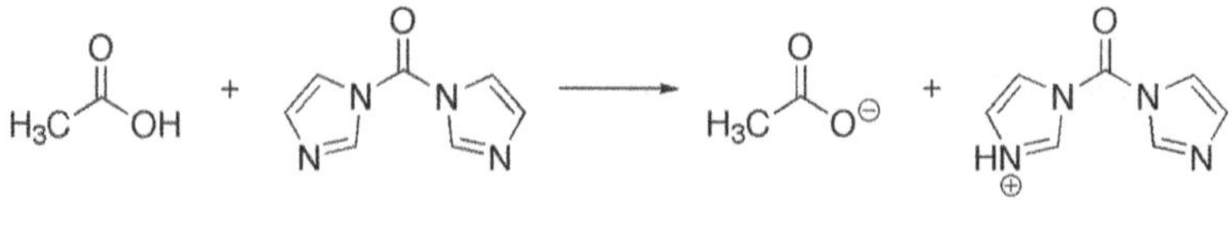

Following proton transfer, the carboxylate anion participates in an addition–elimination reaction where the carboxylate anion adds to the carbonyl of carbonyl diimidazole and imidazole is eliminated. Intermediate **C** then results from a second addition–elimination step where imidazole adds to the resulting anhydride species, and the group being eliminated decomposes to carbon dioxide and imidazole. This sequence of events is illustrated in the following using arrow-pushing.

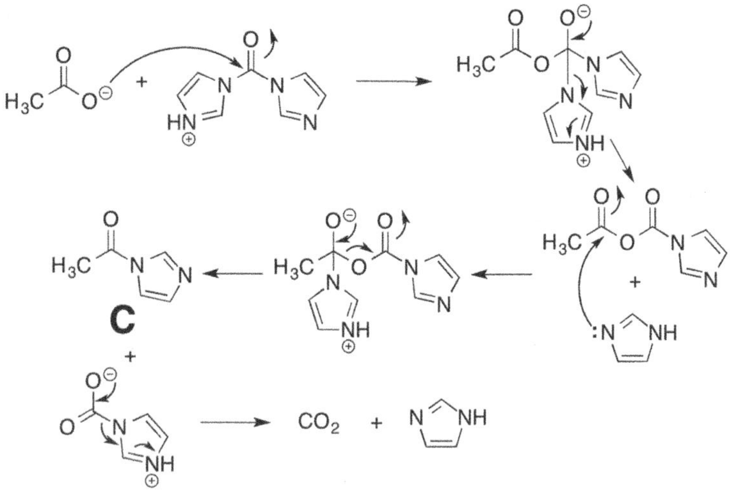

Like the mixed carbonic anhydride (intermediate **A** from 7a), the intermediate imidazolide (intermediate **C**) is an activated carboxy group that can react with amines to form amides. The addition–elimination mechanism, illustrated in the following using arrow-pushing, involves addition of an amine followed by elimination of imidazole.

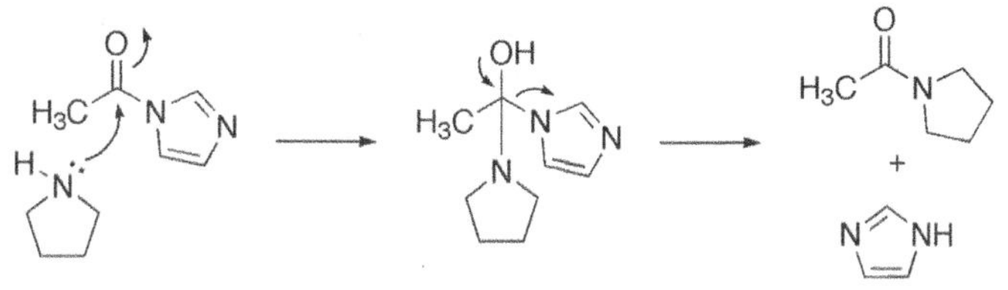

CHAPTER 9 SOLUTIONS

1. *Schemes 9.1 and 9.2 illustrate carbene formation through α-elimination. Predict the carbene species formed from the following compounds. Explain your answer using arrow-pushing.*

a. $CHBr_3$

On reaction with a base, bromoform undergoes deprotonation, forming the illustrated anion. A bromide anion then leaves through α-elimination giving dibromocarbene.

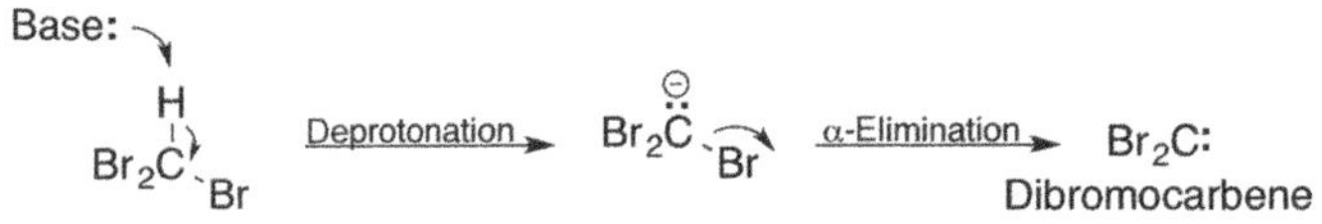

b. CH_2Cl_2

On reaction with a base, dichloromethane (methylene chloride) undergoes deprotonation forming the illustrated anion. A chloride anion then leaves through α-elimination, giving chlorocarbene.

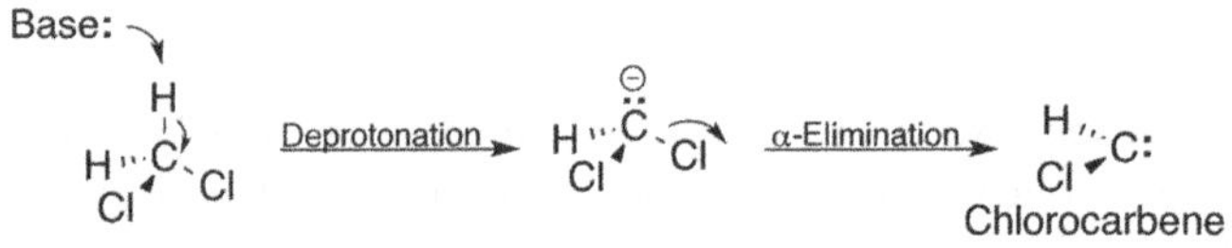

c. CH_2I_2

On reaction with a base, diiodomethane (methylene iodide) undergoes deprotonation forming the illustrated anion. An iodide anion then leaves through α-elimination giving iodocarbene.

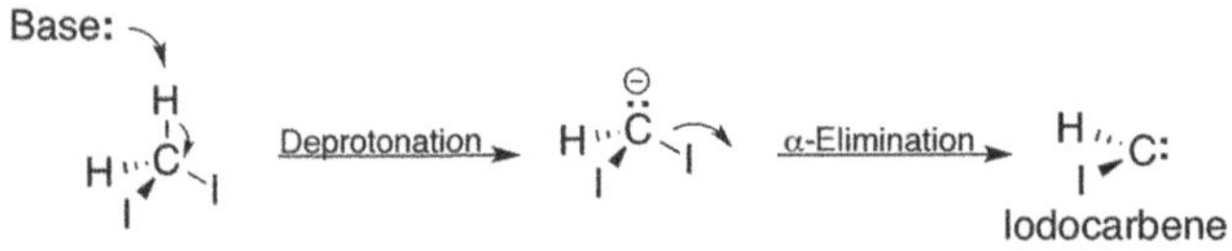

d. CH_2N_2

On exposure to heat or light, diazomethane undergoes α-elimination giving carbene and nitrogen gas.

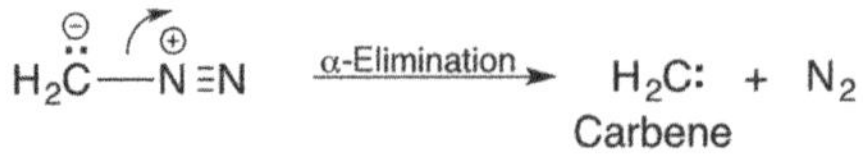

e. 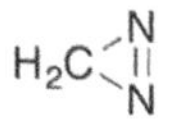 $H_2C \overset{N}{\underset{N}{\overset{|}{|}}}$

Diazirines are known to decompose to carbenes. One possible mechanism involves cleavage of a carbon–nitrogen bond through assistance of a nitrogen lone pair forming diazomethane. Subsequent α-elimination gives carbene and nitrogen gas.

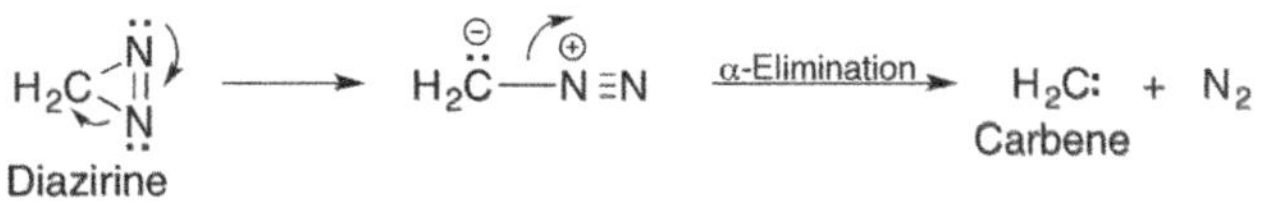

$$H_2C \overset{N}{\underset{N}{\overset{|}{|}}} \longrightarrow H_2\overset{\ominus}{C}—\overset{\oplus}{N} \equiv N \xrightarrow{\alpha\text{-Elimination}} H_2C\colon + N_2$$

Diazirine   Carbene

2. *Predict all of the products of the following reactions using arrow-pushing. Show stereochemistry where applicable.*

While this exercise may seem repetitive, the lesson actually focuses on reaction outcomes. When planning synthetic strategies, it is important to recognize both the mechanisms associated with given transformations as well as the likely products or product mixtures. For many of the following examples, reaction outcomes are based on the spatial orientations associated with how reactants and reactive species approach one another.

a.  $\diagup\!\!\diagup$ + $CHCl_3$ $\xrightarrow{^t BuOK}$

Chloroform, on reaction with potassium *tert*-butoxide, forms dichlorocarbene.

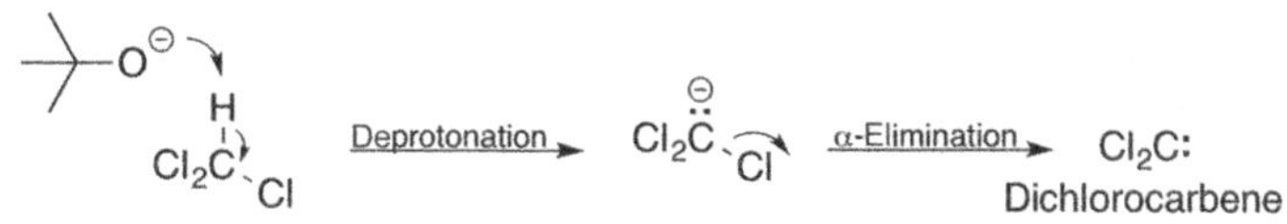

$$\xrightarrow{\text{Deprotonation}} Cl_2\overset{\cdot\cdot}{C} \cdot_{Cl} \xrightarrow{\alpha\text{-Elimination}} Cl_2C\colon$$

Dichlorocarbene

Dichlorocarbene inserts into the 1-butene double bond forming a cyclopropane ring.

$$Cl_2C\colon \longrightarrow$$

Because this reaction involves approaches from both the top and the bottom faces of 1-butene, a mixture of two enantiomeric products is formed.

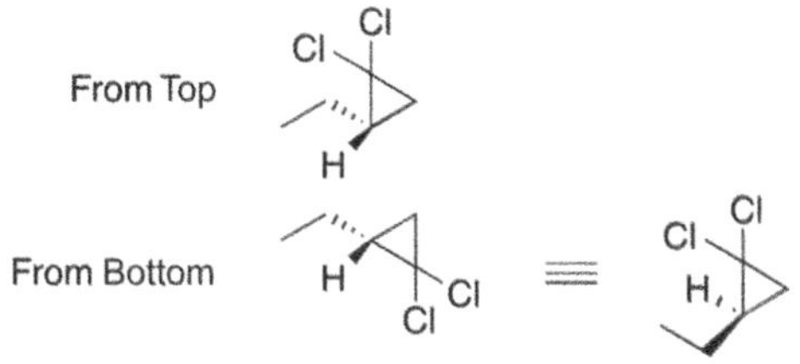

From Top

From Bottom

b.

+ CHCl₃ $\xrightarrow{\text{\textit{t}BuOK}}$

As with Problem 2a, the reactive species is dichlorocarbene that inserts into the 2-methyl-1-butene double bond forming a cyclopropane ring.

Because this reaction involves approaches from both the top and the bottom faces of 2-methyl-1-butene, a mixture of two enantiomeric products is formed.

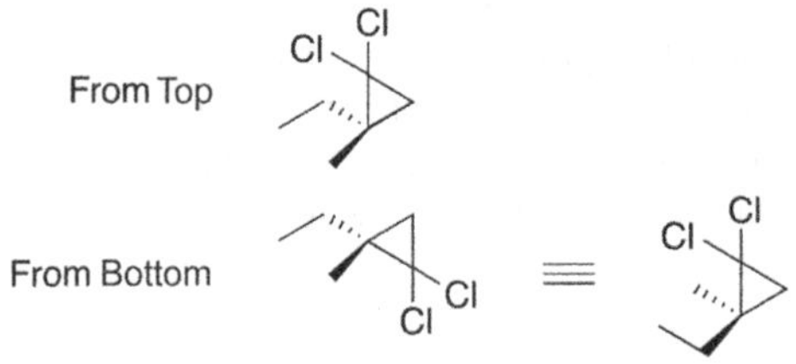

c.

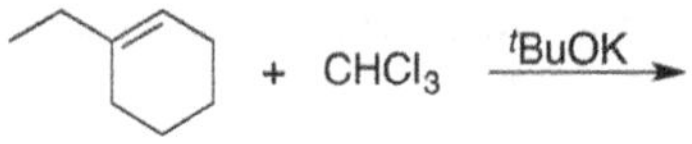

+ CHCl₃ $\xrightarrow{\text{\textit{t}BuOK}}$

As with Problem 2a, the reactive species is dichlorocarbene that inserts into the 1-ethyl-1-cyclohexene double bond forming a cyclopropane ring.

Because this reaction involves approaches from both the top and the bottom faces of 1-ethyl-1-cyclohexene, a mixture of two enantiomeric products is formed.

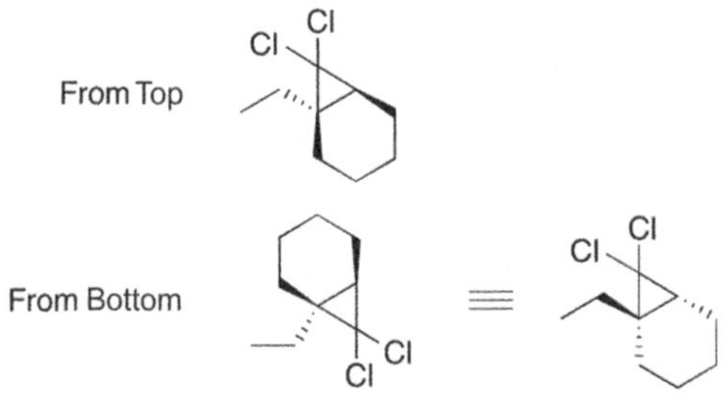

d. 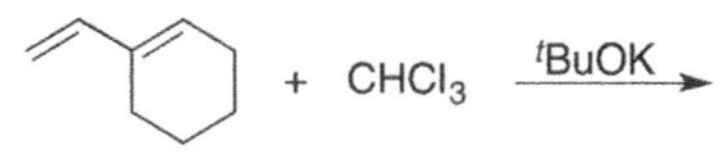   + CHCl₃ —*t*BuOK→

As with Problem 2a, the reactive species is dichlorocarbene that inserts into both of the 1-ethylenyl-1-cyclohexene double bonds forming two cyclopropane rings.

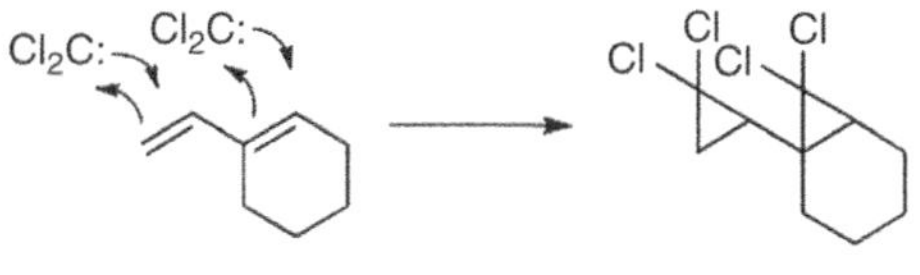

Because this reaction involves approaches from both the top and the bottom faces of the 1-ethylenyl-1-cyclohexene double bonds, there are four possible trajectory combinations:

- Trajectory combination 1—both double bonds are approached from the top face.
- Trajectory combination 2—both double bonds are approached from the bottom face.
- Trajectory combination 3—the ethylene double bond is approached from the top face and the cyclohexene double bond is approached from the bottom face.
- Trajectory combination 4—the ethylene double bond is approached from the bottom face, and the cyclohexene double bond is approached from the top face.

Due to the multiple trajectory combinations, a mixture of four products is formed. This mixture comprises two sets of enantiomers. It is important to note that these sets of enantiomers are diastereomers of one another.

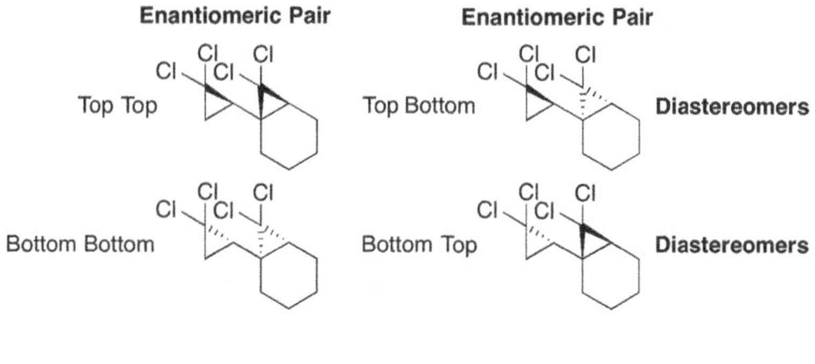

e. 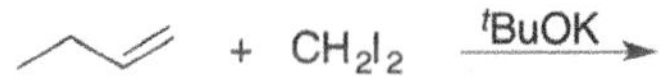   + CH₂I₂ —*t*BuOK→

Methylene iodide, on reaction with potassium *tert*-butoxide, forms iodocarbene.

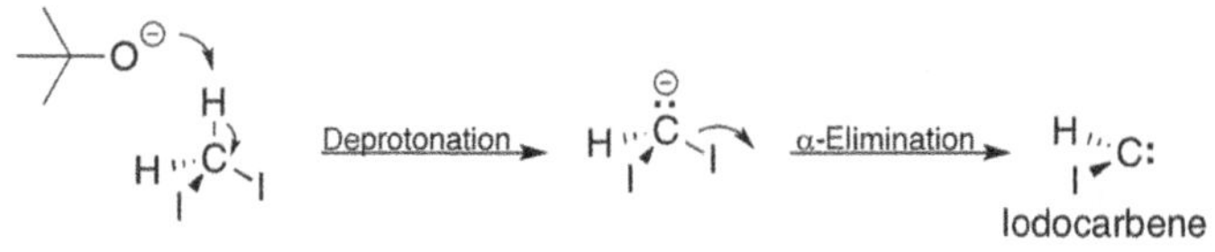

Iodocarbene inserts into the 1-butene double bond forming a cyclopropane ring.

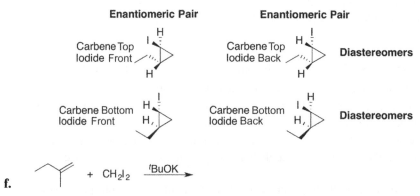

Because this reaction involves approaches from both the top and the bottom faces of 1-butene as well as spatial orientations of iodocarbene where the iodide is either in front or in back, a mixture of four products is formed. This mixture comprises two sets of enantiomers. It is important to note that these sets of enantiomers are diastereomers of one another.

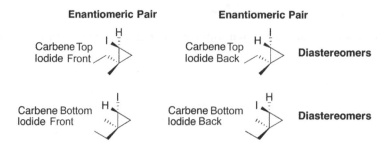

f.

As with Problem 2e, the reactive species is iodocarbene that inserts into the 2-methyl-1-butene double bond forming a cyclopropane ring.

Because this reaction involves approaches from both the top and the bottom faces of 2-methyl-1-butene as well as spatial orientations of iodocarbene where the iodide is either in front or in back, a mixture of four products is formed. This mixture comprises two sets of enantiomers. It is important to note that these sets of enantiomers are diastereomers of one another.

g.

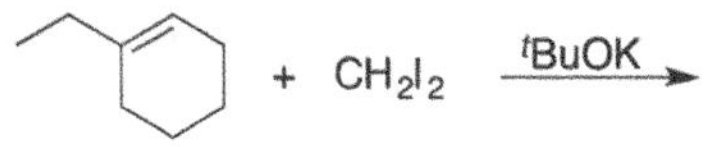

As with Problem 2e, the reactive species is iodocarbene that inserts into the 1-ethyl-1-cyclohexene double bond forming a cyclopropane ring.

Because this reaction involves approaches from both the top and the bottom faces of 1-ethyl-1-cyclohexene as well as spatial orientations of iodocarbene where the iodide is either in front or in back, a mixture of four products is formed. This mixture comprises two sets of enantiomers. It is important to note that these sets of enantiomers are diastereomers of one another.

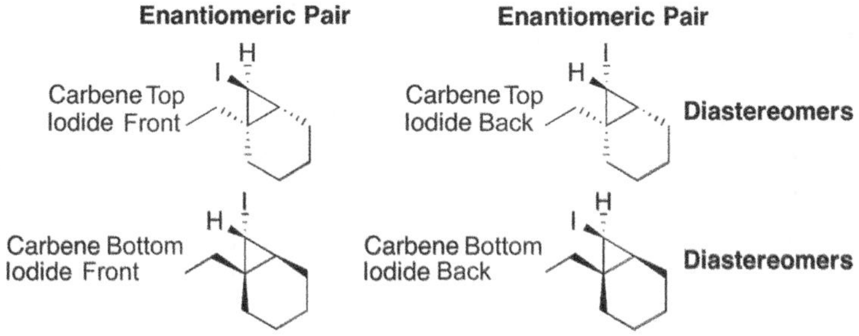

h.

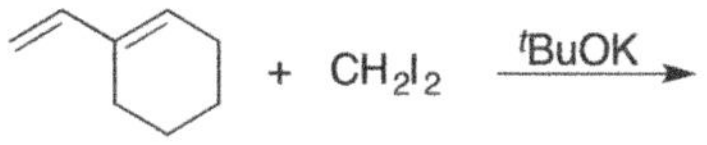

As with Problem 2e, the reactive species is iodocarbene that inserts into both of the 1-ethylenyl-1-cyclohexene double bond forming two cyclopropane rings.

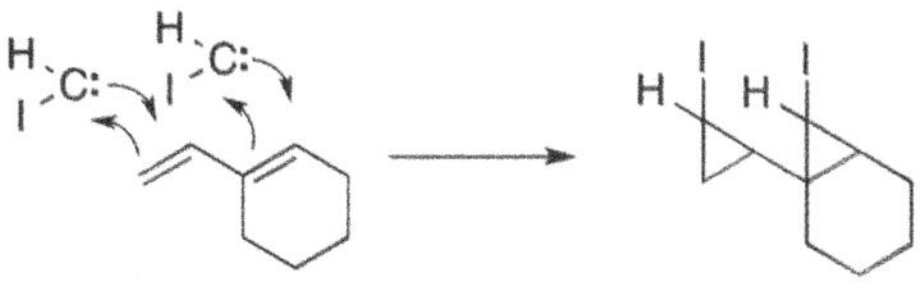

Because this reaction involves approaches from both the top and the bottom faces of the 1-ethylenyl-1-cyclohexene double bonds as well as spatial orientations of iodocarbene where the iodide is either in front or in back, there are eight possible trajectory combinations:

- Trajectory combination 1—both double bonds are approached from the top face with both iodides in front.
- Trajectory combination 2—both double bonds are approached from the top face with both iodides in back.

- Trajectory combination 3—both double bonds are approached from the top face with the iodide associated with the ethylene double bond in front and the iodide associated with the cyclohexene double bond in back.
- Trajectory combination 4—both double bonds are approached from the top face with the iodide associated with the ethylene double bond in back and the iodide associated with the cyclohexene double bond in front.
- Trajectory combination 5—both double bonds are approached from the bottom face with both iodides in front.
- Trajectory combination 6—both double bonds are approached from the bottom face with both iodides in back.
- Trajectory combination 7—both double bonds are approached from the bottom face with the iodide associated with the ethylene double bond in front and the iodide associated with the cyclohexene double bond in back.
- Trajectory combination 8—both double bonds are approached from the bottom face with the iodide associated with the ethylene double bond in back and the iodide associated with the cyclohexene double bond in front.
- Trajectory combination 9—the ethylene double bond is approached from the top face with the iodide in front, and the cyclohexene double bond is approached from the bottom face with the iodide in front.
- Trajectory combination 10—the ethylene double bond is approached from the top face with the iodide in front, and the cyclohexene double bond is approached from the bottom face with the iodide in back.
- Trajectory combination 11—the ethylene double bond is approached from the top face with the iodide in back, and the cyclohexene double bond is approached from the bottom face with the iodide in front.
- Trajectory combination 12—the ethylene double bond is approached from the top face with the iodide in back, and the cyclohexene double bond is approached from the bottom face with the iodide in back.
- Trajectory combination 13—the ethylene double bond is approached from the bottom face with the iodide in front, and the cyclohexene double bond is approached from the top face with the iodide in front.
- Trajectory combination 14—the ethylene double bond is approached from the bottom face with the iodide in front, and the cyclohexene double bond is approached from the top face with the iodide in back.
- Trajectory combination 15—the ethylene double bond is approached from the bottom face with the iodide in back, and the cyclohexene double bond is approached from the top face with the iodide in front.
- Trajectory combination 16—the ethylene double bond is approached from the bottom face with the iodide in back, and the cyclohexene double bond is approached from the top face with the iodide in back.

Due to the multiple trajectory combinations, a mixture of 16 products is formed. This mixture comprises eight sets of enantiomers. It is important to note that these sets of enantiomers are diastereomers of one another. Use of molecular models will help in understanding the illustrated stereochemistry.

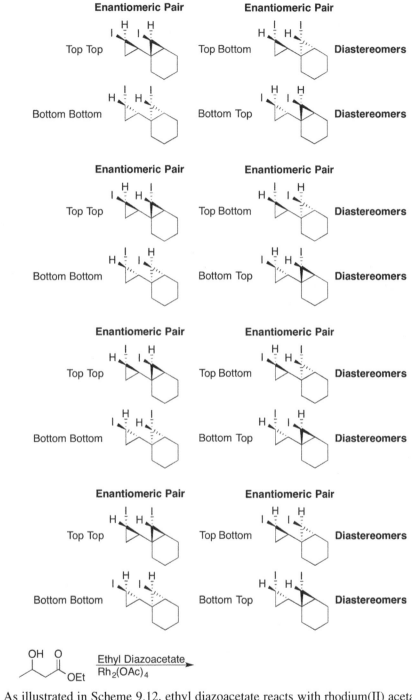

i.

As illustrated in Scheme 9.12, ethyl diazoacetate reacts with rhodium(II) acetate to form a reactive carbenoid species. In the interest of simplifying structures, the mechanisms associated with this reaction are shown using the ethyl acetate carbene. In a manner similar to that shown in Scheme 9.11, a lone electron pair from the hydroxyl oxygen adds to the carbene. Proton transfer results in charge neutralization and completion of the reaction.

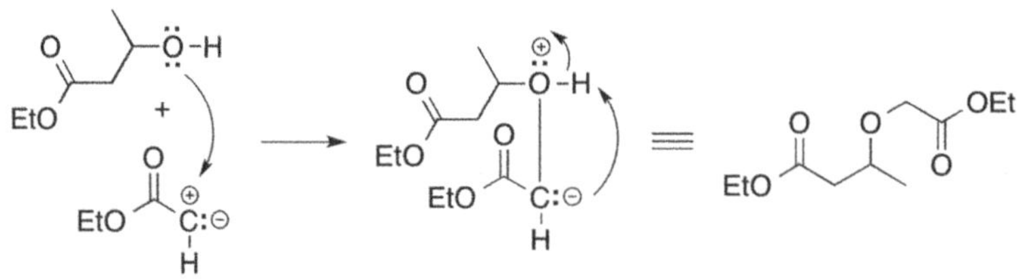

j.

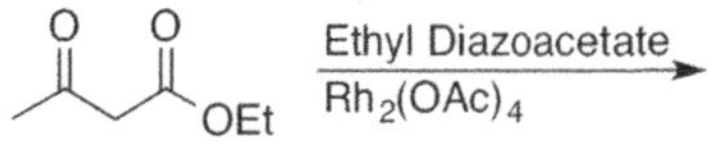

Initial examination of ethyl acetoacetate does not reveal a hydroxyl group leading one to assume that no reaction will take place. However, ethyl acetoacetate can enolize forming a relatively stable species bearing a hydroxyl group. Stabilization of the enol form is due to hydrogen bonding between the hydrogen of the hydroxyl group and the carbonyl oxygen of the ethyl ester as well as through conjugation of the double bond with the ester carbonyl.

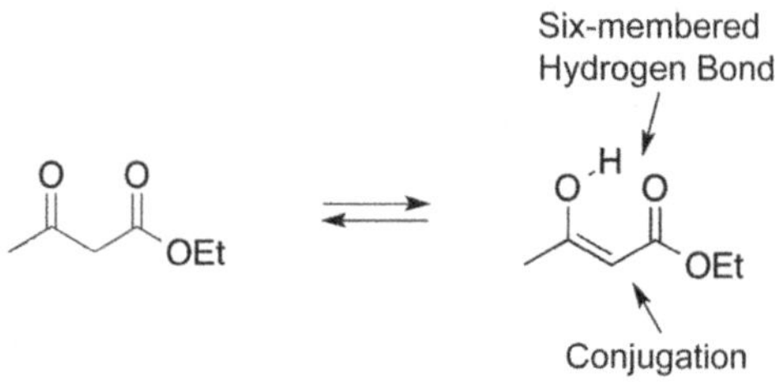

In its enol form, ethyl acetoacetate can react with the ethyl acetate carbene by initial addition of the hydroxyl oxygen to the carbene followed by proton transfer.

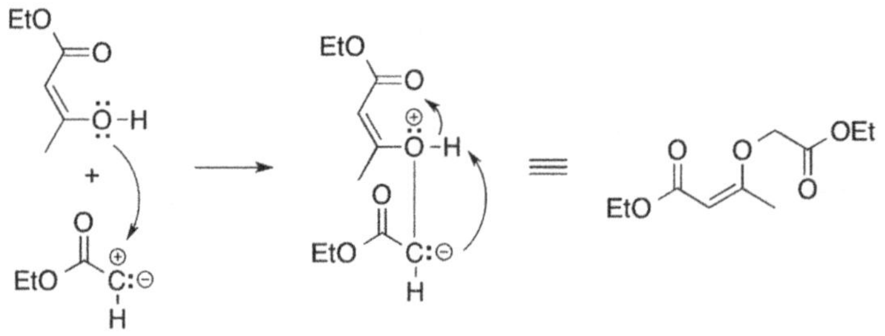

k.

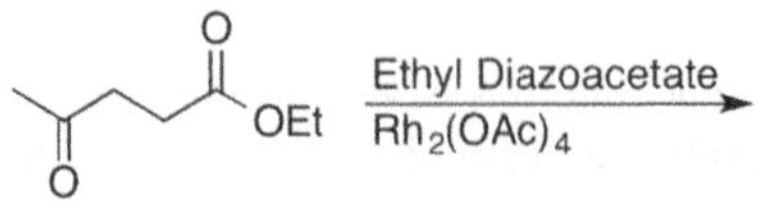

Unlike example 2i, 4-oxy-ethyl pentanoate does not possess a hydroxyl group. Furthermore, unlike example 2j, this compound cannot enolize because the resulting double bond cannot conjugate to the carbonyl group of the ethyl ester. Additionally, if enolization was possible, the resulting hydrogen bond to the carbonyl oxygen would represent a difficult-to-form seven-membered ring.

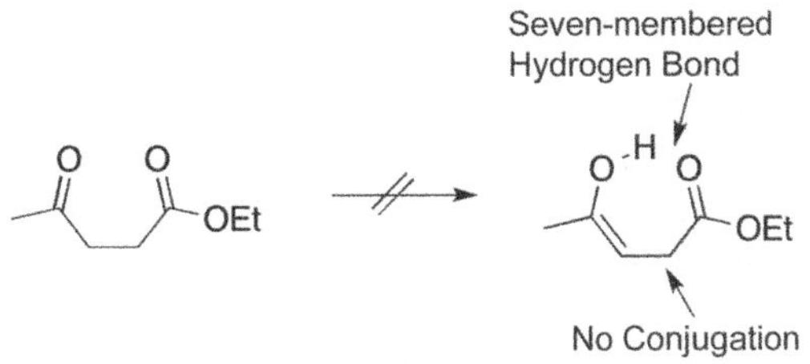

Therefore, ethyl diazoacetate in the presence of rhodium(II) acetate will not react with 4-oxy-ethyl pentanoate.

3. *The following transformations involve both carbenes and free radicals. Using arrow-pushing, explain the products of the following reactions.*

a.

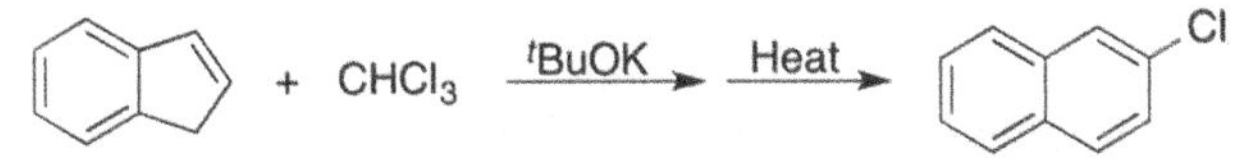

Chloroform, on reaction with potassium *tert*-butoxide, forms dichlorocarbene.

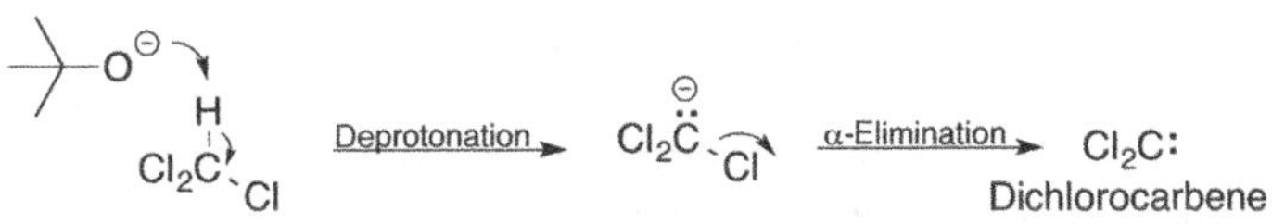

Once dichlorocarbene forms, it inserts into the illustrated double bond forming a cyclopropane ring. On heating, one of the cyclopropane bonds undergoes homolytic cleavage forming a diradical. With loss of a chlorine atom, one radical forms a new carbon–carbon double bond. The released chlorine atom then abstracts a hydrogen atom from a benzylic carbon with simultaneous formation of a second carbon–carbon double bond.

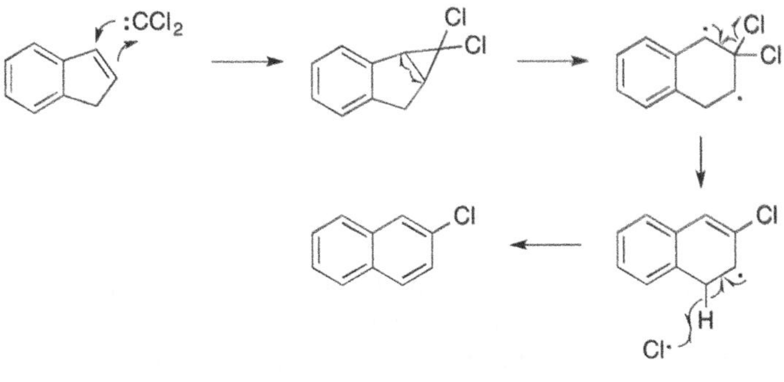

b.

Understanding that rhodium(II) acetate forms carbenoids on reaction with azides, the first step in approaching this problem is to recognize that a carbene-like species is generated. Because carbenoids have similar reactivities to carbenes, the explanation to this problem focuses on the simpler carbene model. As described in this chapter, carbenes react with double bonds to form cyclopropane rings. However, in this case, formation of a cyclopropane ring through intramolecular addition of the carbene into the double bond forms a fairly complex and highly strained ring system. Therefore, direct cyclopropanation will not occur.

Exploring the mechanism in which double bonds react with carbenes, initial addition of the π-bond to the empty p-orbital of the carbene generates a new carbon–carbon bond. This bond formation places a positive charge on one carbon atom formerly associated with the double bond. Because negative charges can delocalize into adjacent carbonyl groups, the next step in the mechanism allows incorporation of the carbene nonbonded electron pair into a new carbon–carbon double bond with placement of the negative charge on the oxygen atom. When the negative charge moves back to the carbon atom, breaking a carbon–carbon bond allows neutralization of all charges on this molecule with formation of a new carbon–carbon double bond and a ketene functional group. As shown, the ketene functional group comprises both a carbon–carbon double bond and a carbon–oxygen double bond.

Ketenes readily react with protic species such as water, amines, and alcohols. When this ketene species is exposed to methanol, a lone electron pair from the methanol oxygen adds to the ketene carbonyl. Proton transfer results in charge neutralization and formation of an ester enol. Tautomerization of the ester enol to a methyl ester completes the transformation.

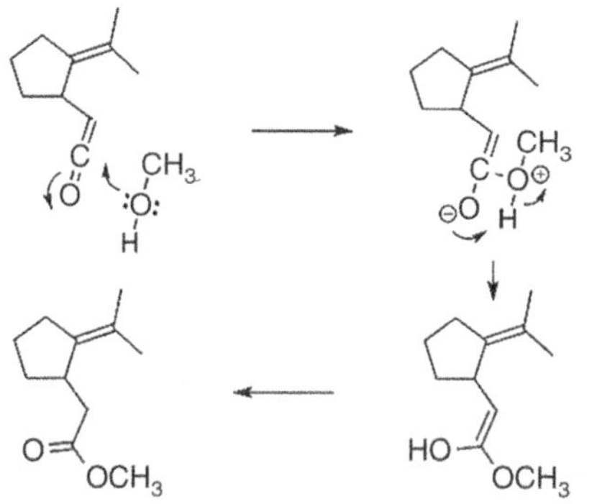

4. *Referring to Figure 9.7, show the product resulting from reaction of each starting material with ethyl diazoacetate and rhodium(II) acetate. Show all reaction pathways using arrow-pushing, and illustrate why the side reaction products shown in Figure 9.7 are not likely to form.*

a.

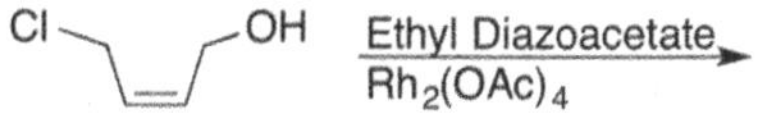

Using the ethyl acetate carbene as the reactive species, the first step is addition of a lone electron pair from the alcohol oxygen to the carbene empty *p*-orbital. Proton transfer from the oxygen to the carbon completes the carbene insertion reaction into the alcohol.

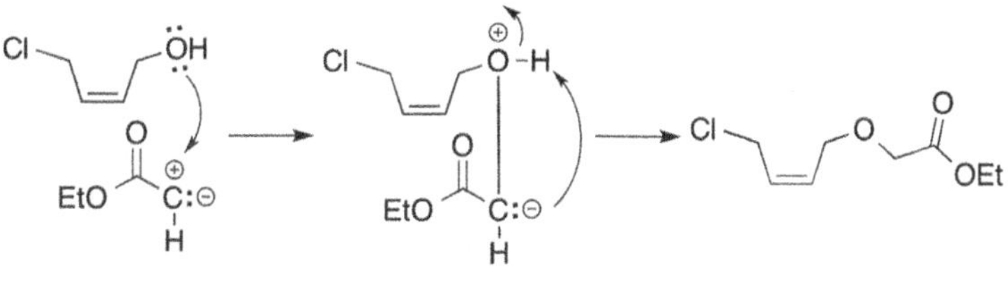

The S_N2 cyclization illustrated in Figure 9.7 is not likely to occur because the ethyl acetate carbene will not deprotonate the hydroxyl group, leading to oxygen anion formation.

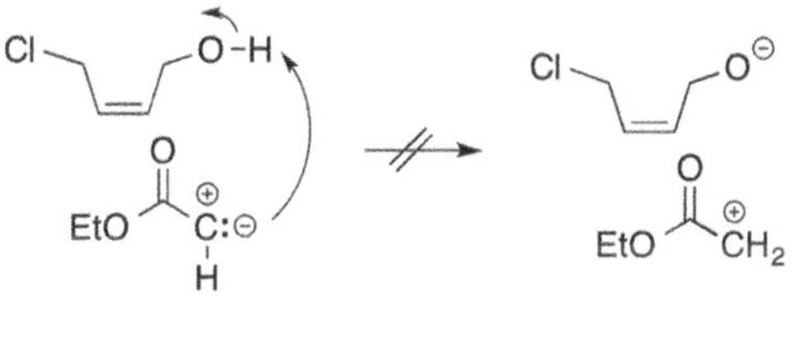

b.

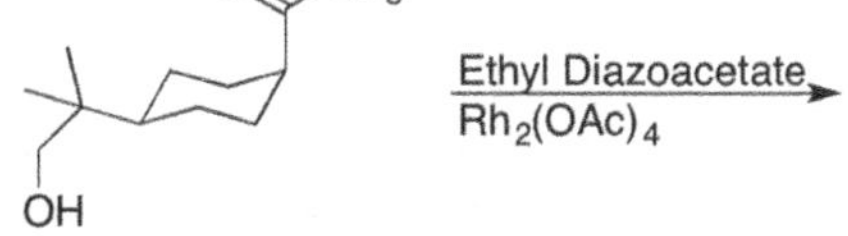

Using the ethyl acetate carbene as the reactive species, the first step is addition of a lone electron pair from the alcohol oxygen to the carbene empty *p*-orbital. Proton transfer from the oxygen to the carbon completes the carbene insertion reaction into the alcohol.

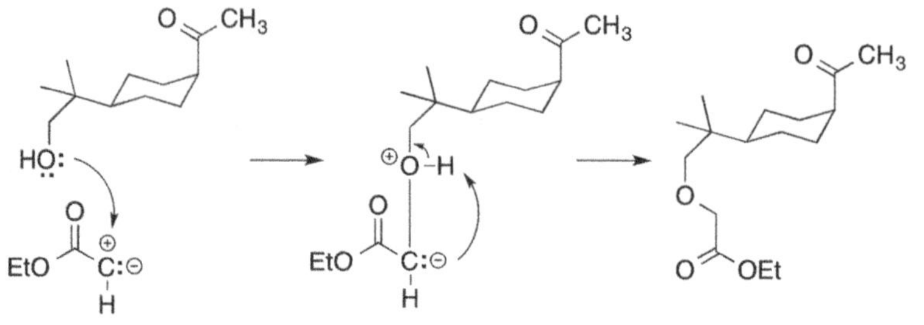

The epimerization illustrated in Figure 9.7 is not likely to occur because the ethyl acetate carbene will not deprotonate the carbon adjacent to the ketone carbonyl, leading to enolate formation.

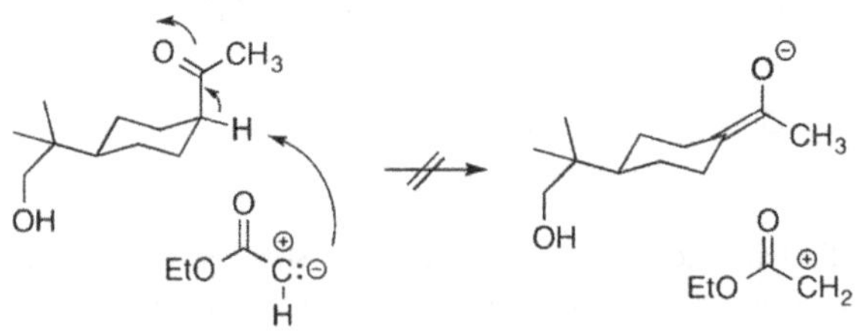

c.

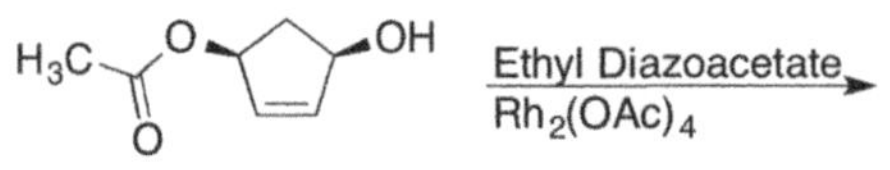

Using the ethyl acetate carbene as the reactive species, the first step is addition of a lone electron pair from the alcohol oxygen to the carbene empty *p*-orbital. Proton transfer from the oxygen to the carbon completes the carbene insertion reaction into the alcohol.

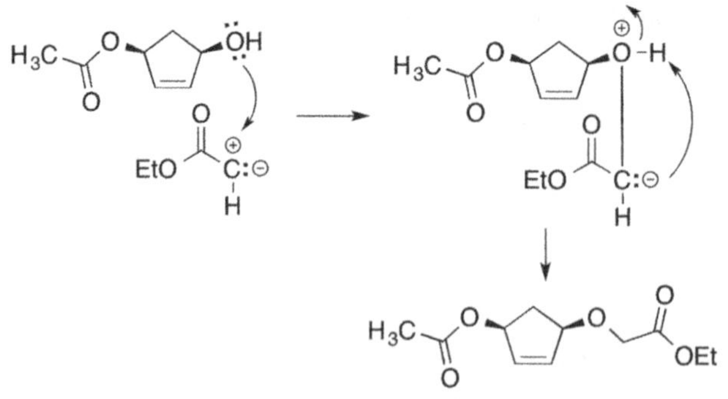

The acetate migration illustrated in Figure 9.7 is not likely to occur because the ethyl acetate carbene will not deprotonate the hydroxyl group, leading to oxygen anion formation.

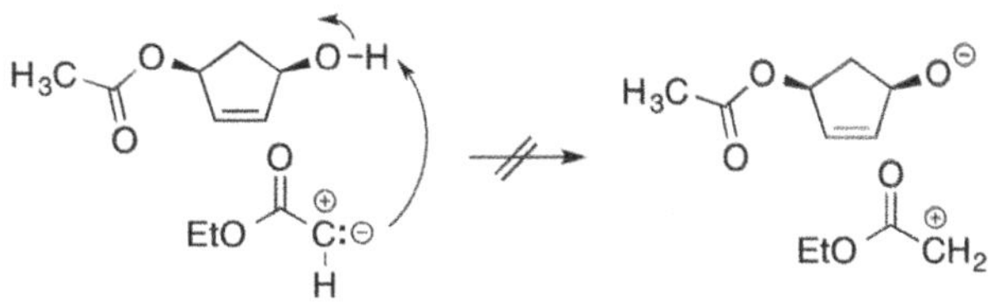

d.

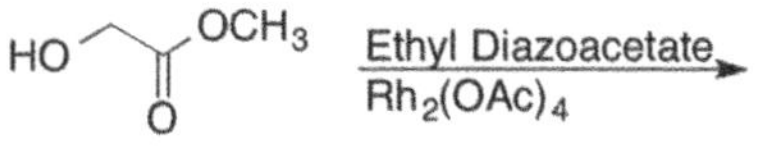

Using the ethyl acetate carbene as the reactive species, the first step is addition of a lone electron pair from the alcohol oxygen to the carbene empty *p*-orbital. Proton transfer from the oxygen to the carbon completes the carbene insertion reaction into the alcohol.

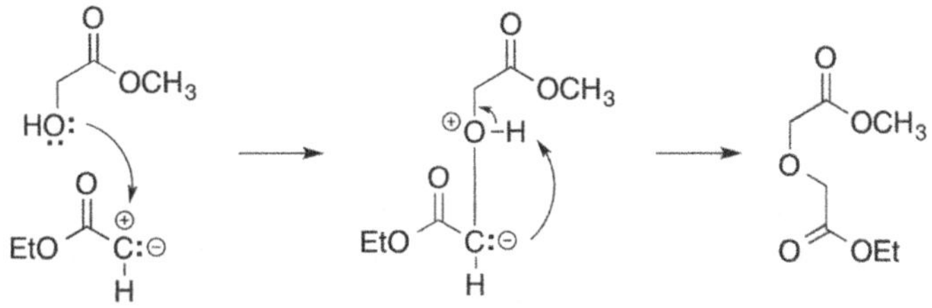

The ester hydrolysis illustrated in Figure 9.7 is not likely to occur because formation of the ethyl acetate carbene does not depend upon bases such as sodium hydroxide or potassium hydroxide. In fact, because the illustrated carbene reaction does not require the presence of water or any protic solvent, the illustrated ether product will be the only product formed.

5. *The thermal or photolytic decomposition of azides results in the formation of nitrenes. Nitrenes are chemically similar to carbenes due to their electronic nature and reactivity. One important nitrene reaction is the Curtius rearrangement in which an acyl azide is converted to a carbamate via an isocyanate.*

   a.   *Using arrow-pushing, show the mechanism for conversion of benzoyl azide to its corresponding nitrene.*

In the first stage of the Curtius rearrangement, an acyl azide undergoes α-elimination forming a nitrene. The α-elimination step is not obvious from the illustrated acyl azide structure. However, on consideration of the acyl azide resonance form comprising a nitrogen–nitrogen triple bond, the α-elimination step becomes more recognizable.

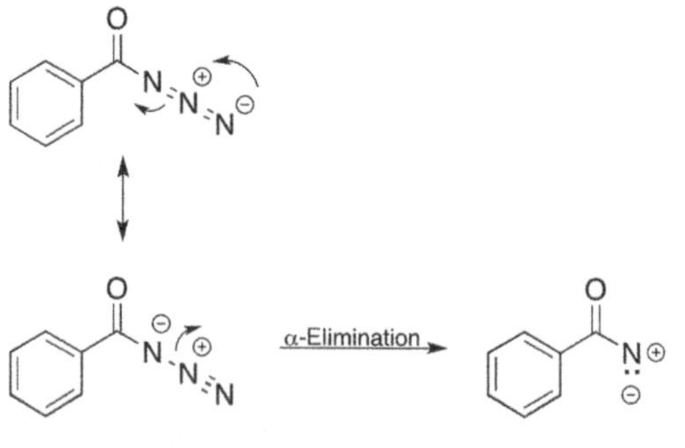

b.   *The Curtius rearrangement of benzoyl nitrene is illustrated in the following. Using arrow-pushing, draw the mechanism for formation of the intermediate isocyanate.*

In the second stage of the Curtius rearrangement, stabilization of the acyl nitrene through conjugation into the carbonyl must be considered. Return of the electrons from the anionic oxygen atom results in migration of the phenyl ring to the nitrogen neutralizing all charges. The species formed from this transformation is an isocyanate functional group comprising both a nitrogen–carbon double bond and a carbon–oxygen double bond.

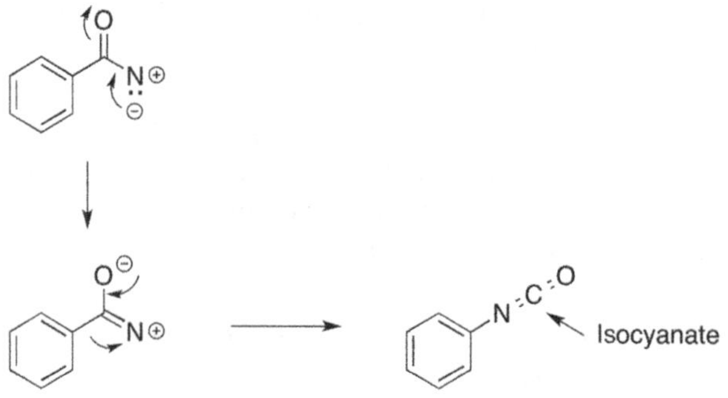

c.   *Using arrow-pushing, draw the mechanism for conversion of the isocyanate earlier to the illustrated tert-butyl carbamate.*

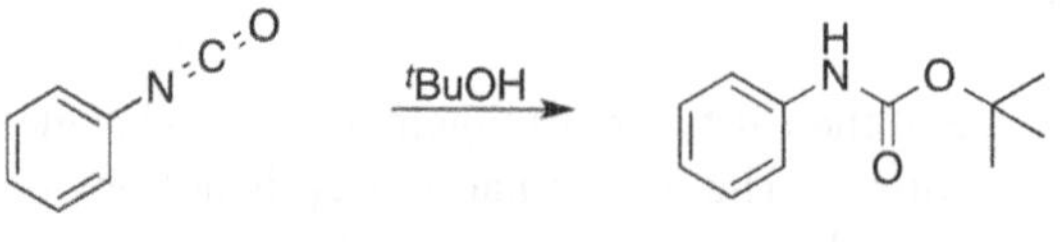

Isocyanates are reactive species similar in nature to ketenes described in Problem 3b. As such, they readily react with protic species such as water, amines, and alcohols.

When this isocyanate species is exposed to *tert*-butanol, a lone electron pair from the *tert*-butanol oxygen adds to the isocyanate carbonyl. Proton transfer results in charge neutralization, and tautomerization generates the final *tert*-butyl carbamate functional group.

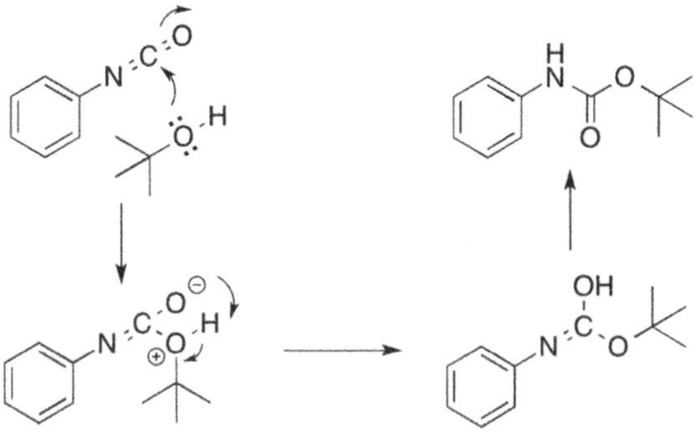

Through the steps described in this problem, the Curtius rearrangement can be recognized as a useful tool for converting readily available carboxylic acids and acid halides into less readily available amines conveniently protected as hydrolyzable carbamates.

CHAPTER 10 SOLUTIONS

In all previous chapters and solution sets, reaction mechanisms were illustrated using lone electron pairs when appropriate. Understanding that lone electron pairs are present without actually showing them is a common shortcut in describing reactions. By now, readers should be familiar with the octet rule and should also be able to recognize when and where lone electron pairs should be present. In this solution set, there are instances where lone electron pairs are not illustrated. In such cases, readers are encouraged to work without them or to draw them in as needed.

1. *The Carroll rearrangement is mechanistically similar to a variation of the Claisen rearrangement. Using arrow-pushing, propose a mechanism for the conversion of allyl acetoacetate to 5-hexene-2-one.*

Allyl Acetoacetate   5-Hexene-2-one

As discussed in the explanation for Problem 2j in Chapter 9, β-ketoesters such as allyl acetoacetate can enolize forming a stable six-membered ring hydrogen bond and a carbon–carbon double bond conjugated to the ester. The rationale leading to enolization of the ketone can also be applied to enolization of the ester. When the ester enolizes, the carbon–carbon double bonds present in allyl acetoacetate can arrange themselves into a configuration suitable for a [3,3] sigmatropic rearrangement to occur. This transformation breaks the carbon–oxygen σ-bond, leaving a carboxylic acid. β-Ketoacids readily decarboxylate under thermal conditions through a mechanism similar to the pericyclic reactions discussed in this chapter. Decarboxylation of the illustrated 2-allyl acetoacetic acid results in formation of the final products, 5-hexene-2-one and carbon dioxide.

Allyl Acetoacetate

Decarboxylation   [3,3] Sigmatropic
Rearrangement

2-Allyl Acetoacetic Acid

Tautomerization

5-Hexene-2-one Carbon Dioxide   + O=C=O

2. *The Fischer indole synthesis brings together phenylhydrazine and acetone to form 2-methylindole. The mechanism involves a sigmatropic rearrangement. Using arrow-pushing, propose a mechanism for this reaction.*

Phenylhydrazine   Acetone   2-Methylindole

The first stage of the Fischer indole synthesis involves the formation of a hydrazone—a structure in which a hydrazine unit condenses with a ketone. In this example, phenylhydrazine condenses with acetone according to mechanisms discussed in earlier chapters. As illustrated using arrow-pushing, the lone electron pair of the phenylhydrazine terminal amino group adds to the acetone carbonyl. Proton transfer from the cationic nitrogen to the anionic oxygen neutralizes all charges. A lone electron pair from the newly formed hydroxyl group then interacts with the adjacent nitrogen–hydrogen bond, leading to elimination and formation of a new nitrogen–carbon double bond. This structure is a hydrazone.

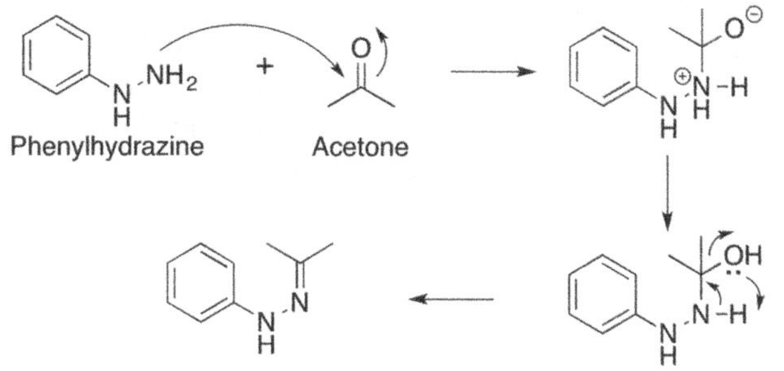

In the second stage of the Fischer indole synthesis, the hydrazone tautomerizes to form an enamine. Under acidic conditions, the enamine is protonated and then undergoes a [3,3] sigmatropic rearrangement. The resulting species possesses an iminium ion (protonated imine) and a non-protonated imine conjugated with two carbon–carbon double bonds. This second imine undergoes tautomerization, leading to aromatization of the six-membered ring and formation of an aniline. The lone electron pair of the aniline then adds to the iminium ion forming a five-membered ring. Following proton transfer from the protonated aniline nitrogen to the charge-neutral amine, ammonia leaves along with a proton adjacent to the phenyl ring, forming the final indole structure.

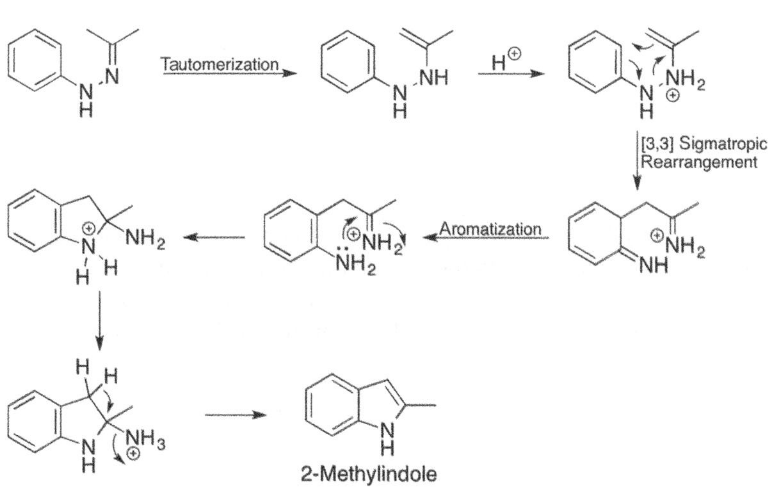

2-Methylindole

3. *As demonstrated by the Fischer indole synthesis, sigmatropic rearrangements can occur with heteroatoms such as nitrogen and oxygen. Furthermore, the use of heteroatoms is*

broadly applicable to pericyclic reactions. Predict the products of the following pericy-
clic reactions. For each reaction, show the mechanism and predict the major product
formed. Justify your answers using arrow-pushing and electronic/steric influences. For
each example, state the type of pericyclic reaction (electrocyclic, cycloaddition, or sig-
matropic rearrangement). Predict the intermediate structures when not shown.

a.

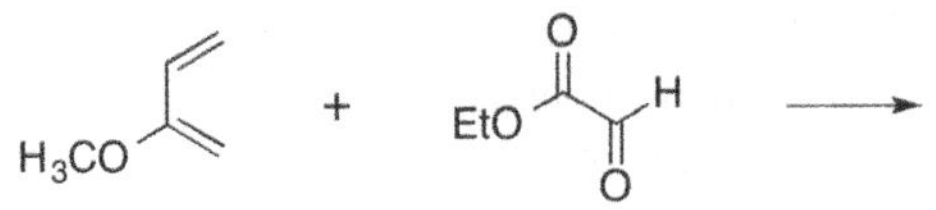

The oxygen atom present on 2-methoxy-1,3-butadiene possesses two lone electron
pairs that are conjugated into the adjacent double bond. By virtue of this conjugation,
partial charges can be placed around this molecule as illustrated in the following.
Furthermore, carbonyl groups are polar in nature with the partial charges present on
ethyl glyoxylate as illustrated in the following:

Aligning the aldehyde of ethyl glyoxylate with 2-methoxy-1,3-butadiene such that all
partial charges alternate rationalizes the orientation with which these compounds
come together. The observed cycloaddition is an oxo-Diels–Alder reaction. The illustrated stereochemistry is expected from the endo approach of the aldehyde to the diene.

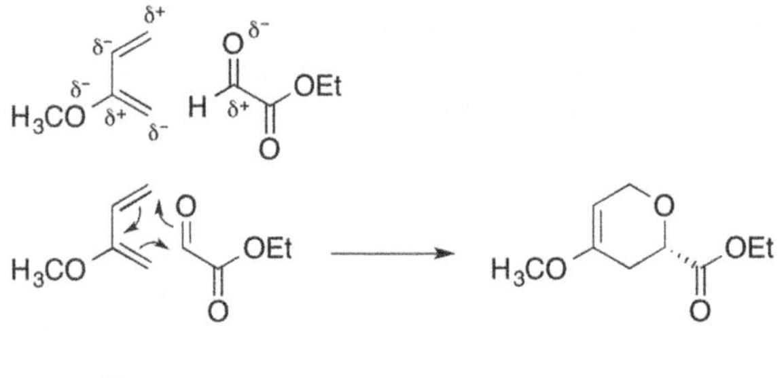

b.

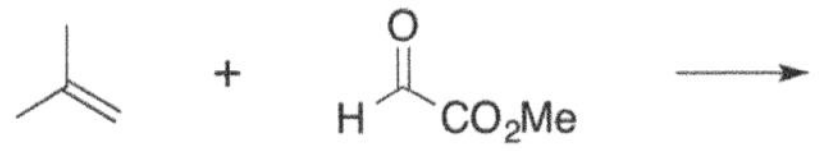

The reaction between an aldehyde and an allylic olefin is a cycloaddition known as
the carbonyl ene reaction. In order to predict the product of this reaction, it is necessary to recognize that hydrogen atoms carry partial positive charges. Thus, the partial
charges of the reactive atoms for isobutylene and methyl glyoxylate are illustrated in
the following:

Aligning the aldehyde of methyl glyoxylate with isobutylene such that all partial charges alternate rationalizes the orientation with which these compounds come together. The observed carbonyl ene reaction produces the illustrated product.

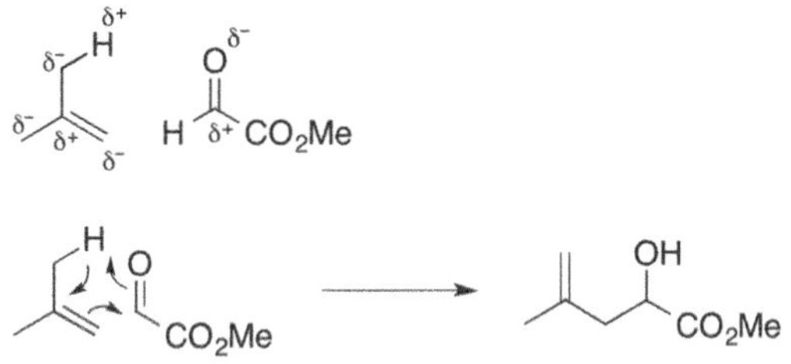

c.

This two-step sequence begins with the reaction of benzylamine with formaldehyde according to mechanisms discussed in earlier chapters. As illustrated, the initial step is addition of the amine to the aldehyde followed by dehydration to form the illustrated imine.

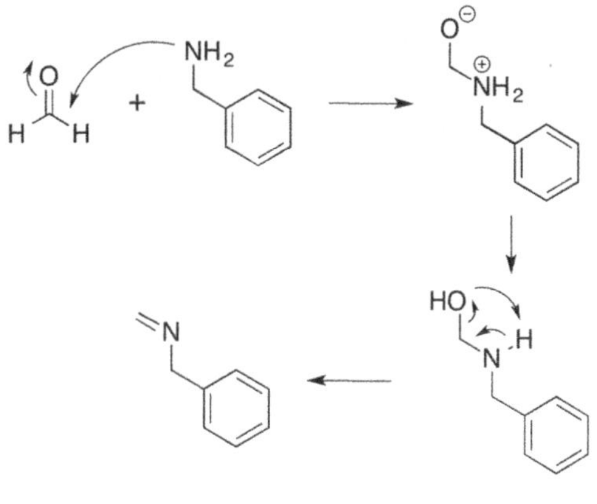

Once the imine is formed, it is available to participate in an aza-Diels–Alder cycloaddition reaction with cyclopentadiene. Because there are no substitutions on the cyclopentadiene, there are no isomers formed based on the orientation of approach of the imine to the diene.

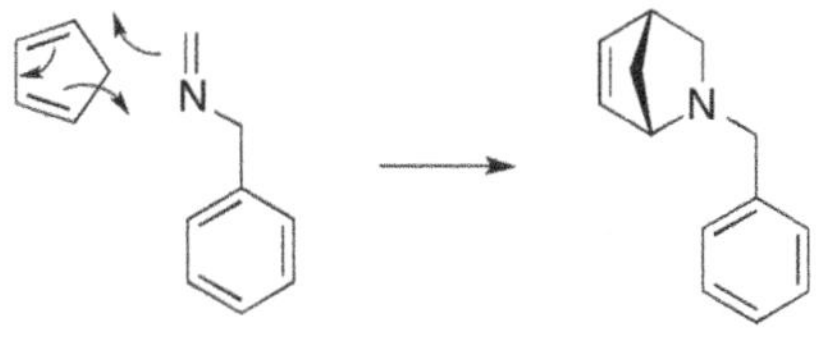

d.

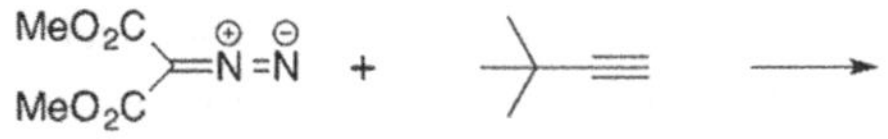

This reaction is a 1,3-dipolar cycloaddition between dimethyl diazomalonate and *tert*-butyl acetylene. Due to steric interactions between the *tert*-butyl group and the methyl esters, the orientation with which these compounds approach one another will maximize the distance between these groups. Thus, the illustrated reaction will proceed as shown in the following:

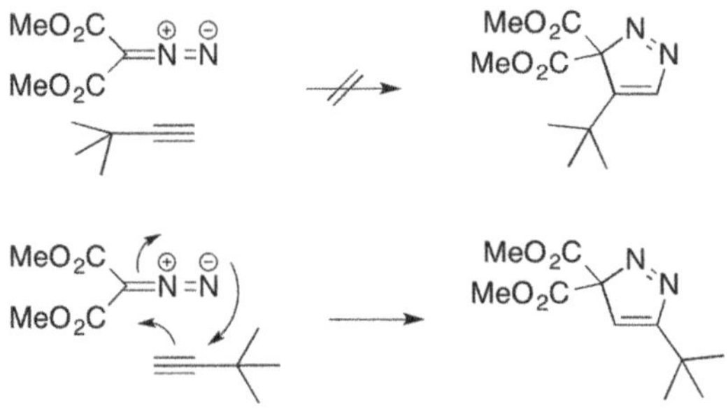

e.

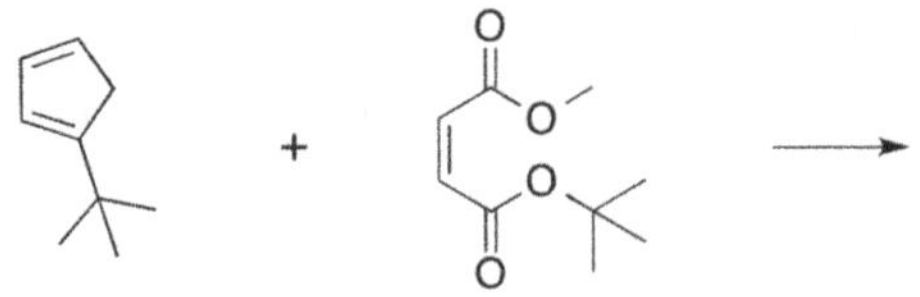

The outcome of this Diels–Alder cycloaddition is influenced by the endo approach of methyl *tert*-butyl maleate with *tert*-butyl cyclopentadiene. In addition, minimization of steric effects between the two *tert*-butyl groups will direct these groups to opposite sides of the product. Thus, the illustrated reaction will proceed as shown in the following:

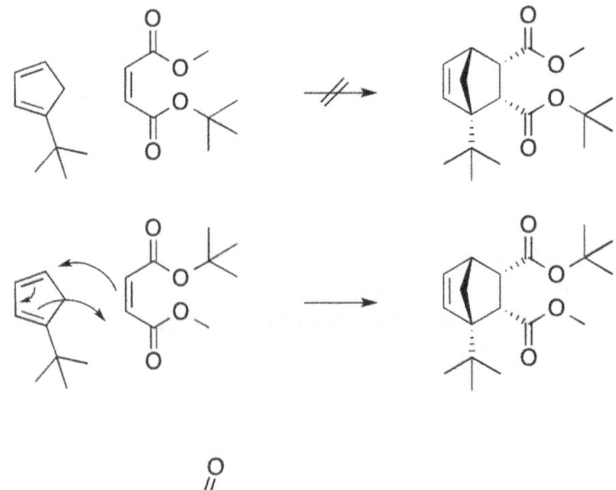

f.

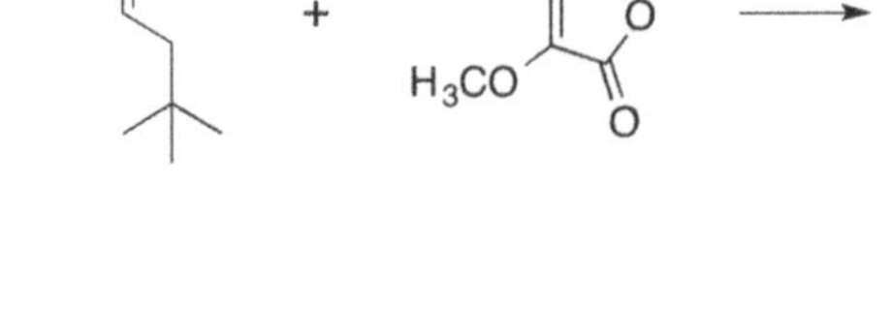

The outcome of this ene reaction is influenced by steric factors that develop between the methoxy group of methoxy maleic anhydride and the *tert*-butyl group attached directly to the double bond of *trans*-2,2,6,6-tetramethyl-3-heptene. Evaluation of the prospective products in the following reveals one product with much less steric congestion. Thus, this cycloaddition proceeds with the formation of a product with substitutions on separate carbon atoms instead of two substitutions on the same carbon atom.

4. *The aza-Cope rearrangement follows the same general mechanistic pathway as the standard Cope rearrangement. Using principles discussed throughout this book, predict the structures formed at each stage of the illustrated reaction and leading to the formation of 3-butenylamine and benzaldehyde.*

The first stage of the aza-Cope rearrangement is the formation of an imine. The mechanism for imine formation is illustrated in the solution for Problem 3c applied to benzylamine—a very similar starting material. Because of the similarities between benzylamine and the amine of this problem, only product **A** is shown in the following:

The second stage of this reaction sequence is the aza-Cope [3,3] sigmatropic rearrangement. The mechanism is illustrated in the following using arrow-pushing.

Compound **B**, being a benzylidene imine, is subject to hydrolysis according to mechanisms discussed in previous chapters. The reaction of compound **B** with water is illustrated in the following. As shown, a lone electron pair from water adds to the carbon–nitrogen double bond. Proton transfer from the oxygen to the nitrogen neutralizes all charges forming a hemiaminal. Under acidic conditions, the nitrogen is protonated. As a lone electron pair from the oxygen forms a new carbon–oxygen double bond, 3-butenylamine is released.

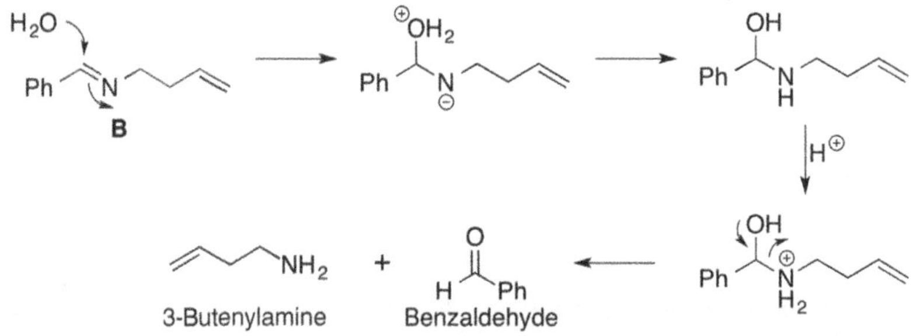

3-Butenylamine   Benzaldehyde

5. *As stated in this chapter, orthoesters are readily hydrolyzed to esters and carboxylic acids. Propose a mechanism for the hydrolysis of trimethyl orthoacetate to methyl acetate. Justify your answer using arrow-pushing. Explain why trimethyl orthoacetate may be useful in keeping reaction mixtures anhydrous.*

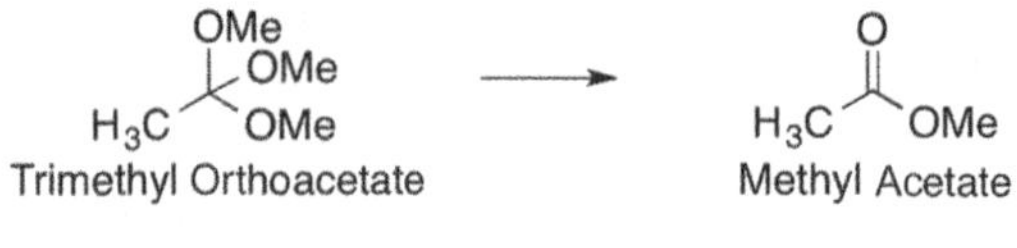

Trimethyl Orthoacetate   Methyl Acetate

As illustrated in the following using arrow-pushing, a lone electron pair from the water oxygen atom displaces a methoxide ion through an S_N2 mechanism. Proton transfer neutralizes all charges. Next, a lone electron pair from the hydroxyl group displaces a second methoxide ion through an elimination process. Proton transfer again neutralizes all charges, leading to the formation of methyl acetate.

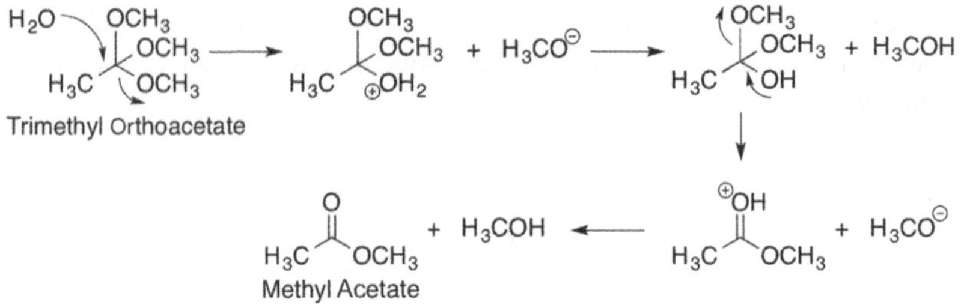

Trimethyl Orthoacetate

Methyl Acetate

Because trimethyl orthoacetate readily reacts with water to form methyl acetate, trimethyl orthoacetate is useful for eliminating residual water from reaction mixtures either as water forms or if any exists prior to starting a reaction. All of this is predicated upon the requirement that trimethyl orthoacetate and methyl acetate are inert to the specific reaction conditions utilized.

6. *Scheme 10.10 illustrates how sigmatropic rearrangements are classified by counting the atoms between the broken σ-bond and the newly formed σ-bond. Specify the type of sigmatropic rearrangement for the following reactions. Show your work by numbering the relevant atoms.*

a.

In this example, a carbon–hydrogen σ-bond is broken and a new carbon–hydrogen σ-bond is formed. Counting the atoms between the broken bond and the formed bond, we find that the σ-bond shifts from the [1,1] position to the [1,5] position. Because the newly formed σ-bond is in the [1,5] position relative to the broken σ-bond, this is a [1,5] sigmatropic rearrangement.

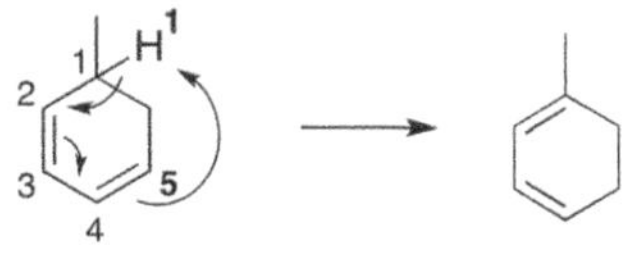

b.

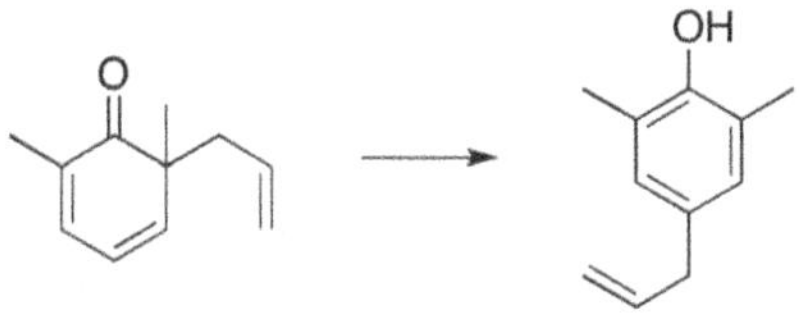

In this example, a carbon–carbon σ-bond is broken and a new carbon–carbon σ-bond is formed. Counting the atoms between the broken bond and the formed bond, we find that the σ-bond shifts from the [1,1] position to the [3,3] position. Because the newly formed σ-bond is in the [3,3] position relative to the broken σ-bond, this is a [3,3] sigmatropic rearrangement. In this case, the initial sigmatropic rearrangement is followed by aromatization, forming the illustrated phenol.

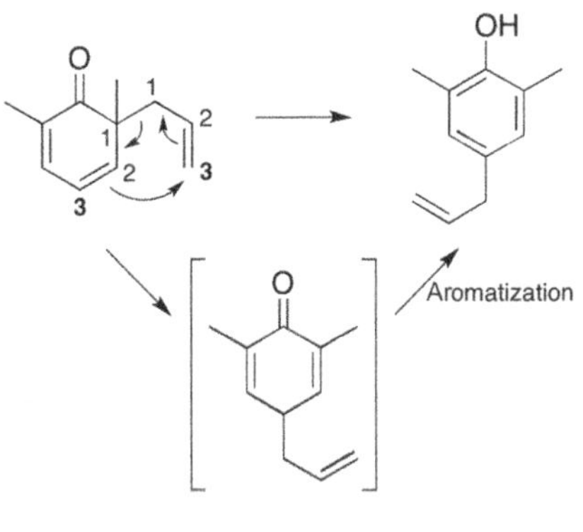

c.

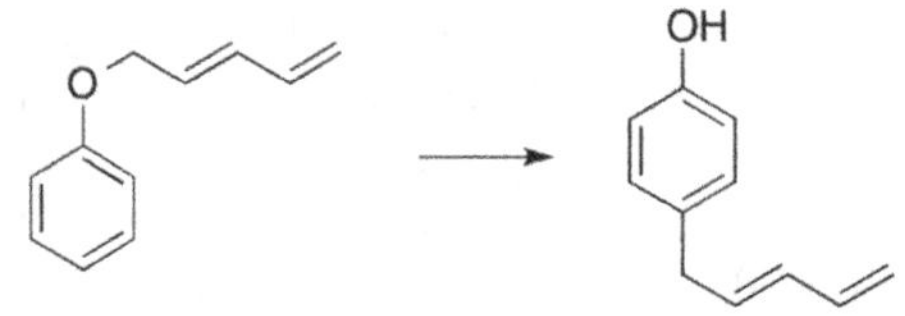

In this example, a carbon–oxygen σ-bond is broken and a new carbon–carbon σ-bond is formed. Counting the atoms between the broken bond and the formed bond, we find that the σ-bond shifts from the [1,1] position to the [5,5] position. Because the newly formed σ-bond is in the [5,5] position relative to the broken σ-bond, this is a [5,5] sigmatropic rearrangement. In this case, the initial sigmatropic rearrangement is followed by aromatization, forming the illustrated phenol.

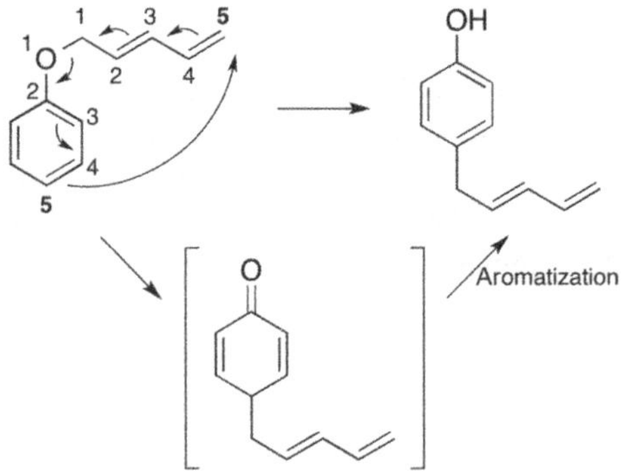

d.

In this example, a carbon–sulfur σ-bond is broken and a new carbon–oxygen σ-bond is formed. Counting the atoms between the broken bond and the formed bond, we find that the σ-bond shifts from the [1,1] position to the [2,3] position. Because the newly formed σ-bond is in the [2,3] position relative to the broken σ-bond, this is a [2,3] sigmatropic rearrangement. In this case the sigmatropic rearrangement results in neutralization of all charges.

e.

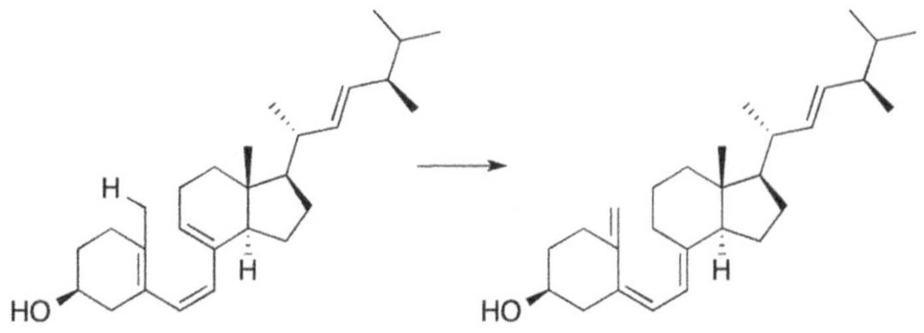

In this example, a carbon–hydrogen σ-bond is broken and a new carbon–hydrogen σ-bond is formed. Counting the atoms between the broken bond and the formed bond, we find that the σ-bond shifts from the [1,1] position to the [1,7] position. Because the newly formed σ-bond is in the [1,7] position relative to the broken σ-bond, this is a [1,7] sigmatropic rearrangement.

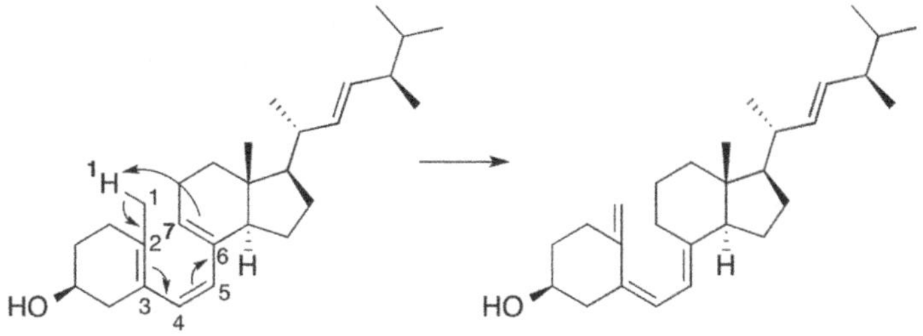

7. *Pericyclic reactions are useful for the generation of more complex ring systems. Propose a pericyclic reaction strategy for the formation of the following structures from completely acyclic starting materials. Show product formation using arrow-pushing, and specify the type of pericyclic reaction applied. In some cases, multiple pericyclic reactions are required.*

This set of exercises serves as an introduction to multistep organic syntheses. When approaching problems of this type, retrosynthetic analysis is a useful tool. Retrosynthetic analysis involves evaluation of the target structure and identifying bonds that can be broken in order to reveal simpler starting materials. This process of working backwards serves all organic chemists in the design of rationally designed synthetic strategies. As illustrated in the examples in the following, retrosynthetic analyses of individual reaction steps are represented by large block arrows.

a.

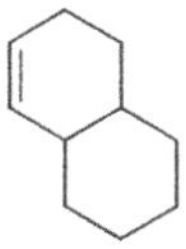

Recognizing that cyclohexene rings are available from Diels–Alder reactions, the target ring system is accessible from *trans*-1,3,9-decatriene. Because this Diels–Alder brings together a diene and a dienophile present in the same starting material, this is an intramolecular Diels–Alder reaction. Finally, this reaction produces a 6,6-*trans* fused ring system that is a very stable form of two fused six-membered rings. Use of molecular models will help in understanding the stereochemical outcome of this reaction.

b.

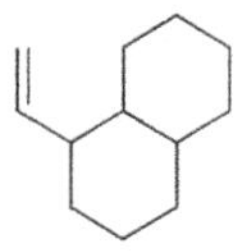

Recognizing that olefin substitutions are available from ene reactions, the first step in evaluating a possible synthetic route to this target structure is to identify potential ene reaction sites. By drawing one of the hydrogen atoms associated with the ring system, a potential six-membered ene reaction can be identified.

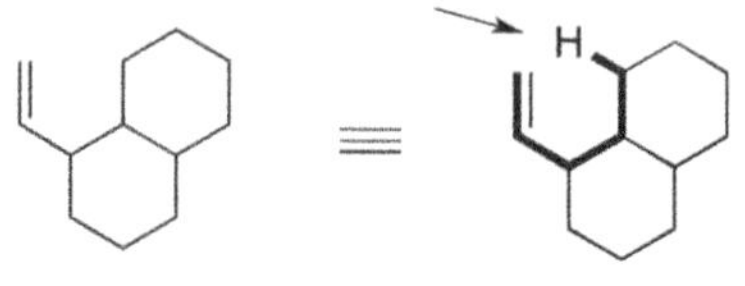

Once the ene reaction site is identified, reversing the ene reaction reveals a substituted cyclohexene structure.

As in Problem 7a, cyclohexene rings are accessible through the Diels–Alder reaction. Thus, the illustrated cyclohexene structure can be assembled from the illustrated diene and ethylene.

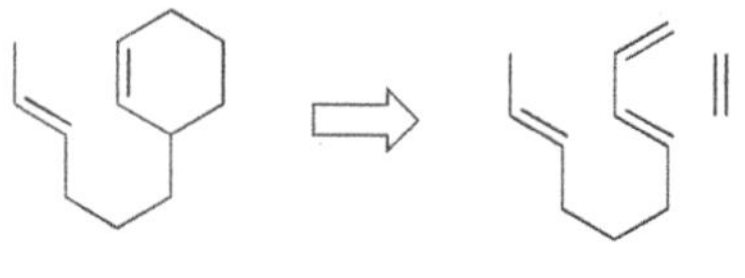

It is important to note that *trans-trans*-1,3,8-decatriene is similar to *trans*-1,3, 9-decatriene from Problem 7a. As such, one might expect *trans-trans*-1,3, 8-decatriene to undergo the intramolecular Diels–Alder reaction illustrated in the following. In fact, this reaction does not take place because the resulting product would be a 5,6-*trans* fused ring system. Due to ring strain, a five-membered ring fused to a six-membered ring in a *trans* configuration is relatively unstable. Use of molecular models will help in understanding the strain associated with 5,6-*trans* fused ring systems.

Based on the aforementioned evaluation, the target molecule can be prepared as illustrated in the following using arrow-pushing. Initially, *trans-trans*-1,3,8-decatriene undergoes a Diels–Alder reaction with ethylene. Next, the resulting cyclohexene structure undergoes an intramolecular ene reaction forming the final product.

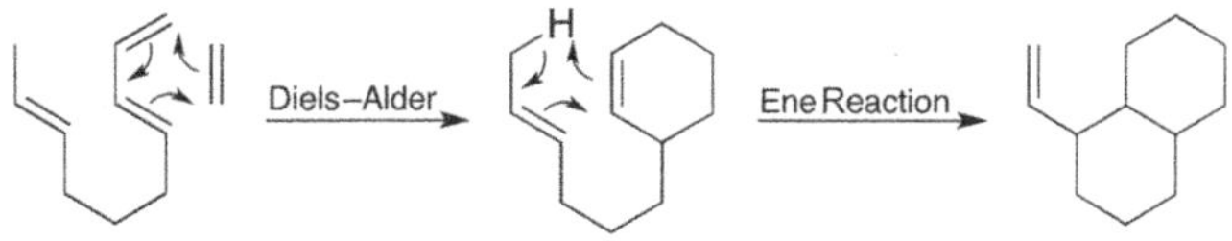

c.

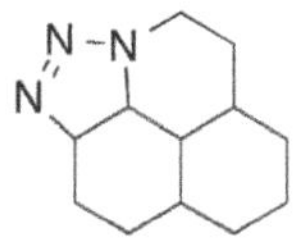

Recognizing that triazoline rings are available from 1,3-dipolar cycloadditions between azides and olefins, the first step in evaluating a possible synthetic route to this target structure is to identify an appropriate structure bearing both an azide and an olefin. Such a structure is illustrated in the following:

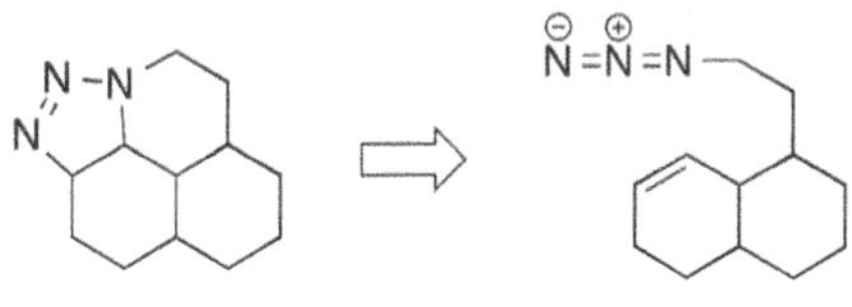

Having dissected the 1,3-dipolar cycloaddition, a 6,6-fused cyclohexene ring system similar to that in Problems 7a and 7b is revealed. Like previous examples, this structure is available from an intramolecular Diels–Alder reaction as illustrated in the following:

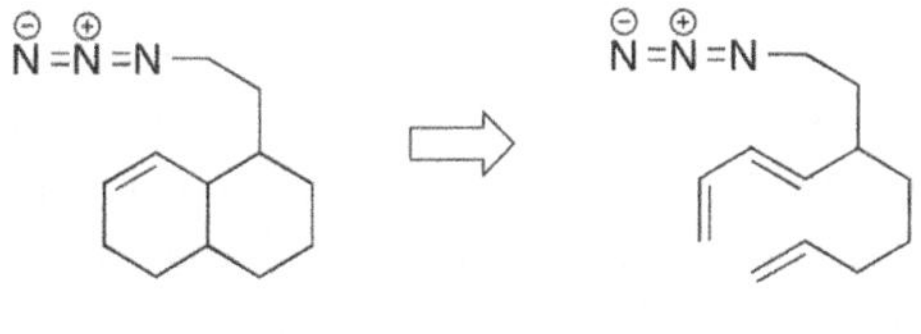

Assuming that the illustrated acylic azide structure aforementioned is readily available, the target compound can be prepared by simply subjecting this molecule to thermal conditions. Under these conditions, an intramolecular Diels–Alder reaction will be followed by an intramolecular 1,3-dipolar cycloaddition. This sequence is illustrated in the following using arrow-pushing.

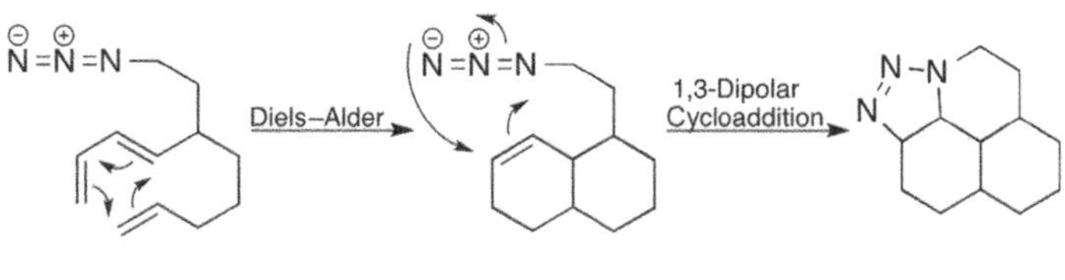

d.

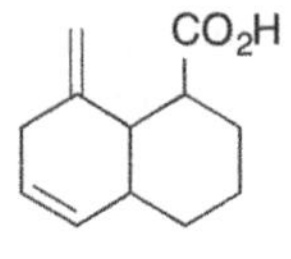

With a carboxylic acid and an olefin present in a 4-pentenoic configuration, the Ireland–Claisen rearrangement becomes a transformation of interest. Reversing the Ireland–Claisen rearrangement reveals a 10-membered lactone.

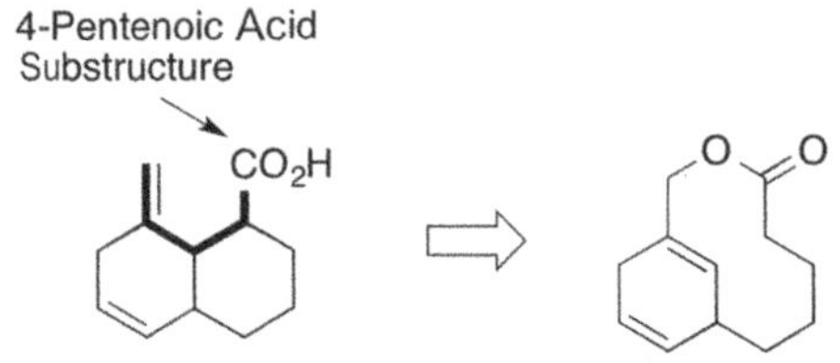

4-Pentenoic Acid
Substructure

While the lactone may seem more complex than the starting material, recognizing that a 1,4-cyclohexadiene is present simplifies the problem. 1,4-Cyclohexadienes are available from Diels–Alder reactions between butadienes and acetylenes. Reversing the intramolecular Diels–Alder reaction reveals the propargylic ester of *trans*-6,8-nonadieneoic acid.

1,4-Cyclohexadiene
Substructure

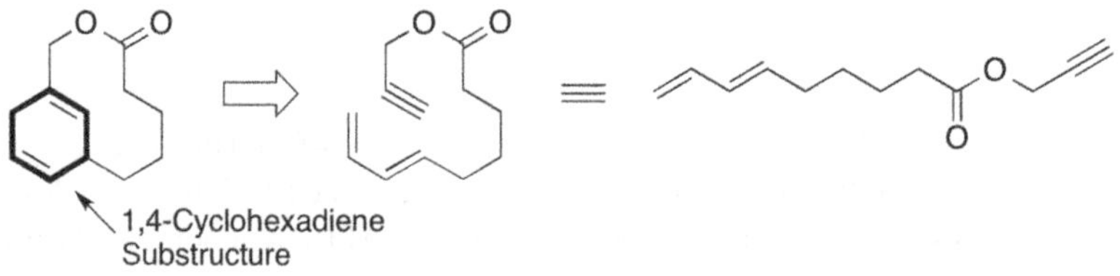

Based on the aforementioned evaluation, the target molecule can be prepared as illustrated in the following using arrow-pushing. Initially, the propargylic ester of *trans*-6,8-nonadieneoic acid undergoes an intramolecular Diels–Alder reaction forming the illustrated lactone. Following deprotonation of the lactone and formation of the illustrated silyl ketene acetal, the Ireland–Claisen rearrangement generates the target carboxylic acid upon hydrolytic workup.

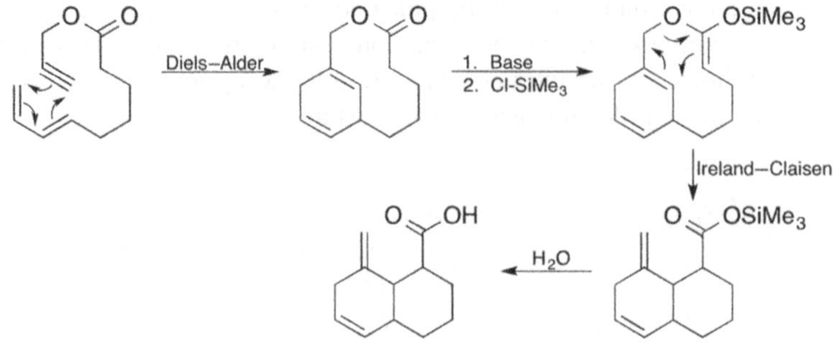

CHAPTER 11 SOLUTIONS

1. *Describe the following functional group transformation in mechanistic terms. Show arrow-pushing.*

a.

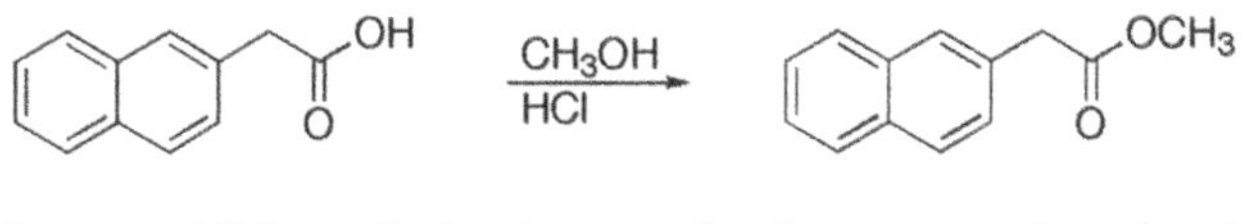

This is an addition–elimination reaction between methanol and a protonated carboxylic acid. As illustrated in the following, hydrochloric acid protonates the carboxylic acid. Methanol then adds to the protonated carboxylic acid. Elimination of water liberates the methyl ester.

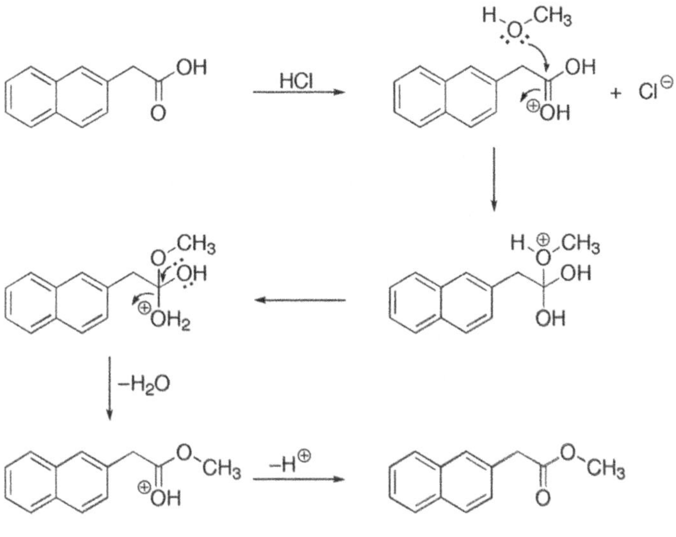

Please note that this reaction is generally run with methanol as the solvent. Under these circumstances, the reverse reaction, ester hydrolysis, does not proceed because the water being liberated during the reaction is so dilute in the methanol that water molecules never interact with the forming ester.

b.

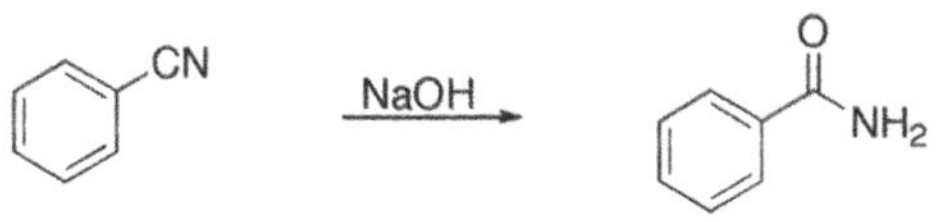

This is a hydrolysis reaction where a hydroxide anion adds to a nitrile. As illustrated in the following, the hydroxide anion adds to the nitrile carbon atom. Proton transfer from the hydroxyl group to the nitrogen anion is followed by charge transfer through resonance. This charge transfer results in the formation of a carbonyl and a nitrogen anion. The nitrogen anion is neutralized when the reaction is quenched with acid.

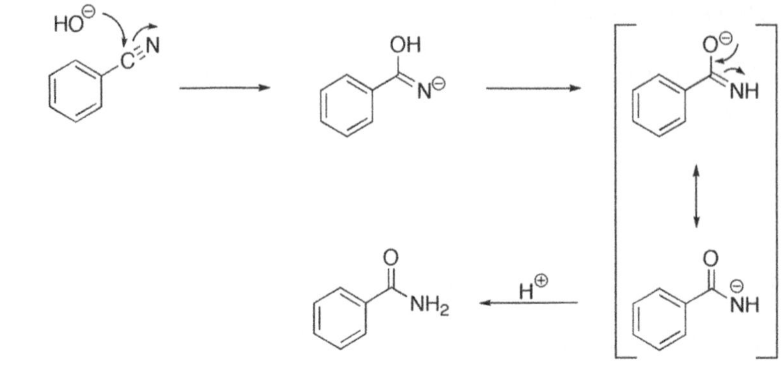

c.

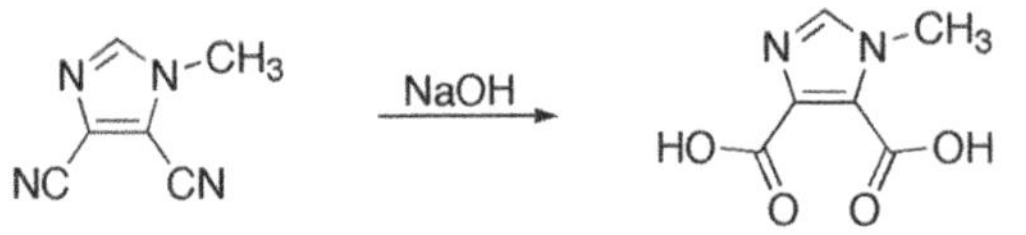

The first step in this reaction is the hydrolysis of two nitrile groups to form amides. The mechanism for the amide formation is identical to that illustrated in the previous example. Continuing from the amides, hydroxide anions add to the carbonyls generating negative charges on each functional group. Following the addition–elimination mechanistic sequence, the negative charges residing on the oxygen atoms displace amine anions (amide ions) liberating the illustrated carboxylic acids. However, since ammonia is less acidic than a carboxylic acid, the amine anions deprotonate the carboxylic acids generating carboxylate anions and ammonia. These carboxylate anions become neutralized on treatment with acid.

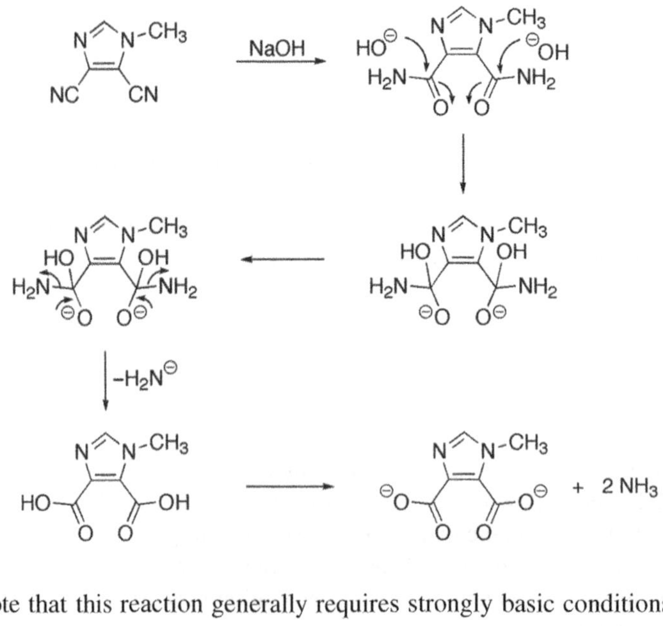

Please note that this reaction generally requires strongly basic conditions and high temperatures.

2. *Explain the following reactions in mechanistic terms. Show arrow-pushing and describe the reaction as a name reaction.*

a.

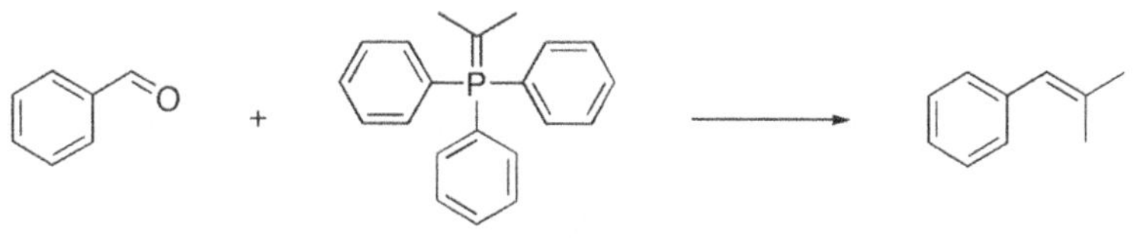

This is an example of the Wittig reaction. Wittig reactions occur when a phosphorus ylide reacts with an aldehyde or a ketone. An ylide is a molecule in which there exists a natural state of charge separation. In this case, the ylide is dimethyl-methylene triphenylphosphorane illustrated in the following. Note that the phosphorus possesses a positive charge and is electrophilic, while the negative charge resides on a carbon atom rendering it nucleophilic.

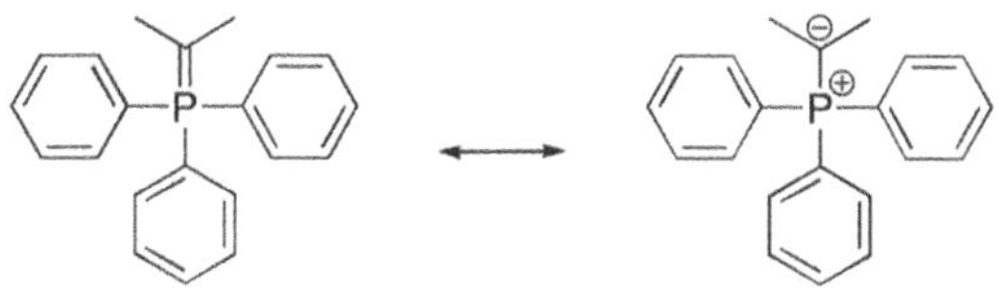

The Wittig reaction mechanism involves addition of the anionic carbon atom to the carbon atom of an aldehyde. As illustrated in the following, the now negatively charged oxygen atom adds to the positively charged phosphorus atom forming a four-membered ring. This ring, known as a betaine, then decomposes to form an olefin and triphenylphosphine oxide.

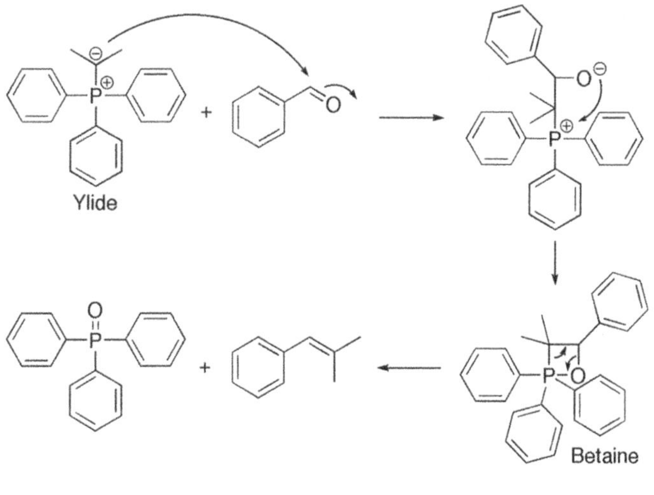

b.

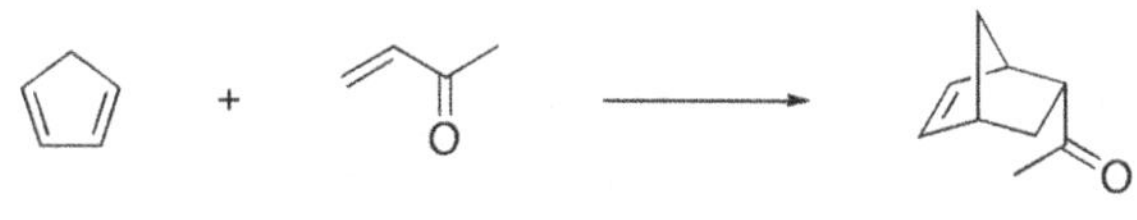

This is an example of a Diels–Alder reaction. This is an electrocyclic reaction where no charges are involved. While no charges are involved, electron pairs do move and their movement can be illustrated using arrow-pushing. The mechanism, illustrated in the following, involves aligning cyclopentadiene (a diene) with methyl vinyl ketone (a dienophile) such that all three double bonds define a six-membered ring. Once the six-membered ring is defined, the electrons simply move to form two new carbon–carbon bonds with a net conversion of two carbon–carbon double bonds to carbon–carbon single bonds. It is important to recognize that in electrocyclic reactions, the total number of bonds never changes. Specifically, seven bonds are involved in the reaction where six of the seven bonds are incorporated in double bonds. During the course of the reaction, these seven bonds are composed of five carbon–carbon single bonds and one carbon–carbon double bond.

The Diels–Alder reaction was discussed in detail in Chapter 10.

c.

This is an example of a Horner–Emmons reaction. The mechanism, illustrated in the following, is similar to that discussed for Problems 7n and 7o from Chapter 5. As shown, the first step involves deprotonation of triethyl phosphonoacetate with sodium hydride. The resulting anion then participates in an addition reaction with acetone. The product of this addition reaction possesses a negatively charged oxygen. This negative charged oxygen adds into the phosphorus–oxygen double bond forming a four-membered ring known as an oxaphosphetane. The oxaphosphetane, on decomposition as illustrated with arrow-pushing, liberates the product.

Oxaphosphetane

When considering the Horner–Emmons reaction, it is important to recognize that the mechanism and products are similar to those observed during a Wittig reaction. In fact, the Horner–Emmons reaction is a recognized and viable alternative to the Wittig reaction.

d.

This is an example of a Claisen rearrangement. This is an electrocyclic reaction where no charges are involved. While no charges are involved, like the Diels–Alder reaction, electron pairs do move, and their movement can be illustrated using arrow-pushing. The mechanism, illustrated in the following, involves moving a lone pair of electrons from the oxygen into the aromatic ring. The aromatic ring then adds electrons to the double bond. The double bond then migrates and the carbon–oxygen bond is cleaved. While the expected product may be the illustrated ketone, spontaneous conversion to the enol form is facilitated by the stability of the resulting aromatic ring. Thus the illustrated product is formed.

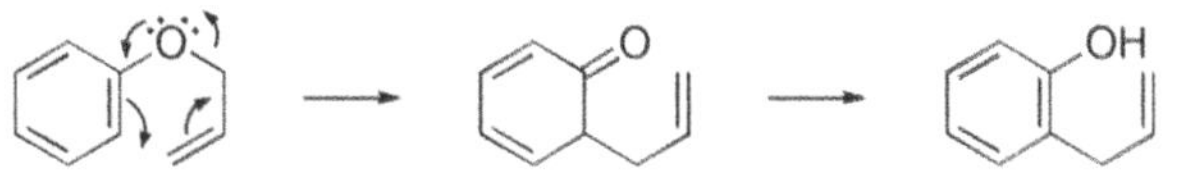

When considering the aforementioned mechanistic description, it is important to recognize that all of these steps occur concurrently. Furthermore, like the Diels–Alder reaction (and all electrocyclic reactions), there is no net loss or gain of bonds.

The Claisen rearrangement was discussed in detail in Chapter 10.

e.

This is an example of a Robinson annulation. The mechanism for the Robinson annulation involves a sequence of conjugate addition reactions and aldol condensations. As illustrated, the first step is deprotonation of cyclohexanedione with sodium hydride. The resulting anion then participates in a 1,4-addition to methyl vinyl ketone. The resulting enolate anion then tautomerizes through resonance, placing the anion adjacent to a carbonyl. Proton transfer migrates this negative charge to the terminal methyl group.

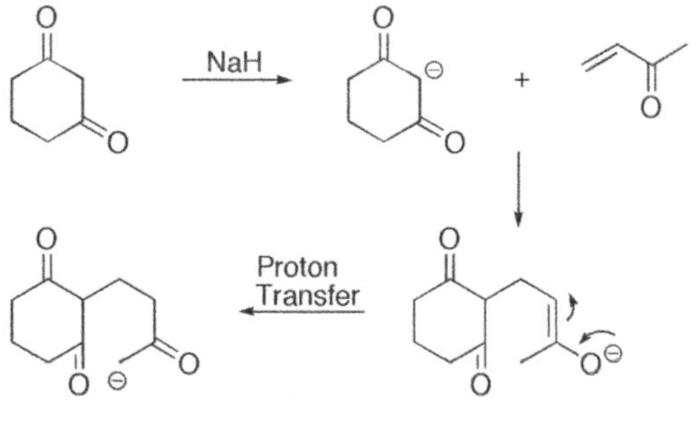

Following the formation of a negative charge at the terminal methyl group, the terminal methyl group participates in an aldol condensation with one of the cyclohexanedione carbonyl groups. This aldol condensation involves initial addition of the anion to the carbonyl followed by subsequent dehydration of the resulting alkoxide. This dehydration usually occurs under acidic conditions during isolation of the product and through mechanistic pathways already presented (consider protonation of a hydroxyl group followed by an E1 elimination under solvolytic conditions).

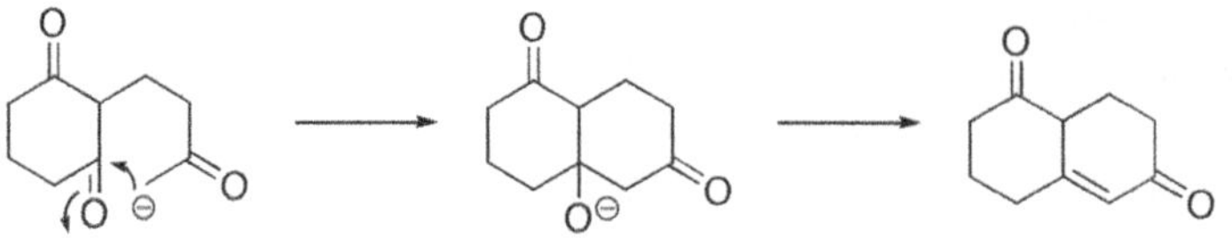

f.

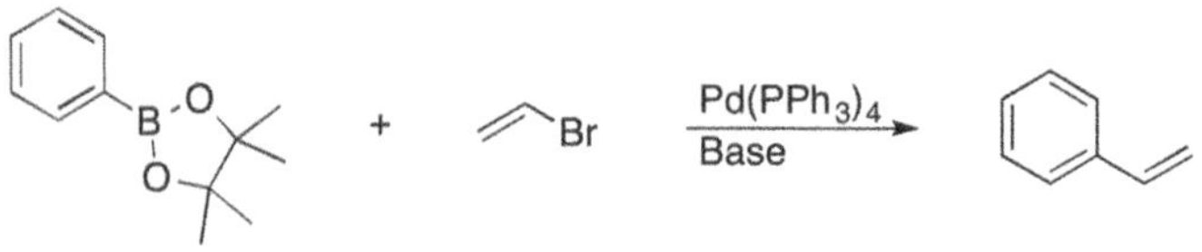

This is an example of a Suzuki reaction. The mechanism for the Suzuki reaction, illustrated in the following, involves insertion of the palladium atom into the carbon–bromine bond of vinyl bromide. This oxidative addition step where the neutral palladium atom takes on a +2 charge is followed by transmetallation where the boronic ester of phenylboronic acid pinacolate is replaced by the oxidized palladium atom. Following transmetallation, the vinyl group forms a new bond with the phenyl ring and expels the palladium atom through reductive elimination where the palladium atom returns to a neutral state.

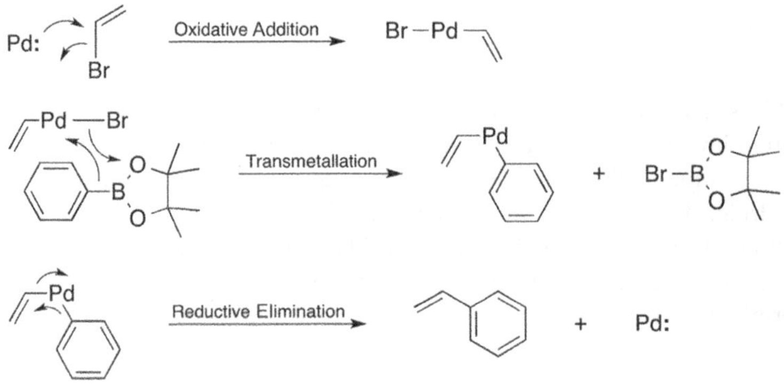

3. *Explain the following name reactions in mechanistic terms. Show arrow-pushing.*

a. *The ene reaction*

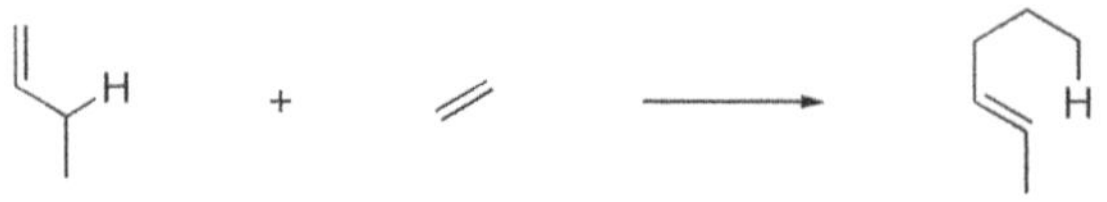

Note: Only the hydrogen involved in the reaction is shown.

The ene reaction is an electrocyclic reaction similar to the Diels–Alder reaction and the Claisen rearrangement. In this reaction, a hydrogen atom is participating in the electrocyclic process. The mechanism, illustrated in the following using arrow-pushing, involves no charges. Note that there is no net gain or loss of bond count between the starting materials and the product.

The ene reaction was discussed in detail in Chapter 10.

b.   *The McLafferty rearrangement*

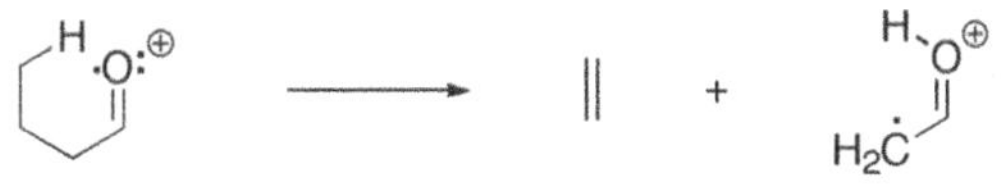

Note: The radical cation present in the starting material is the result of the carbonyl oxygen losing a single electron. This reaction is generally observed during electron impact mass spectrometry.

The McLafferty rearrangement is a reaction generally seen as part of the fragmentation processes observed during mass spectrometry. It is, in fact, during electron impact mass spectrometry that the illustrated starting radical cation is formed. Since this is a radical-mediated process, there are no charges involved in the progression of the reaction mechanism other than the positive charge that remains on the oxygen atom. As shown in the following, using arrow-pushing, the first step of this rearrangement involves transfer of a hydrogen atom to the carbonyl oxygen. This occurs through homolytic bond cleavage and bond formation. The second step, also progressing through a homolytic process, involves cleavage of a carbon–carbon bond and liberation of ethylene.

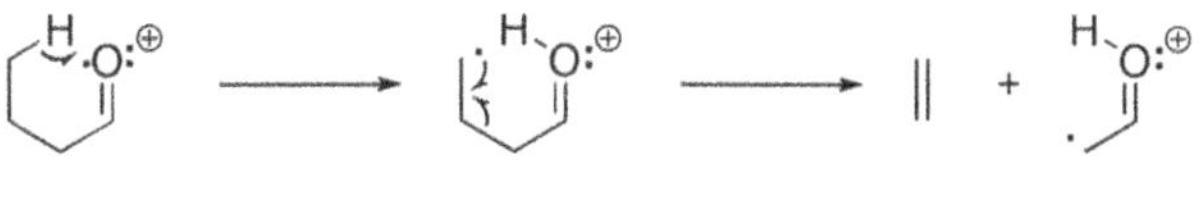

c.   *1,3-Dipolar cycloaddition*

1,3-Dipolar cycloadditions are electrocyclic reactions where one of the starting materials is charged. In fact, it is the charges on the starting material that define the dipole. Like all electrocyclic reactions, there is no net gain or loss of bond count. However, in this case, while the starting material is charged, there are no charges present on the product. The mechanism of this reaction is illustrated in the following using arrow-pushing.

1,2-Dipolar cycloadditions were discussed in detail in Chapter 10.

d.   *The Swern oxidation*

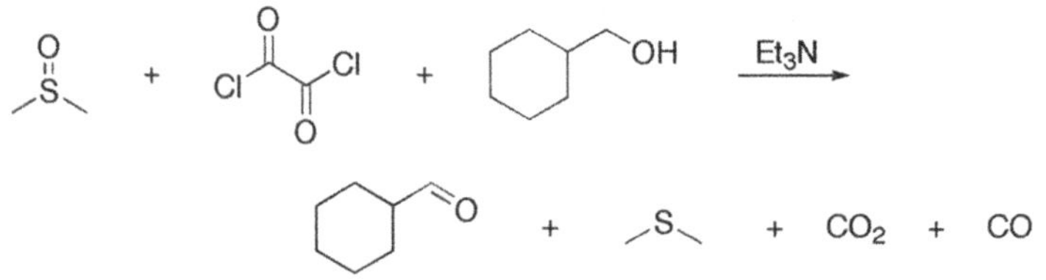

Hint: The oxygen atom in dimethyl sulfoxide is nucleophilic.

In this book, there have been many references to oxidation and reduction reactions. While these reactions are not within the scope of the discussions of this book, their mechanisms do involve the processes presented herein. In the case of the Swern oxidation, the first step is an addition–elimination reaction between dimethyl sulfoxide and oxalyl chloride. This process, illustrated in the following using arrow-pushing, involves addition of the sulfoxide oxygen to a carbonyl with subsequent elimination of a chloride anion.

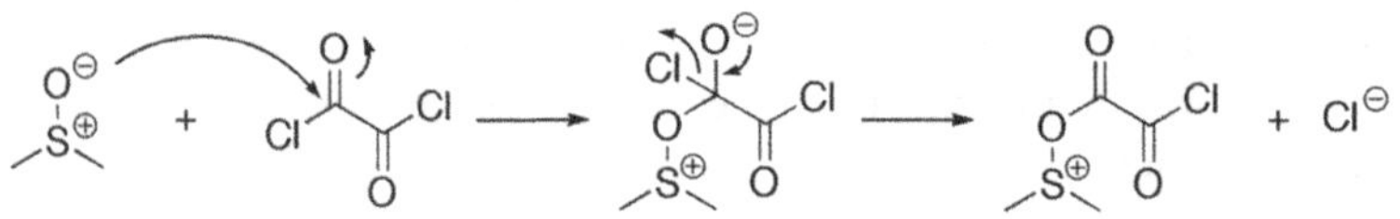

The second stage of the Swern oxidation, illustrated in the following, involves a nucleophilic displacement of the oxalyl group from the sulfur. In this step, the nucleophile is a chloride anion, and the reaction is facilitated by the decomposition of the leaving group into carbon dioxide gas, carbon monoxide gas, and a chloride anion.

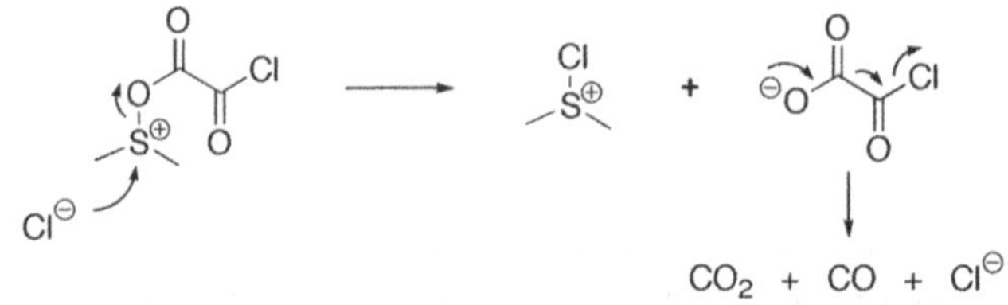

The third stage of this reaction involves another nucleophilic displacement. In this step, the nucleophile is an alcohol and the leaving group is a chloride anion. This step, illustrated in the following, involves protonation of the leaving chloride anion forming hydrochloric acid.

The final stage of this reaction involves an E2 elimination. In this step, illustrated in the following, a proton adjacent to the oxygen is removed by a base such as triethylamine. The negative charge then forms a double bond with the oxygen, and dimethylsulfide is eliminated. The overall oxidation process converts an alcohol into an aldehyde.

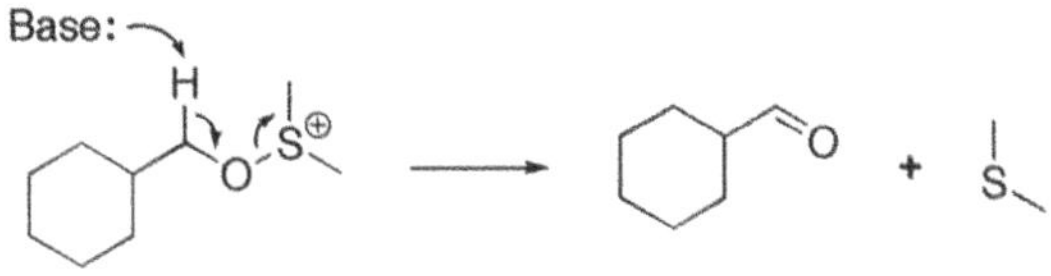

e.   *The Sonogashira reaction*

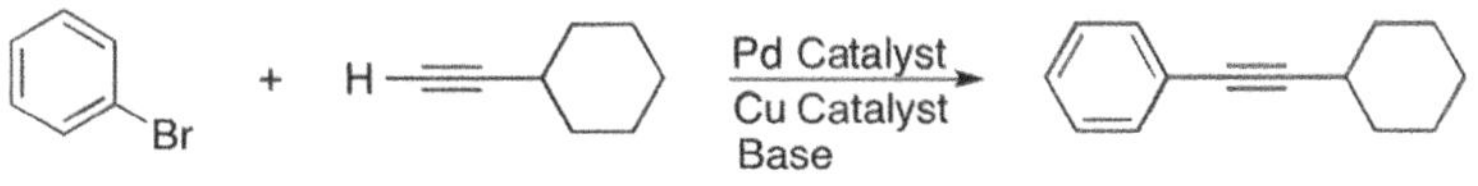

Hint: The copper catalyst metallates the acetylene.

As suggested by the hint, the Sonogashira reaction involves two catalytic cycles. For simplification purposes, only the palladium cycle is addressed due to similarities to the Suzuki reaction. The copper catalytic cycle is represented only by metallation of the acetylene reactant.

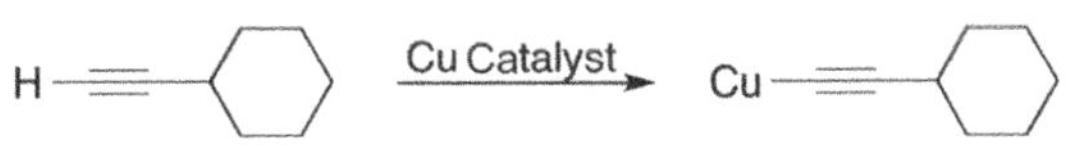

Following the general mechanistic pathway for the Suzuki coupling, the palladium catalyst inserts into the carbon–bromine bond of bromobenzene through oxidative addition.

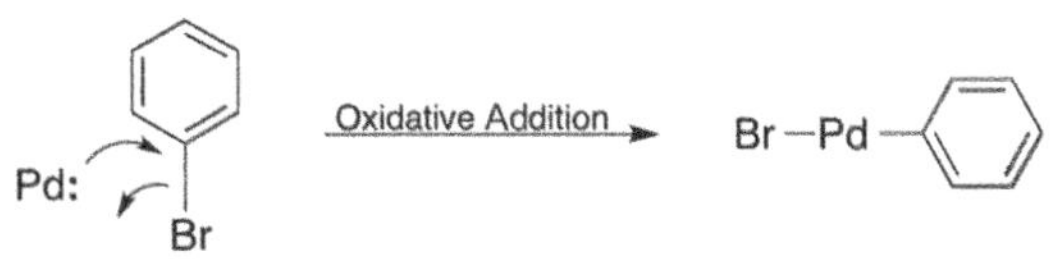

Following oxidative addition, a transmetallation step replaces the copper with palladium.

Finally, reductive elimination results in formation of the final product with regeneration of the palladium catalyst.

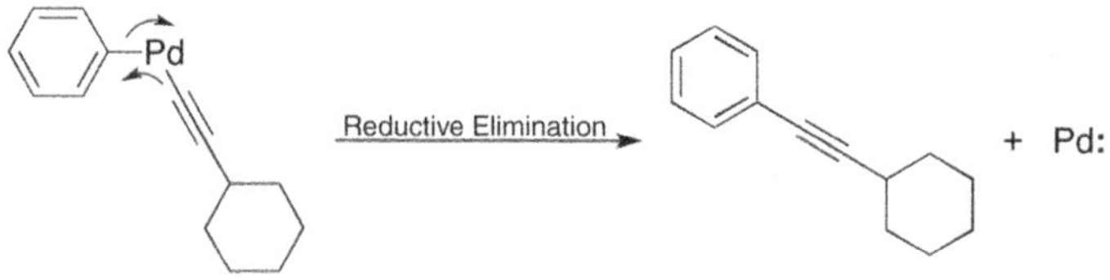

It is important to note that while base is mentioned in the illustrated reaction, its role is omitted from this discussion for simplicity. The role of the base in the Sonogashira reaction is similar to its role in the Suzuki reaction.

f.   *The Buchwald–Hartwig amination*

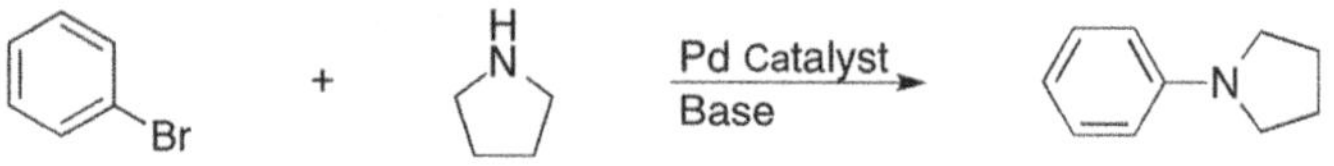

The Buchwald–Hartwig amination follows the general mechanistic pathway for the Suzuki reaction. Initially, the palladium catalyst inserts into the carbon–bromine bond of bromobenzene through oxidative addition.

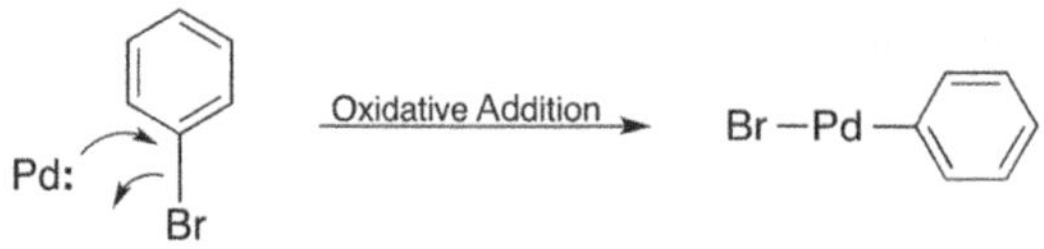

Following oxidative addition, the lone electron pair associated with the amine nitrogen coordinates with the palladium atom.

Finally, reductive elimination results in formation of the final product with regeneration of the palladium catalyst and formation of hydrobromic acid.

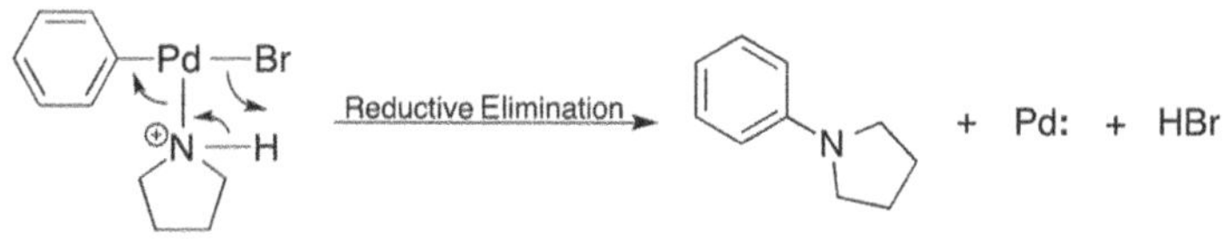

In the Buchwald–Hartwig amination, base is required to facilitate reductive elimination and to absorb the acid that is formed during the reaction.

4. *The Friedel–Crafts acylation, illustrated in Scheme 11.12, shows the formation of one product. However, the reaction, as illustrated, actually forms a mixture of two products. Using the arguments presented in the solution set for Chapter 8, identify the second product. Show partial charges and arrow-pushing.*

Just as double bonds possess nucleophilic characteristics, so do aromatic rings. By analyzing the charge distribution around an aromatic ring, sites of partial positive charge and sites of partial negative charge can be identified. The sites of partial positive charge are electrophilic in nature, and the sites of partial positive charge are nucleophilic in nature. The partial charge distribution for methoxybenzene was the subject of Problem 2h from Chapter 1 and is shown in the following:

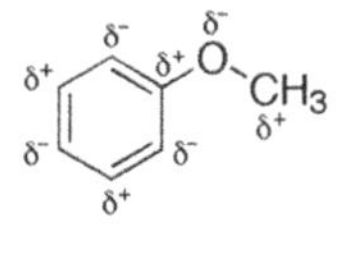

Having identified the nucleophilic sites, this mechanism now becomes an addition–elimination reaction between methoxybenzene and acetyl chloride where methoxybenzene is being added and chloride is being eliminated. As shown in the following, using arrow-pushing, electron movement starts with the methoxy oxygen and moves through the aromatic ring. The addition–elimination steps occur as shown in Chapter 8, Problem 6a. Finally, due to the conjugated and charged system, the proton present on the reactive carbon atom of the phenyl ring becomes acidic. Loss of this proton allows rearomatization and neutralization of the cationic intermediate, thus allowing conversion to the final product.

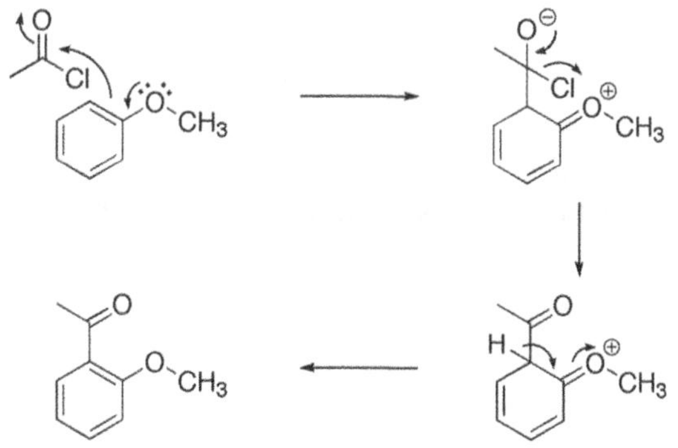

Please note that while the Friedel–Crafts acylation reaction is presented in discussions of addition–elimination reaction mechanisms, this reaction is actually an electrophilic aromatic substitution reaction. The correct mechanism for a Friedel–Crafts acylation was presented in the solution for Problem 6h from Chapter 8.

5. *Predict all products formed from a Friedel–Crafts acylation on the following compounds with acetyl chloride. Rationalize your answers using partial charges.*

a.

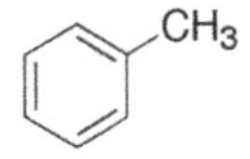

Identification of the partial charges on toluene (methylbenzene), illustrated in the following, was the subject of Problem 2g in Chapter 1.

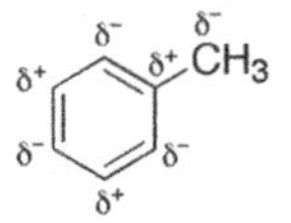

Based on the arguments presented in Chapter 8 and in Problem 4 of this chapter, acylation leads to the formation of the two structures shown in the following.

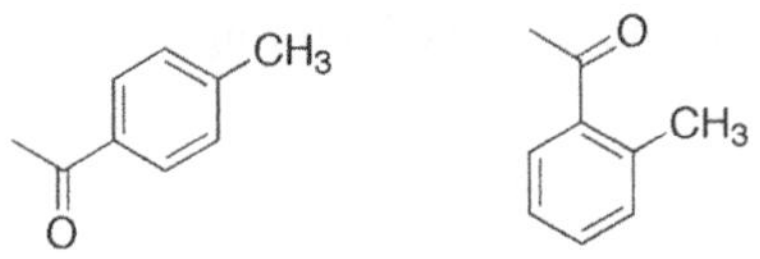

b.

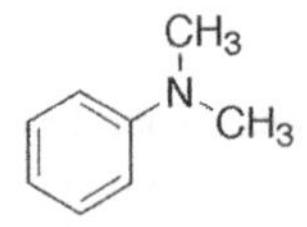

Like methoxybenzene (see Problem 4 and Problem 2h from Chapter 1), the partial charges of dimethylaniline (dimethylaminobenzene) are dependent upon the electron-donating properties of nitrogen. Thus, the partial charges are distributed as shown in the following.

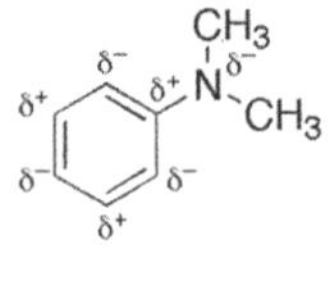

Based on the arguments presented in Chapter 8 and in Problem 4 of this chapter, acylation leads to the formation of the two structures shown in the following.

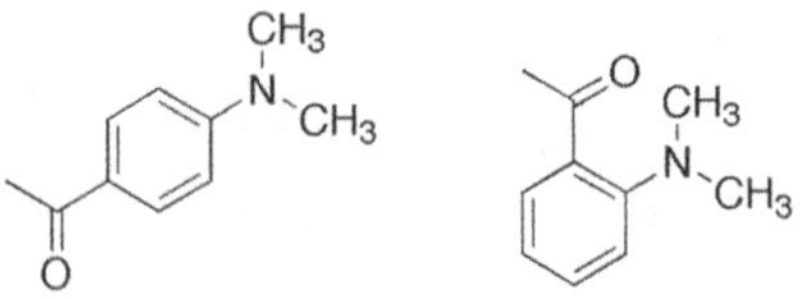

c.

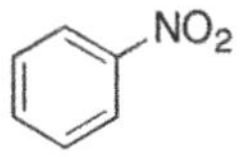

Identification of the partial charges on nitrobenzene, illustrated in the following, was the subject of Problem 2j in Chapter 1.

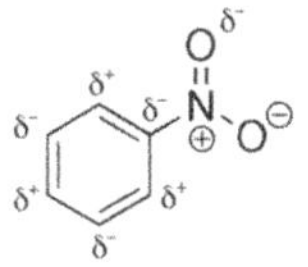

Based on the arguments presented in Chapter 8 and in Problem 4 of this chapter, acylation leads to the formation of the structure shown in the following. Please note that while there are two carbon atoms bearing partial negative charges, acylation of each of these leads to the formation of identical products.

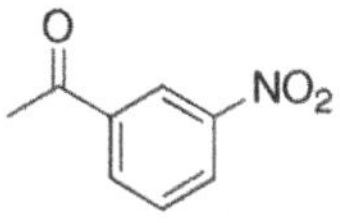

d.

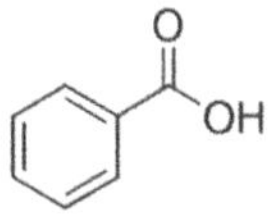

Identification of the partial charges on benzoic acid, illustrated in the following, was the subject of Problem 2k in Chapter 1.

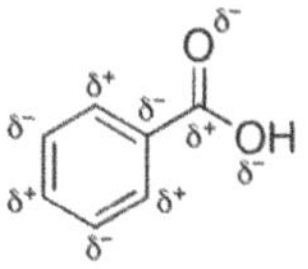

Based on the arguments presented in Chapter 8 and in Problem 4 of this chapter, acylation leads to the formation of the structure shown in the following. Please note that while there are two carbon atoms bearing partial negative charges, acylation of each of these leads to the formation of identical products.

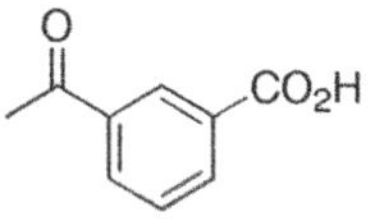

e.

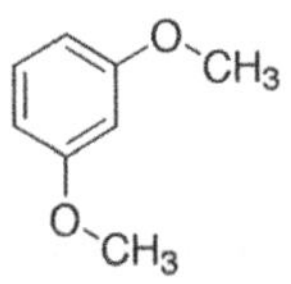

Extrapolating from the arguments presented in Problem 2h of Chapter 1, the partial charge distribution of 1,3-dimethoxybenzene is as shown in the following:

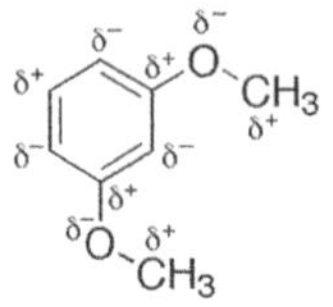

Based on the arguments presented in Chapter 8 and in Problem 4 of this chapter, acylation leads to the formation of the two structures shown in the following. Please note that while there are three carbon atoms bearing partial negative charges, acylation of two of these leads to the formation of identical products.

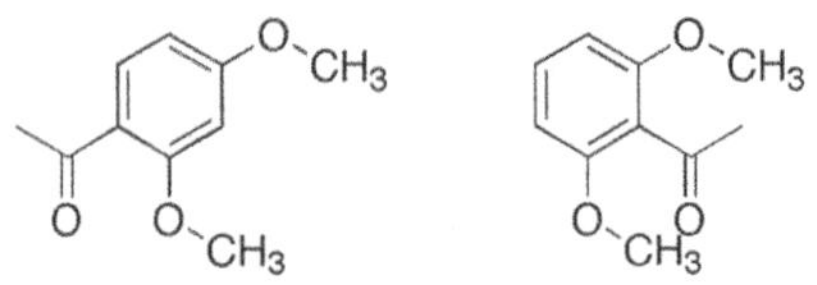

f.

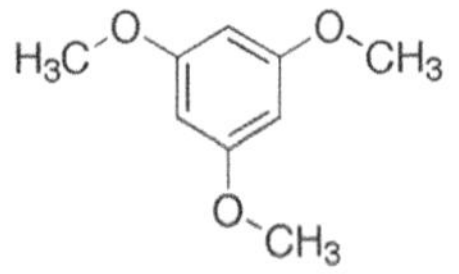

Extrapolating from the arguments presented in Problem 2h of Chapter 1, the partial charge distribution of 1,3,5-trimethoxybenzene is as shown in the following.

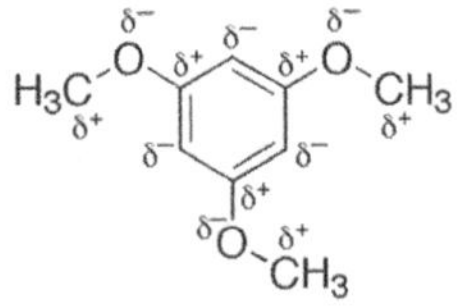

Based on the arguments presented in Chapter 8 and in Problem 4 of this chapter, acylation leads to the formation of the structure shown in the following. Please note that while there are three carbon atoms bearing partial negative charges, acylation of each of these leads to the formation of identical products.

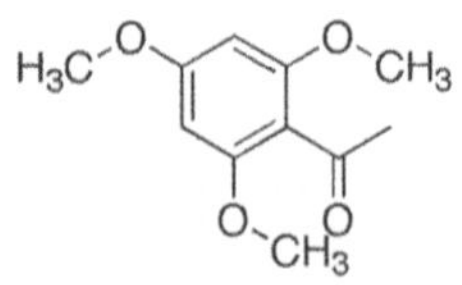

6. *From the following list of compounds, propose a synthetic strategy for the specified compounds. Up to three synthetic steps may be required. Any chemical reagents described in this book or any general organic chemistry text may be used. Show all arrow-pushing.*

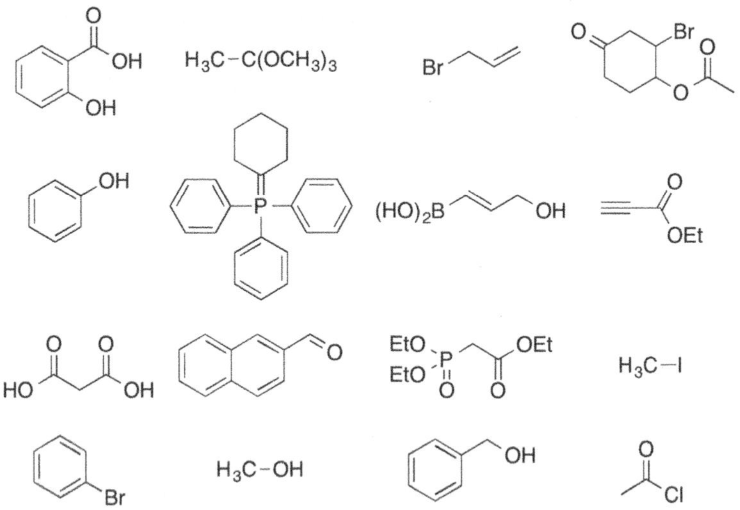

a. 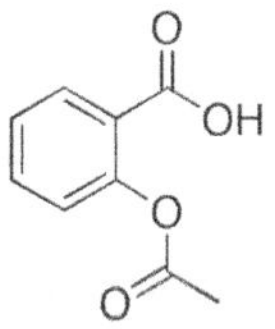   *(acetylsalicylic acid, aspirin)*

Acetylsalicylic acid, a common pain reliever, is composed of two fragments resembling structures from the aforementioned list of compounds. These fragments are illustrated in the following and relate to salicylic acid and acetyl chloride.

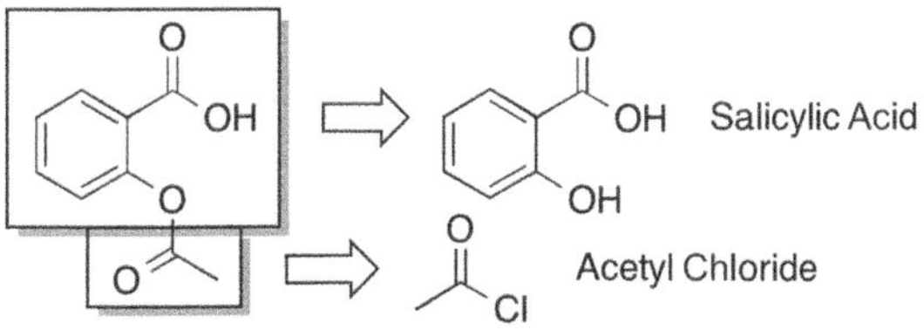

The reaction between salicylic acid and acetyl chloride is an addition–elimination reaction where the hydroxyl group of salicylic acid adds to the carbonyl of acetyl chloride. This addition is followed by the elimination of hydrochloric acid as shown in the following.

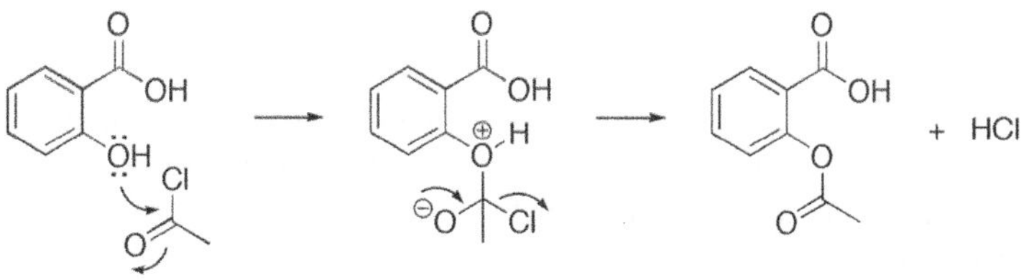

b. 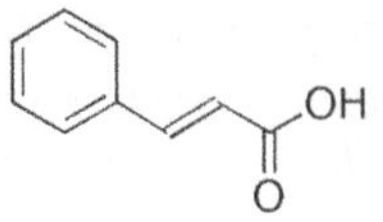 *(cinnamic acid)*

Cinnamic acid, the active flavor compound in cinnamon, is composed of two fragments resembling structures from the aforementioned list of compounds. These fragments are illustrated in the following and relate to benzyl alcohol and triethyl phosphonoacetate.

Triethyl Phosphonoacetate

Benzyl Alcohol

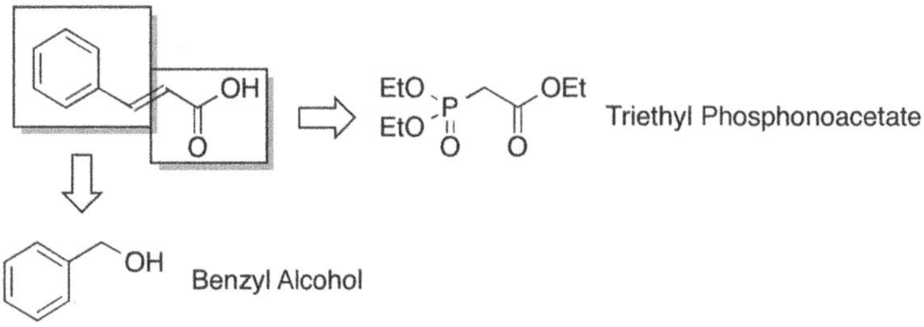

The combination of these compounds will generate cinnamic acid through the synthetic sequence illustrated in the following. As shown, benzyl alcohol is oxidized to benzaldehyde using the Swern oxidation. Next, the aldehyde is reacted with triethyl phosphonoacetate by applying the Horner–Emmons reaction. Finally, the ester is hydrolyzed to a carboxylic acid. With arrow-pushing, the mechanism for the Swern oxidation is shown in Problem 3d of this chapter, the mechanism for the Horner–Emmons reaction is shown in Problem 2c of this chapter, and the mechanism for base-mediated ester hydrolysis was highlighted in Scheme 8.19.

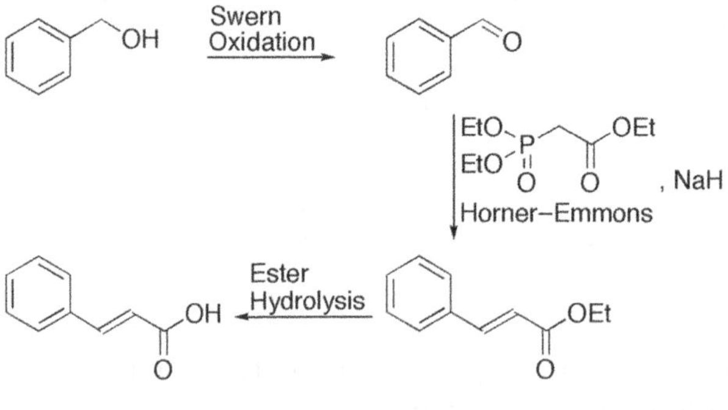

c. 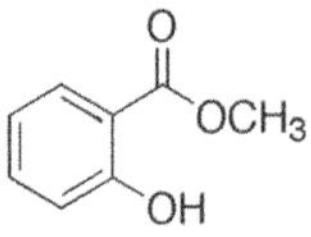 *(methyl salicylate, oil of wintergreen)*

Methyl salicylate, the active flavor compound in wintergreen candy, is composed of two fragments resembling structures from the aforementioned list of compounds. These fragments are illustrated in the following and relate to salicylic acid and methyl alcohol.

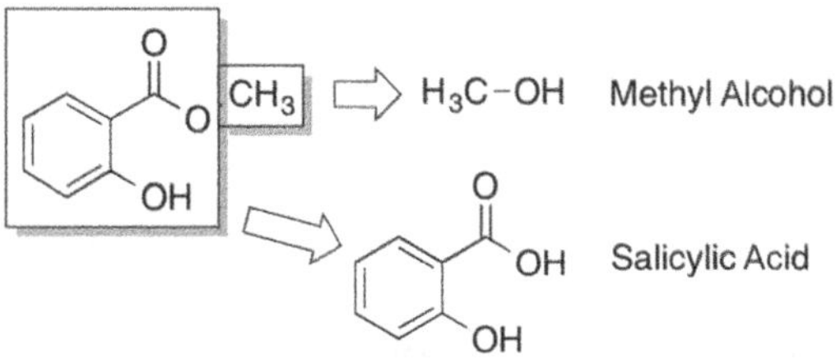

The combination of these compounds will generate methyl salicylate when conditions for an acid-mediated esterification, illustrated in the following, are applied. The mechanism for this type of ester forming reaction is shown in Problem 1a of this chapter.

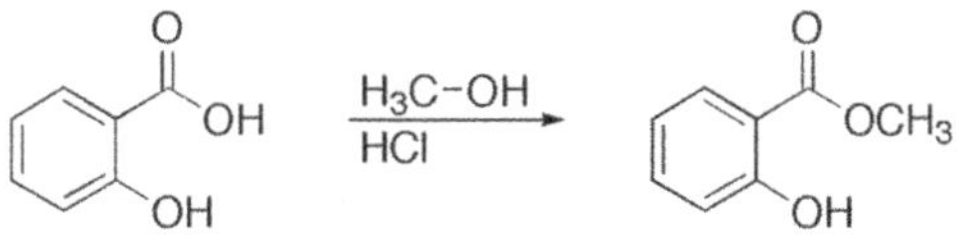

d.

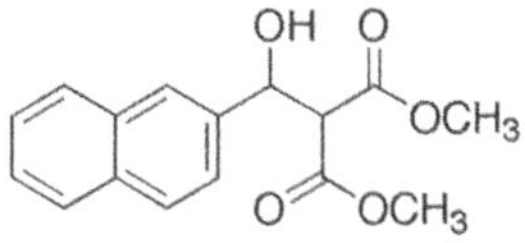

This molecule is composed of three fragments resembling structures from the aforementioned list of compounds. These fragments are illustrated in the following and relate to 2-naphthaldehyde, malonic acid, and methyl iodide.

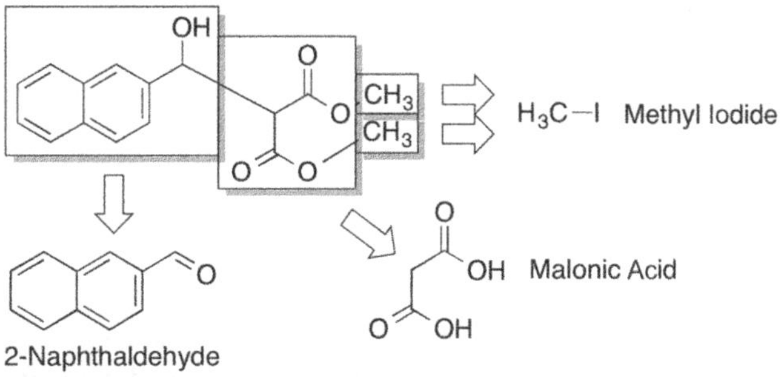

2-Naphthaldehyde

The combination of these compounds will generate the target compound through a two-step synthetic sequence. In the first step, illustrated in the following, malonic acid is converted to dimethyl malonate under mild basic conditions. This esterification reaction proceeds through an S_N2 reaction between a carboxylate anion and methyl iodide. The mechanism for an S_N2 reaction was presented in detail in Chapter 5. In this reaction it is important to use a base that is sufficient to deprotonate a carboxylic acid but not strong enough to remove a proton from the methylene group of malonic acid. Sodium bicarbonate is generally sufficient to affect this deprotonation. Please note that this same esterification can proceed under acidic conditions in methyl alcohol.

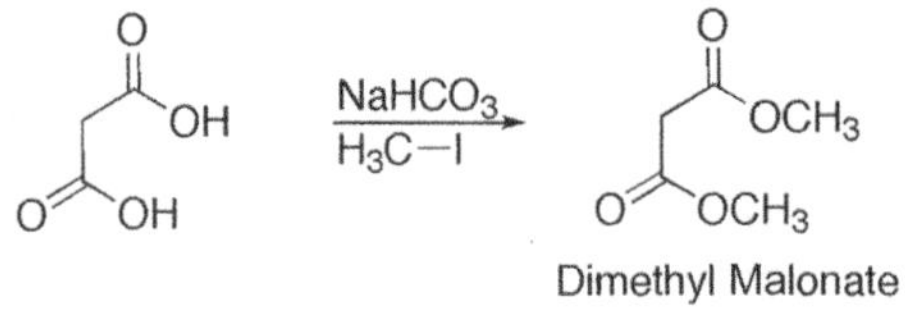

Dimethyl Malonate

The second step in this sequence, illustrated in the following, is an 1,2-addition reaction between a dimethyl malonate anion and 2-naphthaldehyde. The mechanism for 1,2-addition reactions was discussed in detail in Chapter 8. In order for this reaction to proceed, it is important to use a base that is sufficient to deprotonate the methylene group of dimethyl malonate. Furthermore, it is important to use a base that will not hydrolyze the methyl esters. Sodium hydride is generally sufficient to affect this deprotonation.

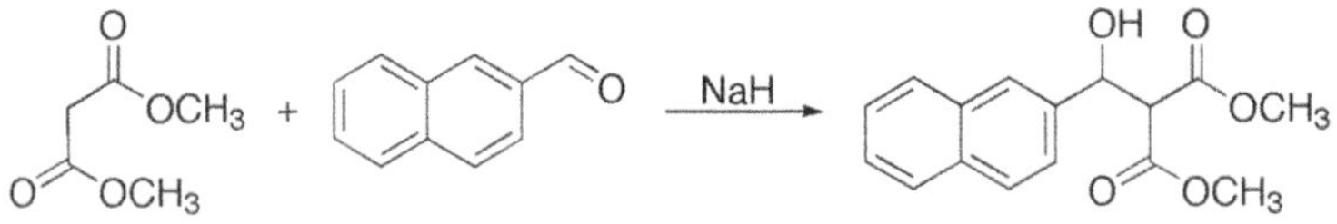

e.

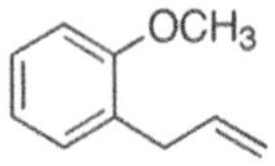

This molecule is composed of three fragments resembling structures from the aforementioned list of compounds. These fragments are illustrated in the following and relate to phenol (hydroxybenzene), allyl bromide, and methyl iodide.

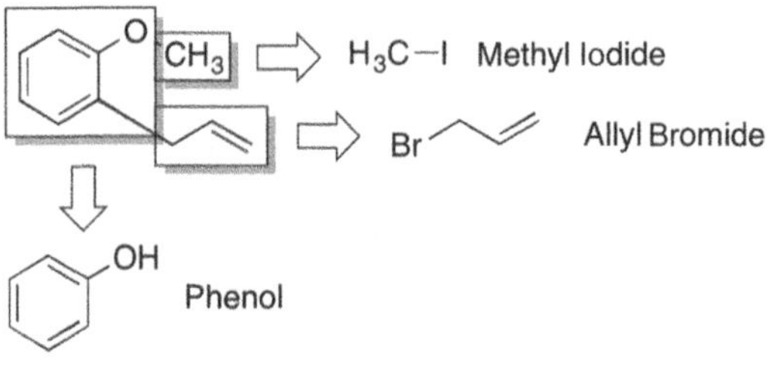

The combination of these compounds will generate the target compound through a three-step synthetic sequence. In the first step, illustrated in the following, phenol is alkylated with allyl bromide through an S_N2' mechanism. The mechanism for an S_N2' reaction was presented in detail in Chapter 5. In this reaction it is important to use a base that is sufficient to deprotonate a phenolic hydroxyl group. Potassium *tert*-butoxide is generally sufficient to affect this deprotonation.

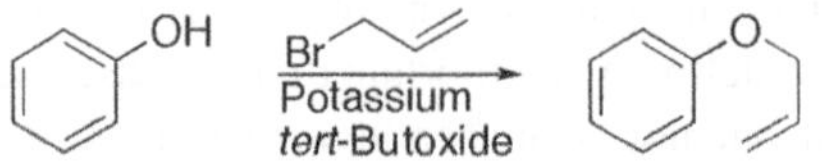

The second step of this sequence, illustrated in the following, is a Claisen rearrangement where the allyl group is migrated from the oxygen onto the aromatic ring. The mechanism for the Claisen rearrangement was presented in Problem 2d of this chapter.

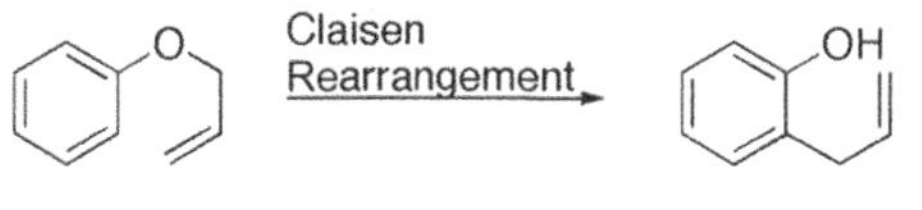

The third step of this sequence, illustrated in the following, is an S_N2 reaction between a phenol anion and methyl iodide. The mechanism for an S_N2 reaction was presented in detail in Chapter 5. In this reaction it is important to use a base that is sufficient to deprotonate a phenolic hydroxyl group. Potassium *tert*-butoxide is generally sufficient to affect this deprotonation.

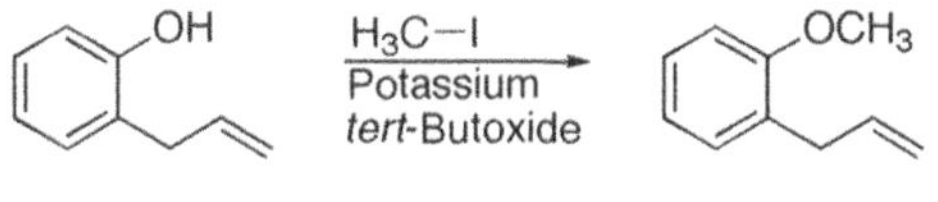

f.

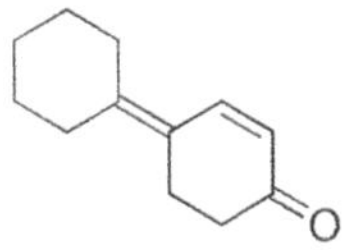

This molecule is composed of two fragments resembling structures from the aforementioned list of compounds. These fragments are illustrated in the following and relate to cyclohexylidine triphenylphosphorane and 3-bromo-4-acetoxycyclohexanone.

3-Bromo-4-acetoxycyclohexanone

Cyclohexylidine Triphenylphosphorane

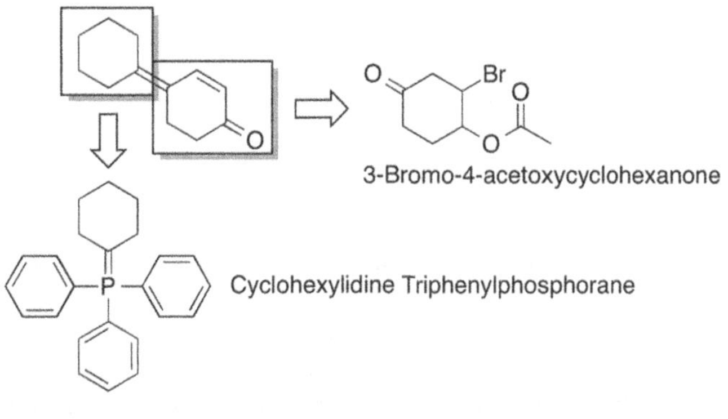

The combination of these compounds will generate the target compound through a four-step synthetic sequence. The first step, illustrated in the following, is a Wittig reaction between cyclohexylidine triphenylphosphorane and 3-bromo-4-acetoxycyclohexanone. The mechanism for the Wittig reaction was presented in Problem 2a of this chapter.

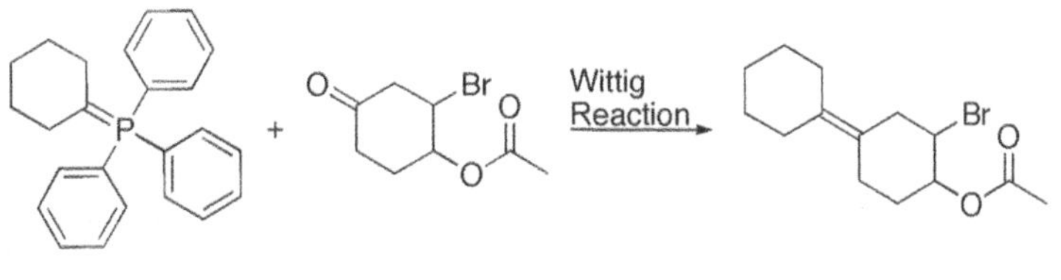

The second step of this sequence is an E2 elimination reaction generating a diene. The mechanism for an E2 elimination was presented in detail in Chapter 7. For this reaction to proceed, it is important to choose a base that is non-nucleophilic and strong enough to remove an allylic proton. LDA is generally sufficient to affect this transformation.

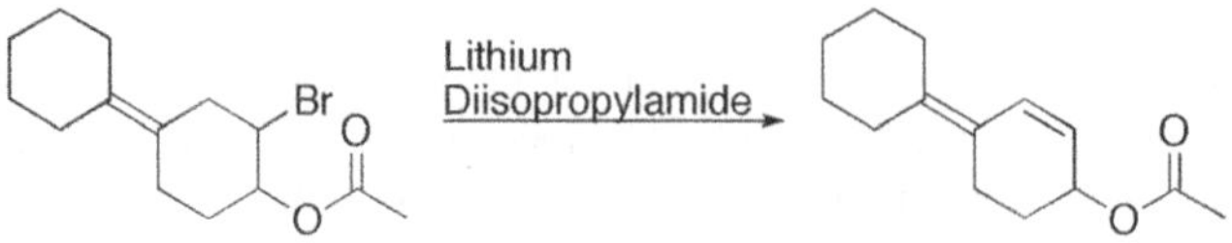

The third step of this sequence, illustrated in the following, is an ester hydrolysis reaction. The mechanism for a base-mediated ester hydrolysis was highlighted in Scheme 8.19.

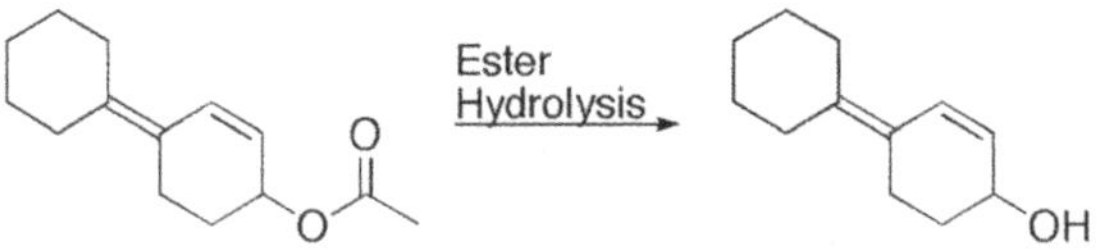

The fourth and final step of this sequence, illustrated in the following, is an oxidation of an alcohol to a ketone. This transformation can be accomplished utilizing the Swern oxidation. The mechanism for the Swern oxidation is shown in Problem 3d of this chapter.

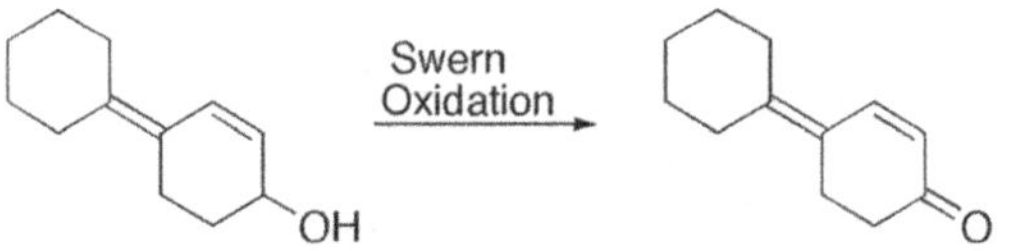

g. 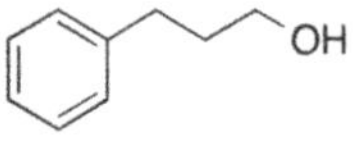 *via a route not related to that used in problem 6b*

This molecule resembles cinnamic acid—the subject of Problem 6b. One objective of this exercise is to demonstrate that there are frequently multiple correct solutions to a given synthetic problem. Therefore, not relying upon the chemistry described for Problem 6b, a reasonable question would be directed at how one may modify a phenyl ring with the required alcohol attachment. From the list of available starting materials, bromobenzene is available to deliver the required phenyl ring. Furthermore, ethyl propiolate, a starting material with three carbon atoms leading to an oxygen atom, is a reasonable starting material for the alcohol unit.

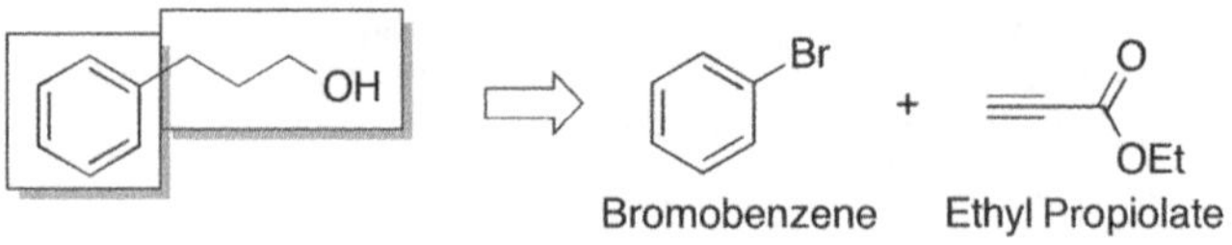

Bromobenzene   Ethyl Propiolate

Bringing bromobenzene and ethyl propiolate together can be accomplished using the Sonogashira reaction generating the illustrated product.

Following the Sonogashira reaction, the final product is available through reduction of the triple bond and reduction of the ethyl ester to an alcohol. Though not covered in this book, general organic chemistry coursework presents many methods for the reduction of alkenes and alkynes to saturated hydrocarbons. Among the most convenient methods is the catalytic hydrogenation in which the reduction takes place in the presence of palladium and hydrogen gas.

Finally, while not covered in this book, general organic chemistry coursework presents many methods for the reduction of esters to alcohols. Among the most convenient methods is the use of lithium aluminum hydride. Therefore, reduction of the ester using lithium aluminum hydride can complete the synthesis of the target structure.

h.   *via a route not related to those used in problems 6b and 6g*

Like the target of Problem 6g, this molecule resembles cinnamic acid—the subject of Problem 6b. One objective of this exercise is to demonstrate that there are frequently multiple correct solutions to a given synthetic problem. Therefore, not relying upon the chemistry described for Problem 6b or 6g, a reasonable question would be directed at how one may modify a phenyl ring with the required attachment. In analyzing such problems, it is often beneficial to study the relationships between the different groups present. In the case of this target molecule, the phenyl ring is attached to a 4-pentenoic acid unit. From Chapter 10, we recall that esters of 4-pentenoic acids are available using the Johnson–Claisen rearrangement. Using this logic, reverting the target molecule to a methyl ester, the Johnson–Claisen rearrangement substrate can be derived.

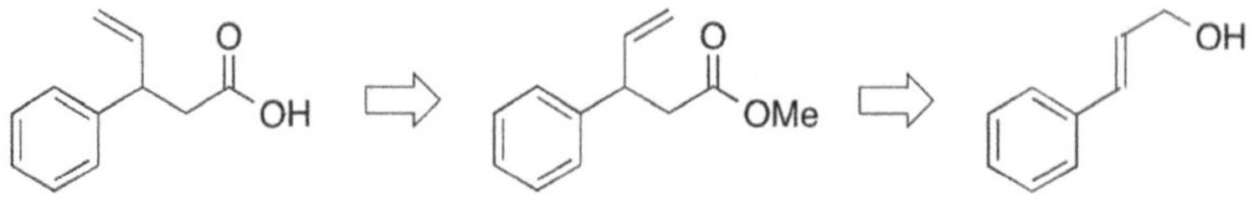

Recognizing that the substrate for the Johnson–Claisen rearrangement is a phenyl-substituted allylic alcohol, bromobenzene and 3-hydroxy-1-propenylboronic acid can be identified as appropriate raw materials from the aforementioned list.

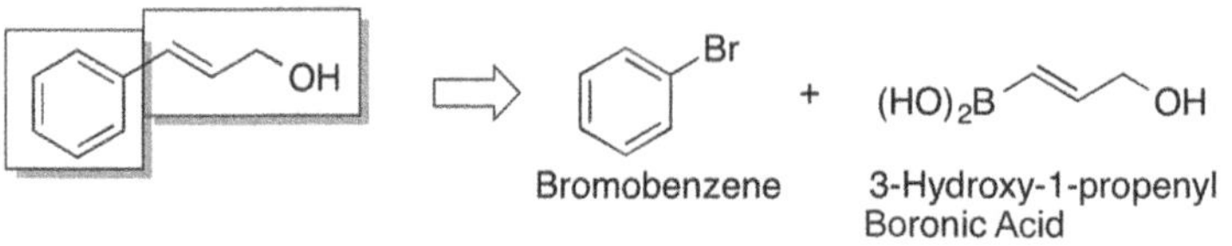

Bromobenzene   3-Hydroxy-1-propenyl
                    Boronic Acid

Having identified appropriate raw materials, the target structure can be assembled using an initial Suzuki reaction to couple bromobenzene with 3-hydroxy-1-propenylboronic acid giving the illustrated product.

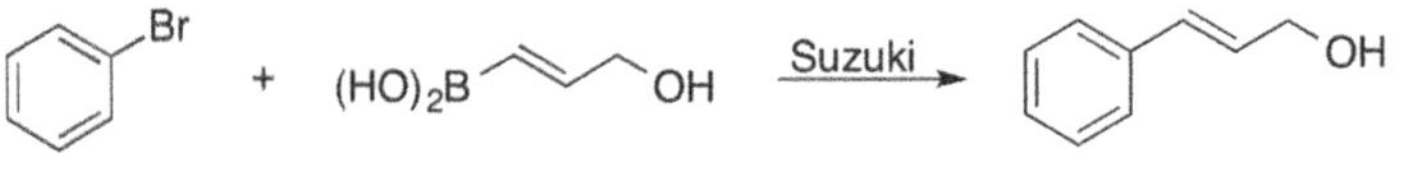

Following the Suzuki reaction, the product can be treated with trimethyl orthoacetate under Johnson–Claisen conditions to generate the target methyl ester.

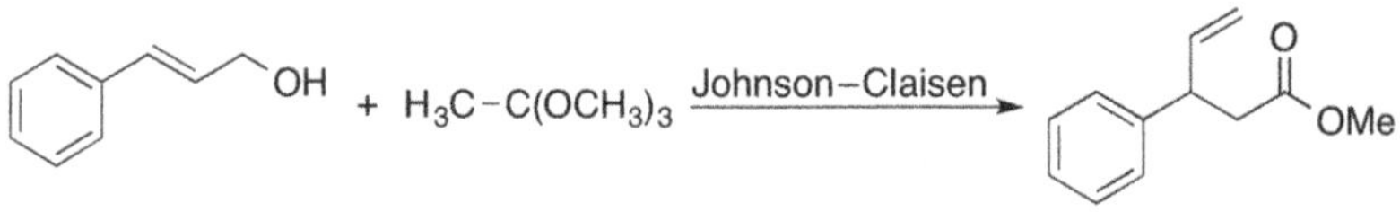

Following the Johnson–Claisen rearrangement, the final product can be completed by simple ester hydrolysis.

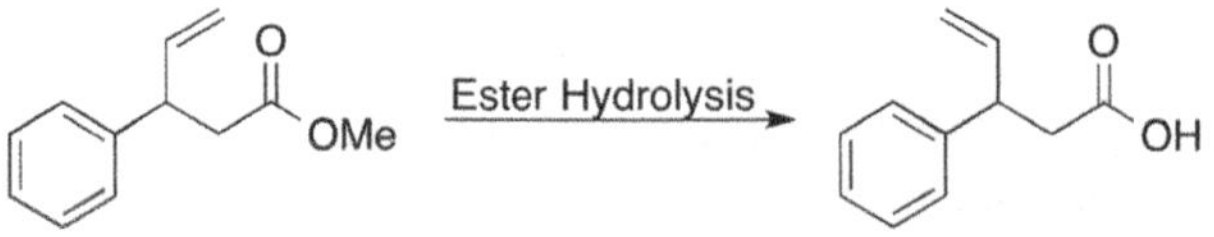

Appendix *3*

Student Reaction Glossary

The premise of this book is based on the presumption that introductory organic chemistry entails very little memorization. As presented in the chapters contained herein, this presumption is valid, provided the student adheres to the philosophy that the study of organic chemistry can be reduced to the study of interactions between organic acids and bases. At this point, use of the principles presented in this book, in conjunction with more detailed coursework, allows students a broader understanding of organic chemistry reactions as described using combinations of fundamental organic mechanistic subtypes.

The mechanistic subtypes presented throughout this book include those related to the acid/base properties of organic molecules. These are protonations, deprotonations, and proton transfers. Mechanistic types based on solvation effects include solvolysis reactions, S_N1, and E1 processes. Additional mechanisms utilizing ionic interactions include S_N2, S_N2', E1cB, E2, 1,2-additions, 1,4-additions, and addition–elimination processes. Finally, those mechanistic types dependent upon the presence of cationic species include alkyl shifts and hydride shifts.

On the following pages, forms are provided, which are designed to aid students in summarizing the various mechanistic components of reactions presented during introductory organic chemistry coursework. The forms are designed to allow students to summarize the name of a reaction in conjunction with its flow from starting material to product and its mechanism. To aid in the description of a reaction's mechanism, mechanistic subtypes are listed at the bottom of the table. Additional spaces are provided for students to add in more advanced mechanistic components such as sigmatropic rearrangements, cycloadditions, oxidative additions, and reductive eliminations presented throughout this book.

As an example, the first form is filled out using the Robinson annulation. In completing this example, each mechanistic step was numbered in order to relate the appropriate mechanistic subtype to those listed in the form. Following this format, students are encouraged to complete additional pages using the reactions described in this book. Students are then encouraged to continue using these forms as an aid in the study of mechanistic organic chemistry.

Arrow-Pushing in Organic Chemistry: An Easy Approach to Understanding Reaction Mechanisms,
Second Edition. Daniel E. Levy.
© 2017 John Wiley & Sons, Inc. Published 2017 by John Wiley & Sons, Inc.

Reaction name: Robinson annulation	Homolytic ☐ Heterolytic ☑ Concerted ☐

Summary:

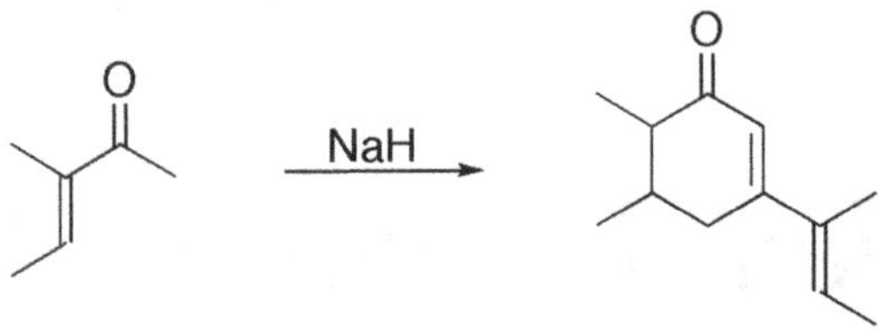

Mechanism:

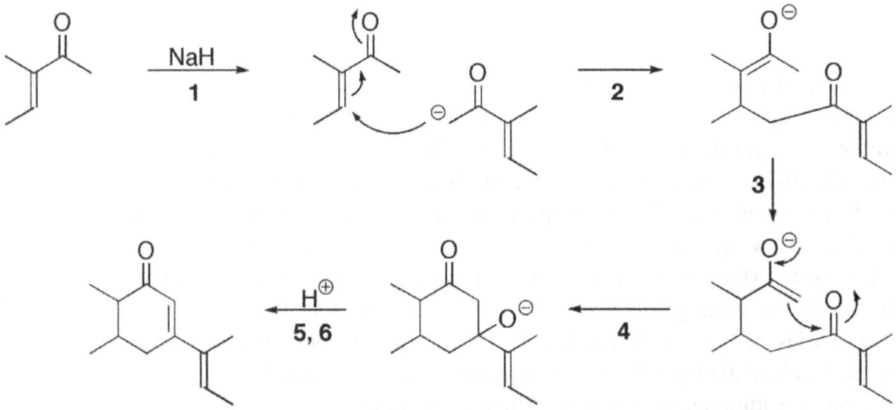

Mechanistic types	Mechanism steps	Mechanistic types	Mechanism steps	Mechanistic types	Mechanism steps
Deprotonation	1	S_N2		Hydride shift	
Protonation	5	S_N2'		Alkyl shift	
Solvolysis		E1	6	1,2-Addition	4
Proton transfer	3	E2		1,4-Addition	2
S_N1		Addition– Elimination			

Reaction name:		Homolytic ☐
		Heterolytic ☐
		Concerted ☐

Summary:

Mechanism:

Mechanistic types	*Mechanism steps*	*Mechanistic types*	*Mechanism steps*	*Mechanistic types*	*Mechanism steps*
Deprotonation		S_N2		Hydride shift	
Protonation		S_N2'		Alkyl shift	
Solvolysis		E1		1,2-Addition	
Proton transfer		E2		1,4-Addition	
S_N1		Addition–Elimination			

*This page may be reproduced as necessary.

Index

Arrow-Pushing in Organic Chemistry: An Easy Approach to Understanding Reaction Mechanisms,
Second Edition. Daniel E. Levy.
© 2017 John Wiley & Sons, Inc. Published 2017 by John Wiley & Sons, Inc.

Printed and bound by CPI Group (UK) Ltd, Croydon, CR0 4YY

05/10/2023

08126532-0002

PERIODIC TABLE OF ELEMENTS

hydrogen 1 **H** 1.0079																	helium 2 **He** 4.0026
lithium 3 **Li** 6.941	beryllium 4 **Be** 9.0122											boron 5 **B** 10.811	carbon 6 **C** 12.011	nitrogen 7 **N** 14.007	oxygen 8 **O** 15.999	fluorine 9 **F** 18.998	neon 10 **Ne** 20.180
sodium 11 **Na** 22.990	magnesium 12 **Mg** 24.305											aluminium 13 **Al** 26.982	silicon 14 **Si** 28.086	phosphorus 15 **P** 30.974	sulfur 16 **S** 32.065	chlorine 17 **Cl** 35.453	argon 18 **Ar** 39.948
potassium 19 **K** 39.098	calcium 20 **Ca** 40.078	scandium 21 **Sc** 44.956	titanium 22 **Ti** 47.867	vanadium 23 **V** 50.942	chromium 24 **Cr** 51.996	manganese 25 **Mn** 54.938	iron 26 **Fe** 55.845	cobalt 27 **Co** 58.933	nickel 28 **Ni** 58.693	copper 29 **Cu** 63.546	zinc 30 **Zn** 65.39	gallium 31 **Ga** 69.723	germanium 32 **Ge** 72.61	arsenic 33 **As** 74.922	selenium 34 **Se** 78.96	bromine 35 **Br** 79.904	krypton 36 **Kr** 83.80
rubidium 37 **Rb** 85.468	strontium 38 **Sr** 87.62	yttrium 39 **Y** 88.906	zirconium 40 **Zr** 91.224	niobium 41 **Nb** 92.906	molybdenum 42 **Mo** 95.94	technetium 43 **Tc** [98]	ruthenium 44 **Ru** 101.07	rhodium 45 **Rh** 102.91	palladium 46 **Pd** 106.42	silver 47 **Ag** 107.87	cadmium 48 **Cd** 112.41	indium 49 **In** 114.82	tin 50 **Sn** 118.71	antimony 51 **Sb** 121.76	tellurium 52 **Te** 127.60	iodine 53 **I** 126.90	xenon 54 **Xe** 131.29
caesium 55 **Cs** 132.91	barium 56 **Ba** 137.33	57-70 *	lutetium 71 **Lu** 174.97	hafnium 72 **Hf** 178.49	tantalum 73 **Ta** 180.95	tungsten 74 **W** 183.84	rhenium 75 **Re** 186.21	osmium 76 **Os** 190.23	iridium 77 **Ir** 192.22	platinum 78 **Pt** 195.08	gold 79 **Au** 196.97	mercury 80 **Hg** 200.59	thallium 81 **Tl** 204.38	lead 82 **Pb** 207.2	bismuth 83 **Bi** 208.98	polonium 84 **Po** [209]	astatine 85 **At** [210]	radon 86 **Rn** [222]
francium 87 **Fr** [223]	radium 88 **Ra** [226]	89-102 **	lawrencium 103 **Lr** [262]	rutherfordium 104 **Rf** [261]	dubnium 105 **Db** [262]	seaborgium 106 **Sg** [266]	bohrium 107 **Bh** [264]	hassium 108 **Hs** [269]	meitnerium 109 **Mt** [268]	ununnilium 110 **Uun** [271]	unununium 111 **Uuu** [272]	ununbium 112 **Uub** [277]	ununtrium 113 **Uut** [284]	ununquadium 114 **Uuq** [289]	ununpentium 115 **Uup** [288]	ununhexium 116 **Lv** [293]	ununseptium 117 **Uus** [292,208]	ununoctium 118 **Uuo** [294]

*Lanthanide series

lanthanum 57 **La** 138.91	cerium 58 **Ce** 140.12	praseodymium 59 **Pr** 140.91	neodymium 60 **Nd** 144.24	promethium 61 **Pm** [145]	samarium 62 **Sm** 150.36	europium 63 **Eu** 151.96	gadolinium 64 **Gd** 157.25	terbium 65 **Tb** 158.93	dysprosium 66 **Dy** 162.50	holmium 67 **Ho** 164.93	erbium 68 **Er** 167.26	thulium 69 **Tm** 168.93	ytterbium 70 **Yb** 173.04

**Actinide series

actinium 89 **Ac** [227]	thorium 90 **Th** 232.04	protactinium 91 **Pa** 231.04	uranium 92 **U** 238.03	neptunium 93 **Np** [237]	plutonium 94 **Pu** [244]	americium 95 **Am** [243]	curium 96 **Cm** [247]	berkelium 97 **Bk** [247]	californium 98 **Cf** [251]	einsteinium 99 **Es** [252]	fermium 100 **Fm** [257]	mendelevium 101 **Md** [258]	nobelium 102 **No** [259]